Modern Petrochemical Technology

Modern Petrochemical Technology

Methods, Manufacturing and Applications

Santi Kulprathipanja
James E. Rekoske
Daniel Wei
Robert V. Slone
Trung Pham
Chunqing Liu

WILEY-VCH

Authors

Dr. Santi Kulprathipanja
Honeywell UOP
25 East Algonquin Road
IL
United States

Dr. James E. Rekoske
Ecolab Inc.
1 Ecolab Place
St. Paul, Minnesota 55102
United States

Dr. Daniel Wei
Honeywell UOP
25 East Algonquin Road
IL
United States

Dr. Robert V. Slone
Honeywell Advanced Mateirals
25 East Algonquin Road
IL
United States

Dr. Trung Pham
Honeywell UOP
25 East Algonquin Road
IL
United States

Dr. Chunqing Liu
Honeywell UOP
25 East Algonquin Road
IL
United States

Cover
Cover Image:
© littlewormy/Shutterstock.com

Library of Congress Card No.:
applied for

British Library Cataloguing-in-Publication Data
A catalogue record for this book is available from the British Library.

Bibliographic information published by the Deutsche Nationalbibliothek
The Deutsche Nationalbibliothek lists this publication in the Deutsche Nationalbibliografie; detailed bibliographic data are available on the Internet at <http://dnb.d-nb.de>.

Print ISBN: 978-3-527-34522-9
ePDF ISBN: 978-3-527-81819-8
ePub ISBN: 978-3-527-81817-4
oBook ISBN: 978-3-527-81816-7

Typesetting SPi Global, Chennai, India
Printing and Binding CPI Group (UK) Ltd, Croydon, CR0 4YY

Printed on acid-free paper

C089595_210321

Contents

Foreword

Over the past 30 years, there has been a dramatic global expansion of the petrochemicals industry. The industrialization and economic growth of countries in Asia, the Middle East, Latin America, and now Africa has raised the purchasing power of billions of people and created surging demand for products that are ultimately derived from petrochemical feedstocks. As a supplier of technology to the petrochemicals industry, Honeywell UOP has been a willing participant in this expansion and has worked with its customers around the world to deploy safe and state-of-the-art processes for making the basic monomers and chemicals that go into making the thousands of products that we all consume every day.

Dr. Santi Kulprathipanja has been at the heart of developing many of Honeywell UOP's technologies. In his very distinguished career, Dr. Kulprathipanja was named as inventor of over a hundred patents, including many of the technologies described in this book. While best known for his many inventions, Dr. Kulprathipanja was always a strong advocate for developing close links between industry and academia, and for many years he has also taught university classes around the world so that he could share his knowledge with the next generation of engineers and chemists.

This book builds on Dr. Kulprathipanja's highly regarded classes in petrochemicals processing and separation technologies and contains additional contributions from colleagues at Honeywell UOP who have joined him in his teaching and provided additional insights from their own technical expertise. It should be a very useful reference for chemical engineers or others who might be starting a career in the petrochemicals industry, or in companies or institutions that serve that industry.

Gavin Towler
Chief Technology Officer and
Vice President of Research and
Development Honeywell UOP

Preface

This book will provide an overview of markets and uses for common petrochemical building blocks and also a survey of the technology used to make the key petrochemical building blocks. The book will be written with a practical chemistry and chemical engineering course work in mind, with the target audience being university professors, students, scientists, and engineers those that are in the field of petrochemicals and petrochemical technology. The book will contain both critical information to help the readers to understand how the technologies are used to produce petrochemicals, how the various petrochemicals are used, and examples and problems designed to reinforce the learning about the basic science, engineering, and use of petrochemicals.

The topics covered by the book include Chapter 1: Refining and Petrochemical Process, Chapter 2: Petrochemical Markets, Chapter 3: Olefins and Synthesis Gas Production Technologies, Chapter 4: Uses of Olefins and Synthesis Gas, Chapter 5: Aromatic Production Technology, Chapter 6: Uses of Aromatics, Chapter 7: Adsorption, Chapter 8: Distillation, Chapter 9: Membranes, Chapter 10: Absorption, and Chapter 11: Extraction. Highlights in each chapter can be found as given in the following content.

Refining and petrochemical process is introduced in Chapters 1 and 2. As we can notice in the last 10 years or so, the combination of high demand for electric cars and higher automobile engine efficiency in the future will mean less conversion of petroleum into fuels. However, the demand for petrochemicals is forecast to rise due to the increase in world population. With this, it is expected that modern and more innovative technologies will be developed to serve the growth of the petrochemical market. Petroleum is refined/converted into intermediate refinery products. The intermediate refinery products are then converted to fuels such as fuel gas, liquefied petroleum gas, gasoline, jet fuel, kerosene, auto diesel, and other heavy products. In addition, these petroleum intermediates can be further processed and separated into products for petrochemical applications.

We start our journey toward a better understanding of petrochemical markets through an exploration of the building blocks of the industry – those commodity chemicals that are utilized in large volume – in Chapter 1. We introduce the sources of these building blocks – predominantly from three separate but important complexes: olefin complexes, aromatic complexes, and synthesis gas complexes.

We also introduce the "interconnected" nature of these building blocks – the fact that their sources and uses are frequently connected in which one building block is co-produced with another. In Chapter 2, we begin to explore this interconnection in more detail, providing some statistics for each building block. Many of the building blocks have many different sources, are made as co-products with other building blocks and chemicals, and the supply of these building blocks depends rarely just on one of the building blocks. Because of this, the markets frequently get out of balance – where either shortages or large surpluses of one building block are noted. The beauty of these markets then comes through, as substitution and movements between the building blocks occur to rebalance the supply and demand. The ability to make many final products (different products) from one building block, as well as the ability to use many building blocks to make one final product (different routes) make the petrochemical markets extremely complex and extremely robust.

Olefins are key building blocks of chemical industry. Light olefins, ethylene and propylene, are the key components used in the manufacture of many polymers and chemicals. Butylene is also in high demand for further processing to produce methyl-*tert*-butyl ether (MTBE) and alkylate for gasoline blending. Heavy normal-olefins (C_{10}–C_{14}) are the main feedstock component of linear alkylbenzene (LAB), the most common raw material in the manufacture of biodegradable household detergents.

Production and uses of olefins are documented in Chapters 3 and 4. They include the olefins productions using steam cracking, paraffin dehydrogenation, and methanol to olefins technologies as well as the uses of the olefins to produce downstream products such as polyolefins, polyvinyl chloride, mono-ethylene glycol, propylene oxide, acrylonitrile, LAB, MTBE, and alkylate. Chapters 3 and 4 also cover the manufacturing technologies for producing syngas, which is used in large volumes for the production of commodity chemicals such as hydrogen, ammonia, and methanol.

Aromatics are important building blocks for petrochemical intermediate and derivative products. Chapters 5 and 6 describe various methods and commercial processes to produce aromatics such as benzene, toluene, and xylene (BTX). Naphtha reforming is still the most efficient and sensible method to produce aromatics since direct dehydrogenation of naphthenes is favorable provided sufficient temperature while valuable hydrogen co-product can be obtained at the same time. Other catalytic processes include fluidified catalytic cracking, aromatization of light naphtha, toluene methylation, aromatic isomerization, transalkylation, and disproportionation. Noncatalytic methods include steam cracking, pyrolysis, and coke-oven benzole production.

Essential materials such as nylons, polyesters, plastics, pesticides, coatings, and resins are primarily derived from basic aromatic building blocks, including BTX. These materials have strong mechanical strength and thermal stability with aromatic ring in the backbone. Electrophilic aromatic substitution allows functional groups to attach to the aromatic molecules that make them flexible monomers for further production or synthesis of polymers for construction and automotive industries or consumer and pharmaceutical products.

The commercially established technologies can be reviewed and leveraged to develop new methods, including new intermediates or advanced materials that can be derived from renewable resources. A circular economy for plastics has started and continues with aims to promote recycle and eliminate wastes; new technologies will emerge to adapt to challenges and opportunities ahead.

Separation processes play a significant role in the chemical industry. These processes include adsorption, distillation, membranes, absorption, extraction, and crystallization. The most utilized technique is distillation, which is used in 90–95% of all processes in the chemical industry. However, due to technological and economic limitations, distillation is not always feasible. For the remaining applications, other techniques must be explored. Adsorption is one of the more desirable options because of its extensive and flexible applications. Particularly, in the bulk industrial liquid adsorption that utilizes zeolite as an adsorbent and counter-current simulated moving-bed technology, which is known as "Sorbex technology." Examples of these bulk liquid industrial adsorption are the separations of *para*-xylene from other xylene-isomer, *meta*-xylene from other xylene-isomer, olefin from paraffin, *n*-paraffin from other paraffin-isomer, fructose from other-sugar. Because zeolite adsorbent and Sorbex technology are critical to the adsorption separation performance processes, Chapter 7 will detail zeolites' synthesis, zeolites' features, and Sorbex technology operation mechanism. Besides, other factors that are critical to the adsorption performance will be detailed.

As mentioned in Chapter 7, the most utilized separation technique is distillation, which is used in 90–95% of all processes in the chemical industry. Distillation is documented in Chapter 8. Chapter 8 covers the commercial development of distillation as well as the theory behind distillation, including vapor–liquid equilibrium (VLE) as well as the construction of McCabe–Thiele diagrams to aid in the design of distillation columns. Our attention then turns to a comparison of both batch and continuous distillation as well as the internals of distillation towers (tray and packed columns).

The strengths and weaknesses of each type of column, tray, and packing are covered in detail. The criteria by which trays are selected vs. packed columns are delineated. The performance and cost aspects of sieve, valve, and bubble-cap trays are compared in terms of the operating range, operating capacity, cost, and pressure drop. We then turn our attention to column-packing materials and discuss the evolution of random column packings from simple Raschig rings to more elaborate Intalox™ designs. Finally, our summary of packing is completed with a survey of structured packing with a discussion of the strengths, challenges, and opportunities within this segment of distillation technology.

Chapter 8 concludes with a survey of commercialization examples starting with crude oil distillation (both atmospheric and vacuum methods). While distillation methods have been developing for thousands of years, innovation continues to drive the introduction of new commercial methods. We finish Chapter 8 with overviews of extractive distillation, azeotropic distillation, dividing wall column technology, and reactive distillation. Each of these methods continues to evolve from a cost–performance perspective.

Membrane technology has been a critical enabler for effective and efficient gas, vapor, and liquid separations for many years. The membrane technology is described in Chapter 9. Polymeric membrane–based industrial separation processes have evolved rapidly since the development of asymmetric integrally skinned polymeric membranes by Loeb and Sourirajan in 1960s and the development of thin-film composite (TFC) membranes by Cadotte in 1980s. Inorganic membranes have the advantages of high thermal stability, high chemical resistance, and resistance to harsh environments over polymeric membranes. Therefore, inorganic membranes not only have been commercially successful in some liquid separation applications, but also have been studied for catalytic membrane reactor, energy storage, and other applications. Further advancement in reducing the cost and improving the processability and reproducibility of the inorganic membranes is still critical for inorganic membranes to be adapted for new commercial applications. Innovation in membrane materials with long-term stability, chemical resistance and specific to applications, membrane fabrication methods, module design, new membrane process design, and process technology integration has been instrumental to the widespread industrial adoption of membrane technology, and it will continue to enable new membrane-based separations and reactions. Chapter 9 will discuss several aspects of membranes, including principles, types and preparation, membrane modules, membrane reactors, and commercial applications.

Absorption technology is given in Chapter 10 to detail method of separation commonly used in the oil and gas and chemical industries for removal of acid contaminants such as H_2S (hydrogen sulfide) and CO_2 (carbon dioxide) from hydrocarbons; this separation strategy is often employed in the purification, transport, and storage of natural gas. As natural gas usage has increased over the past 20 years due to its lower CO_2 emissions vs. coal, the removal of acid gases has become more critical and helps protect large capital investments from corrosion.

There are two primary classes of absorption we cover in Chapter 10: chemical absorption and physical absorption. The strengths and weaknesses of each method are discussed. At low partial pressures of the component (generally low concentration), a chemical solvent is generally preferred while at high partial pressures of the component to be removed, physical solvent has advantage due to its high loading capability. The criteria by which a specific amine is selected are reviewed: (i) Required purity of cleaned gas stream for further processing or end use, (ii) Composition of feed gas, (iii) Utility requirements, process costs, and (iv) Corrosion and solvent degradation/solvent losses.

We then turn our attention to CO_2 removal using an inorganic chemical solvent (K_2CO_3 in the UOP Benfield™ process). The Honeywell UOP Benfield Process is a thermally regenerated cyclical solvent process using an activated, inhibited hot potassium carbonate solution to remove CO_2, H_2S, and other acid gas components.

An overview of the physical solvent options is included: Selexol™ (DMPEG, Dimethyl Ether of Polyethylene Glycol), Rectisol™ (Methanol), and Fluor (PC). Cost and performance factors such as solvent volatility, contaminant solubility, and process conditions such as pressure and temperature often guide the choice of solvent in physical absorption systems.

Finally, we review a couple of key commercialization examples for absorption: Glycol dehydration and the stripping of organic contaminants from air.

Though there are many extraction processes, Chapter 11 will focus on liquid–liquid extraction (LLE) and supercritical fluid extraction (SCFE). LLE is widely used in industry for hydrometallurgy and hydrocarbon extraction. Applications in gas–liquid extraction called absorption are used commercially to extract carbon dioxide and sulfur-species (acid gas) from natural gas. Absorption is covered in Chapter 10.

In Chapter 11, critical materials and processes that impact extraction performance, including solvent selection and type of extractors, will be described. Selecting the right solvent and type of extractors is key to achieving high purity and efficiency in the extraction process. After introducing the fundamentals of extraction, the advantages of extraction over other separation technology will be demonstrated with the following commercial applications: (i) Hydrometallurgy such as extraction of Zr from Hf, and U from Pu; and (ii) Hydrocarbon separation such as aromatics extraction from nonaromatic, *m*-xylene from other xylene-isomers, mercaptan removal, and SCFE of caffeine from coffee.

Honeywell UOP
Des Plaines, Illinois, USA
17 June 2020

Santi Kulprathipanja
James E. Rekoske
Daniel Wei
Robert V. Slone
Trung Pham
Chunqing Liu

Section I

Introduction

1

Refinery and Petrochemical Processes

1.1 Introduction

The combination of high demand for electric cars and higher automobile engine efficiency in the future will mean less conversion of petroleum into fuels. However, the demand for petrochemicals is forecast to rise due to the increase in world population. With this, it is expected that modern and more innovative technologies will be developed to serve the growth of the petrochemical market.

In a refinery process, petroleum is converted into petroleum intermediate products, including gases, light/heavy naphtha, kerosene, diesel, light gas oil, heavy gas oil, and residue. From these intermediate refinery product streams, several fuels such as fuel gas, liquefied petroleum gas, gasoline, jet fuel, kerosene, auto diesel, and other heavy products such as lubricants, bunker oil, asphalt, and coke are obtained. In addition, these petroleum intermediates can be further processed and separated into products for petrochemical applications.

In this chapter, petroleum will be introduced first. Petrochemicals will be introduced in the second part of the chapter. Petrochemicals – the main subject of this book – will address three major areas, (i) the production of the seven cornerstone petrochemicals: methane and synthesis gas, ethylene, propylene, butene, benzene, toluene, and xylenes; (ii) the uses of the seven cornerstone petrochemicals, and (iii) the technology to separate petrochemicals into individual components.

1.2 Petroleum

Petroleum is derived from the Latin words "petra" and "oleum," which means "rock" and "oil," respectively. Petroleum also is known as crude oil or fossil fuel. It is a thick, flammable, yellow-to-black mixture of gaseous, liquid, and solid hydrocarbons formed from the remains of plants and animals. Over millions of years, this organic mixture of plants and animals was subjected to enormous pressure and heat as layers of earth further buried them over time. The organic mixture changed chemically and broke down into hydrocarbons. Because of the time it takes to form petroleum, it is referred to as a nonrenewable energy source.

Modern Petrochemical Technology: Methods, Manufacturing and Applications, First Edition.
Santi Kulprathipanja, James E. Rekoske, Daniel Wei, Robert V. Slone, Trung Pham, and Chunqing Liu.

1.2.1 Forms of Petroleum

Petroleum can be in gas, liquid, and solid forms of hydrocarbons. The forms of petroleum are functions of pressure, temperature, and other surrounding conditions such as rock and soil. Under low surface pressure and temperature conditions, lighter hydrocarbons such as methane, ethane, propane, and butane occur as gases, while heavier hydrocarbons such as octane, benzene, xylenes, and paraffin wax are in the form of liquids or solids. However, in an underground oil reservoir where pressures are high, the proportions of gas, liquid, and solid depend on subsurface conditions. It is expected that there are higher percentages of solids and liquids compared to gas at higher pressures.

1.2.1.1 Gaseous Petroleum: Gaseous Petroleum Can Be Defined as Natural Gas or Shale Gas

Natural gas is a mixture consisting primarily of methane, but also commonly includes varying amounts of other higher alkanes, and small percentages of carbon dioxide, nitrogen, hydrogen sulfide, and helium [1]. Natural gas is found in deep underground reservoirs with sandstone and associated with other hydrocarbon reservoirs.

Shale gas has the same composition as natural gas, except that it is found trapped within shale formations [2]. Shale gas wells depend on hydraulic fractures to allow the gas to flow from the rock and have become an increasingly important source of natural gas in the United States since the start of this century. In 2000, shale gas provided only 1% of U.S. natural gas production; by 2010 it was more than 20%. Today, the United States is the largest shale gas producer in the world [3], while China is estimated to have the world's largest shale gas reserves [4].

1.2.1.2 Liquid Petroleum: Liquid Petroleum Can Be Separated into Light Crude Oil and Heavy Crude Oil

Light crude oil has a low viscosity and is a liquid at room temperature. These properties make light crude oil easy to pump and extract. It is composed of short-chain paraffins, which are straight and branched-chain hydrocarbons. Because virgin light crude oil comprises these short chains, it does not have to be heavily refined to produce gasoline. Typically, the chain-length range in gasoline is 4–12 carbons, making light oil a desirable crude to be processed in a refinery. Approximately 30% of the world's petroleum reserves is light oil crude.

Heavy crude has a higher viscosity, but is still a liquid at room temperature. It usually contains more sulfur, nitrogen, and other contaminants than light oil. To refine heavy oil to gasoline, it must be cracked and treated to remove the contaminants. It requires more energy input and cost. Beyond the need for additional refinement, heavy crude also needs additional extraction techniques to recover oil from the wells. These techniques include stream stimulation to make the oil less viscous and the injection of air into the wells to create fires that burn heavier hydrocarbons and degrade them into lighter, more easily pumped compositions.

The transport of heavy oil requires the addition of diluting agents, particularly in pipelines. The other major drawback to heavy crude is its environmental impact because it contains sulfur and heavy metals, both of which must be removed.

Heavy metals are often toxic and their removal from crude presents disposal issues. Sulfur, which may be as high as 4.5 wt%, is corrosive to pipeline metal and refinery components.

1.2.1.3 Solid Petroleum: Bitumen

Bitumen is a naturally occurring solid hydrocarbon that generally contains a mixture of large polycyclic aromatic with exceptionally low hydrogen-to-carbon ratios. Bitumen crude is found impregnated in sedimentary rock, which is mainly located in Canada. This rock is often referred to as oil sand or tar sand. Bitumen will not flow unless heated or mixed with lighter crudes to reduce viscosity. Most bitumen extraction processes require some level of physical mining as opposed to pump extraction as with gaseous and liquid crude oil.

When crude oil prices increase, it can become profitable to upgrade bitumen to synthetic crude for fuel. Bitumen can also be used for a number of other purposes, including mortar between bricks, base material for statues, and waterproofing. Today, the largest use of bitumen is roofing applications and at a lower concentration in roadways.

1.2.2 Composition of Petroleum

In general, oil wells produce predominantly liquid petroleum (crude oil), with some dissolved natural gas. Because the pressure is lower at the Earth's surface than deep underground, some of the natural gas will come out of the solution and be recovered. Furthermore, at the higher deep underground temperatures and pressures, the dissolved gas may contain heavier hydrocarbons such as pentane, hexane, and heptane. Once these heavier gases reach the lower pressure and temperature surface conditions, they will condense out of the gas phase to form condensate, which has similar appearance to gasoline and is similar in composition to some volatile light crude oils.

Petroleum is a mixture of a very large number of different hydrocarbons. The four most commonly found categories of molecules are alkanes, cycloalkanes, aromatics, and more complicated chemical compounds such as asphaltenes. Petroleum also contains trace organic compounds of sulfur, nitrogen, oxygen, and metals such as iron, nickel, copper, and vanadium. Each petroleum source has a unique mix of molecules that define its physical and chemical properties, such as color and viscosity. The molecular composition of crude oil varies widely depending on the source. The chemical elements [5] in crude oil can vary as shown in Table 1.1. Table 1.2 shows the percent ranges of the four key hydrocarbons in crude oil [6].

Alkanes – also called paraffins – are saturated hydrocarbons with straight or branched chains containing only carbon and hydrogen. They have the general formula C_nH_{2n+2}. Examples of alkanes in crude oil are hexane, decane, and 2,2,4-trimethylpentane (iso-octane).

Cycloalkanes – or, naphthene – are saturated hydrocarbons that have one or more carbon rings to which hydrogen atoms are attached. The general formula of naphthene is C_nH_{2n}. Examples of cycloalkanes are cyclopentane and cyclohexane.

Aromatics are unsaturated hydrocarbons that have one or more planar six-carbon rings called benzene rings to which hydrogen atoms are attached. They have the

Table 1.1 Chemical elements in crude oil.

Element	Percent
Carbon	83–85
Hydrogen	10–14
Nitrogen	0.1–2.0
Oxygen	0.05–1.5
Sulfur	0.05–6.0
Metals	<0.1

Table 1.2 Hydrocarbon in crude oil.

Hydrocarbon	Percent
Alkanes	15–60
Naphthenes	30–60
Aromatics	3–30
Asphaltenes	Remaining

general formula C_nH_{2n-6}. Examples of aromatics are benzene, toluene xylenes, ethylbenzene, 1,2-dimethylbenzene, naphthalene, and biphenyl.

Sulfur compounds in crude oil are typically found as mercaptans such as ethylmercaptan and cyclic sulfides such as tetrahydrothiophen and benzothiophen.

There are two types of nitrogen-based compounds in petroleum. The first type is basic nitrogen molecules that have an extra lone pair of electrons to facilitate reactions. Basic nitrogen compounds are pyridine, quinolone, and phenanthridine. The second type of nitrogen compounds is nonbasic. In this type of nitrogen, the lone pair of electrons with the nitrogen is delocalized in the ring structure making the molecule less reactive. Indole and carbazole are two nonbasic nitrogen compounds.

Some crude oils contain trace organic oxygen compounds such as benzoic acid and phenol.

Metals such as Fe, Ni, Cu, V, Ca, Mg, Hg, As, and Na are generally found in petroleum. Metals can be in the form of organic or inorganic salts. For example, V can be found in an organic porphyrin structure and $CaCl_2$, $MgCl_2$, and NaCl suspended or dissolved as entrained water inorganic salts. Metals in petroleum must be removed or neutralized before processing, as they cause corrosion and fouling to refinery equipment and poison of catalysts.

1.2.3 Petroleum Refinery

An oil refinery is an essential part of the downstream side of the petroleum industry. Petroleum refineries are very large industrial complexes that involve many processing units and auxiliary facilities such as utility units and storage tanks. Some modern petroleum refineries process as much as 900 000 barrels per day of crude oil. According to *Oil and Gas Journal*, a total of 636 refineries operated worldwide at the end of 2014 for a total capacity of 87.75 million barrels per day of crude oil. Reliance Industries' Jamnagar Refinery in Gujarat, India, currently is the world's largest oil refinery. Each refinery has its own unique arrangement and combination of refining processes largely determined by the refinery location, desired products, and economic considerations.

Conversion of heavy oils to useful products requires breaking large molecules into smaller ones. The breaking or cracking of large molecules can be accomplished by one or more combinations of heat, pressure, and chemical reaction. In the process, harmful or unwanted compounds such as metals, sulfur, nitrogen, and oxygen are also removed. The conversion of heavy oils into useful products requires many separation and chemical reaction processes. These processes are briefly characterized as follows:

- *Desalination*: This is the first unit in the refinery complex. Salts from crude oil are extracted before it enters the atmospheric distillation unit.
- *Crude oil distillation (atmospheric distillation)*: The desalted crude oil is separated into various fractions for processing in downstream units.
- *Vacuum distillation*: The residue oil from the bottom of the crude oil distillation unit is further distilled at a vacuum pressure well below atmospheric.
- *Naphtha hydrotreater*: The hydrotreater desulfurizes, denitrogenizes, and deoxygenizes naphtha using hydrogen from the atmospheric distillation unit. The naphtha must be treated before sending the stream to the catalytic reformer unit.
- *Catalytic reformer*: The reformer converts the hydrotreated naphtha into reformate, which has a higher content of aromatics and cyclic hydrocarbons. The end products are high-octane gasoline and *para*-xylene aromatics, which is a critical petrochemical in making PET.
- *Distillate hydrotreater*: This unit, similar to the naphtha hydrotreater, removes sulfur, nitrogen, and oxygen from distillates (such as diesel) and other units within the refinery after atmospheric distillation.
- *Fluid catalytic cracker (FCC)*: The FCC process is used to upgrade the heavier, higher boiling-point fractions from the crude oil distillation by converting them into more valuable lighter and lower boiling-point products.
- *Hydrocracker*: The hydrocracker uses hydrogen to upgrade heavy residual oils from the vacuum distillation unit by thermally cracking them into more valuable lighter and lower-viscosity products.
- *Merox*: The Merox process desulfurizes LPG, kerosene, and jet fuel by oxidizing mercaptans to organic disulfides.
- *Coking*: Delayed cokers, fluid cokers, and flexicokers are used to crack very heavy residual oils into gasoline and diesel fuel leaving petroleum coke as a residual product.
- *Alkylation*: Alkylation is a process that uses sulfuric or hydrofluoric acid or an ionic liquid as a catalyst to produce high-octane components for gasoline blending.

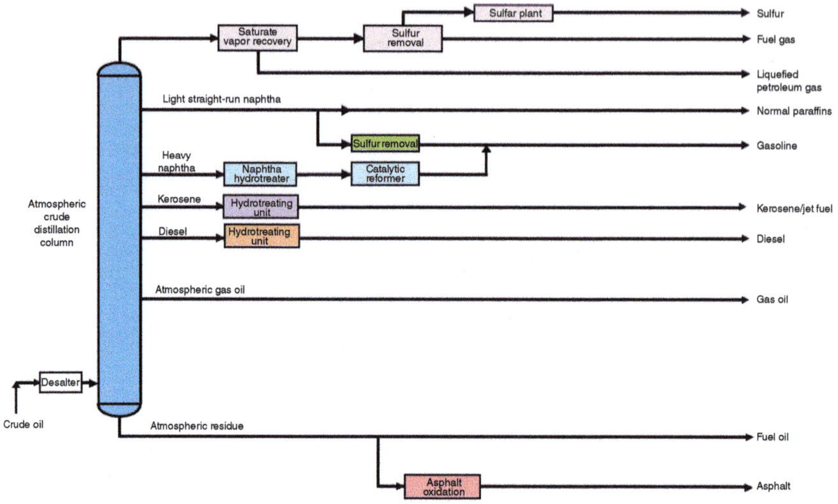

Figure 1.1 Topping and hydroskimming refinery.

For example, iso-butane and butylenes can be converted into iso-paraffin alkylate, which is a very high-octane gasoline.

- *Dimerization*: Dimerization is used to convert olefins into higher-octane gasoline. For example, butenes are dimerized into isooctene, which is subsequently hydrogenated to form isooctane.
- *Isomerization*: This process converts linear molecules such as normal pentane to higher-octane branched molecules for blending into gasoline. Isomerization units also are used to convert linear molecules such as normal butane into isobutane for use in alkylation units.

Oil refining is the process where crude oil is transformed into more desirable and valuable products using the processes described above. Refineries often are classified by the number and type of process units available for transforming crude oil into petroleum products.

Topping refineries are the least complex refineries and are used to separate the crude oil into its constituent petroleum products by atmospheric distillation; naphtha is produced, but no gasoline. Hydroskimming refineries are also one of the simplest types of refineries used in the petroleum industry and are equipped with atmospheric distillation, naphtha reforming, and additional treating processes to produce gasoline. Important to note, a hydroskimming refinery produces a surplus of fuel with a relatively unattractive price and demand. Figure 1.1 shows schematic topping and hydroskimming refineries.

The next refinery classification in complexity level is a medium-conversion refinery. A medium-conversion refinery is shown in Figure 1.2. To design a medium-conversion refinery, unit operations such as vacuum distillation, thermocracking, fluid catalytic cracking, and asphalt oxidation are added to the hydroskimming refinery. This added level of complexity allows conversion of fuel oil to light distillates and middle distillates.

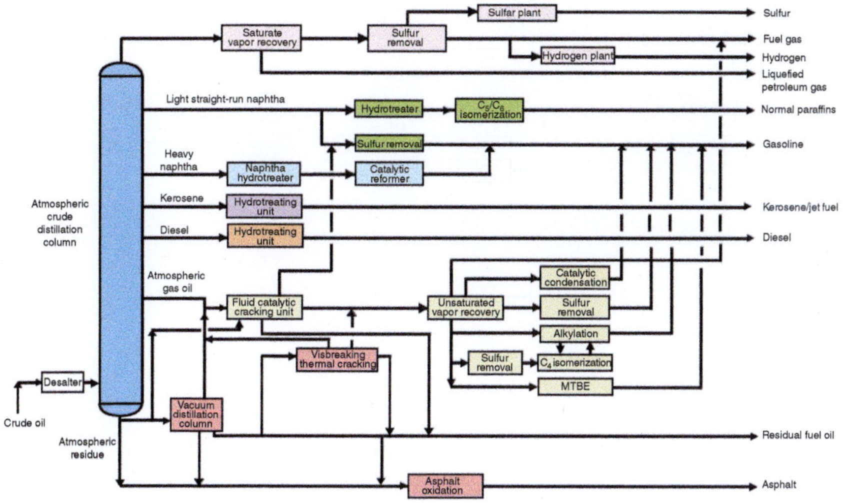

Figure 1.2 Medium-conversion refinery.

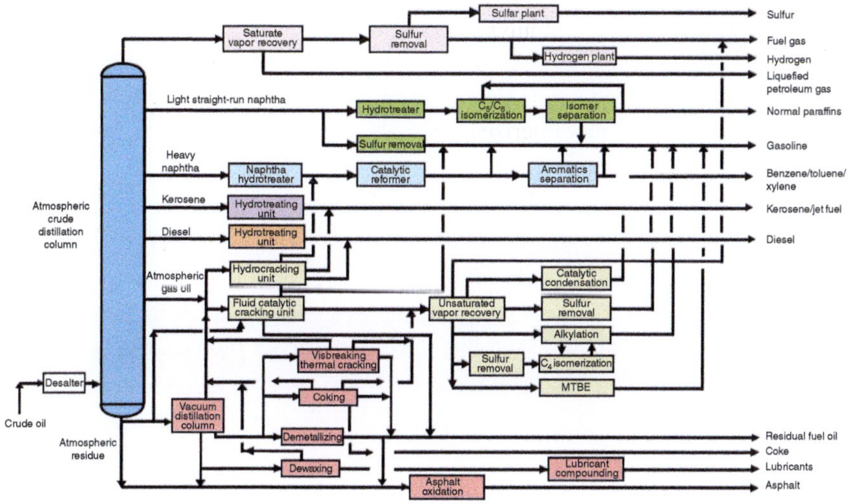

Figure 1.3 High-conversion refinery.

The most complex type of refinery is the high-conversion refinery, which is shown in Figure 1.3. The high-conversion refinery adds more inter-related processes such as hydrocracking, coking, demetallizing, and dewaxing. In particular, a coking unit adds cracking capability for conversion of fuel oil into distillates and petroleum coke, allowing high-efficiency conversion of the crude oil feedstock into higher yields of more valuable products.

1.2.4 Petroleum Products

Petroleum products are complex hydrocarbon mixtures, in contrast to petrochemicals, which are a collection of well-defined usually pure chemical compounds

Table 1.3 A breakdown of the petroleum products made from a typical barrel of U.S. oil.

Petroleum products	Percent
Gasoline	46
Jet fuel	9
Diesel and other fuel	26
Heavy fuel oil	4
Lubricants	1
Asphalt	3
Other products	11

Source: US Energy Information Administration, 31 May 2018.

derived from petroleum refineries. Depending on the composition of the crude oil and the demands of the market, refineries can produce different types and proportions of petroleum products. Petroleum products are usually grouped into four categories: light distillates (LPG, gasoline, naphtha), middle distillates (kerosene, jet fuel, diesel), heavy distillates, and residuum (heavy fuel oil, lubricating oils, wax, asphalt). The largest share of petroleum products is fuels, which include single or blended combinations of hydrocarbons to provide gasoline, jet fuel, diesel fuel, heating oil, and heavier fuel oils. Heavier fractions also can be used to produce asphalt, tar, paraffin wax, lubricating, and other heavy oils. Table 1.3 summarizes a breakdown of the petroleum products made from a typical barrel of U.S. oil.

Refineries also produce chemicals that are used to produce polymers, detergents, and other useful consumer products. Since petroleum contains sulfur-containing molecules, elemental sulfur is often produced as a petroleum byproduct. Carbon, in the form of petroleum coke, and hydrogen gas also are produced as petroleum byproducts. For example, hydrogen often is used as an essential reactant for other oil refinery processes such as hydrocracking and hydrodesulfurization.

1.3 Petrochemical Building Blocks

The industrial use of petrochemicals has focused throughout its history on roughly seven main building blocks, often called intermediates, which are obtained from both natural gas and petroleum processing. These seven basic building blocks are:

- *Ethylene*: Also known as ethene and possessing the formula C_2H_4, it is a colorless flammable gas. Because it contains only two carbon atoms, it is the simplest carbon–carbon double bond building block.
- *Propylene*: Also known as propene or methyl ethylene, propylene has the formula C_3H_6. It also is an excellent C=C double-bond building block allowing unique reactions and functionality because of the presence of both olefinic and aliphatic carbon–carbon bonds.

- *C4 mono- and diolefins*: This grouping is composed of butenes (also called butylenes) having the formula C_4H_8, and butadiene having the formula C_4H_6. While we group these building blocks together for simplicity, they are often used for vastly different chemistry purposes and final products.
- *Benzene*: The simplest of the aromatic ring structures having a formula of C_6H_6 makes benzene a very important organic molecule in the petrochemical industry.
- *Toluene*: Also known as toluol, toluene has the same ring structure as benzene but has one methyl group substituted on the ring. Its formula is C_7H_8.
- *Xylenes*: While strictly defined as the three dimethyl substituted benzene isomers (1,2-, 1,3-, and 1,4-dimethylbenzene), the petrochemical industry often classes ethylbenzene into the same grouping called xylenes. This is because the dimethyl benzenes and ethylbenzene have the same formula (C_8H_{10}) and very similar boiling points. The result is that these compounds are nearly always found together in raw material sources.
- *Synthesis gas*: The oddest grouping in the building blocks, synthesis gas is a mixture of H_2 and carbon monoxide (CO) in a molar ratio typically ranging from around 1.0 to slightly higher than 2.0. Synthesis gas is used to produce many final products and intermediates due to the flexible nature of the building blocks.

Chemical structures of these basic building blocks are shown in Figure 1.4.

From these seven building blocks, and adding abundant, readily available chemicals such as air, pure oxygen, or pure nitrogen, the petrochemical industry has built a wide variety of value-added compounds that have become essential to modern society. For example, in a modern automobile, plastics and petrochemical compounds

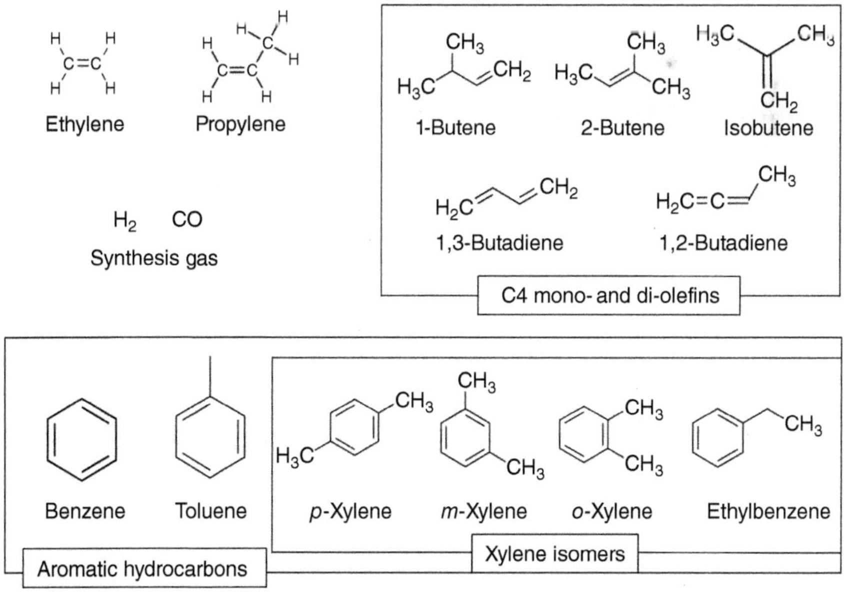

Figure 1.4 Chemical structure of common petrochemical building blocks.

account for more than 50% of the volume of materials used. Despite this large fraction of the volume, the petrochemical parts make up only about 10% of the vehicle weight [7]. Clearly the trend of replacing automotive parts with lighter-weight pieces based on petrochemicals will have to continue as lighter, more energy-efficient, and electric vehicles become more popular.

The seven building blocks are mostly present as components of raw materials (petroleum, natural gas) described earlier. Indeed, most of the building blocks had been identified in the 1800s as components of these raw materials. For example, Faraday [8] was reportedly the first to successfully isolate and characterize benzene in 1825. However, these compounds were needed to be present in much larger quantities, and isolated as intermediates for subsequent chemical reactions, for them to be of use as building blocks. With petroleum and natural gas as starting materials, the industry began to invent methods to obtain these building blocks in vast quantities.

Today, the unit operations (reactions, separations, heat transfer, etc.), which are combined in systematic ways to make petrochemical building blocks in large quantities, often are referred to as *petrochemical complexes*. Although very few petrochemical complexes are exactly alike, we can categorize these complexes into three basic types.

- *Olefin complexes*: The complexes make predominantly the olefins ethylene, propylene, and butenes. It is the primary technology for the product of olefins in steam cracking [9], but other technologies such as dehydrogenation and methanol to olefins technologies [10, 11] are gaining traction. In a steam cracking–based olefin complex, the primary product is ethylene; the other products are formed in far lower quantities. These complexes will sometimes be "fully integrated" – meaning they have full recovery of propylene, butenes, and even the aromatics formed. Others will recover only the ethylene and propylene and sell the remainder of the liquid products to others for recovery. A sample block-flow diagram for an olefins complex using steam cracking of hydrocarbons is shown in Figure 1.5.
- *Aromatics complexes*: These complexes make mostly aromatics hydrocarbons, and often make most of the product as *para*-xylene for use in the synthesis of purified terephthalic acid, PTA [12]. These complexes usually employ a technology to reform the naphtha cut of petroleum into aromatic rings, and then use a series of technologies to interconvert the aromatic rings into the desired products, with

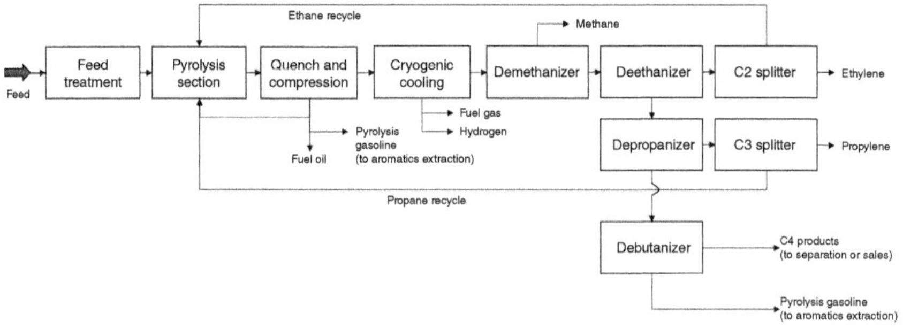

Figure 1.5 Block flow diagram of typical naphtha-based olefins unit.

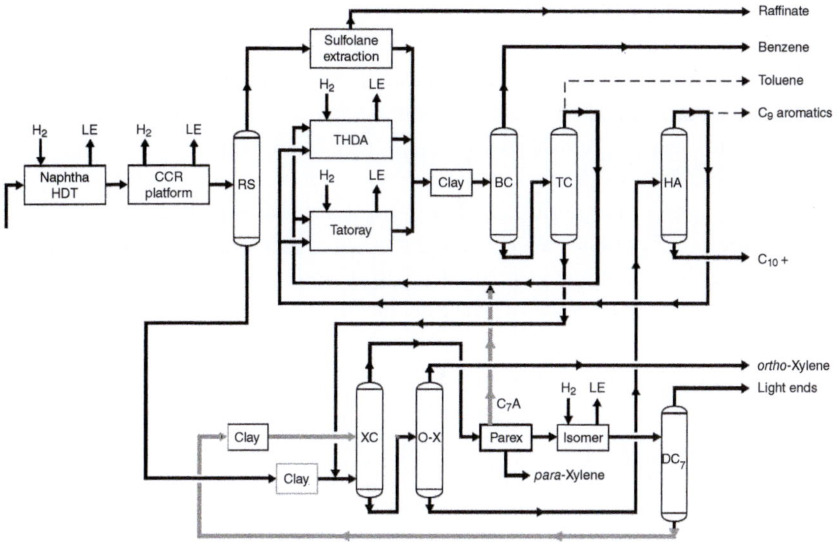

Figure 1.6 Typical aromatics complex.

para-xylene being typically the most desired. A typical plant-flow diagram for an aromatics complex is shown in Figure 1.6.

- *Synthesis gas complexes*: In these complexes, coal, natural gas, or some other hydrocarbon source is steam reformed into a synthesis gas mixture (H_2 and CO). This synthesis gas then is converted into a variety of different products, including methanol or ammonia by reacting the hydrogen with nitrogen recovered from air. While methanol is a very-large-volume commodity chemical, ammonia production is even larger due to the use of ammonia in agricultural fertilizers. While the types of reformers and the layout vary significantly based on feedstock, product ratios, and reformer technology, a typical layout for a synthesis gas plant making methanol from natural gas is shown in Figure 1.7.

Together, these three types of complexes form most of the conversion of raw materials into the building blocks, which support the petrochemical industry.

Many authors have tried to point to the "birth" of petrochemicals – something which has always resulted in controversy. This is because petrochemicals have existed for as long as our natural resources have existed, but were often masked

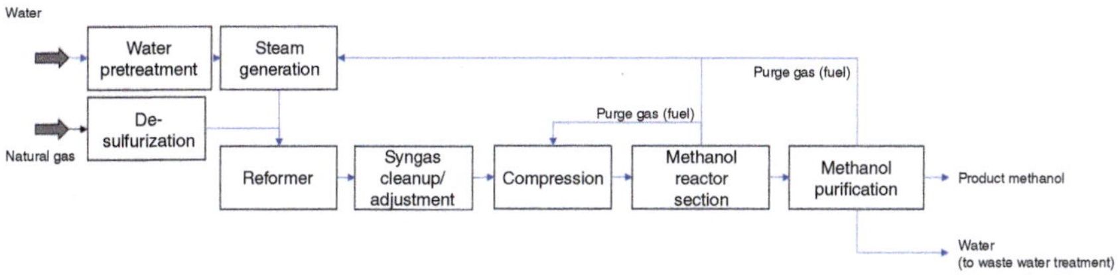

Figure 1.7 Block flow diagram of typical methanol plant.

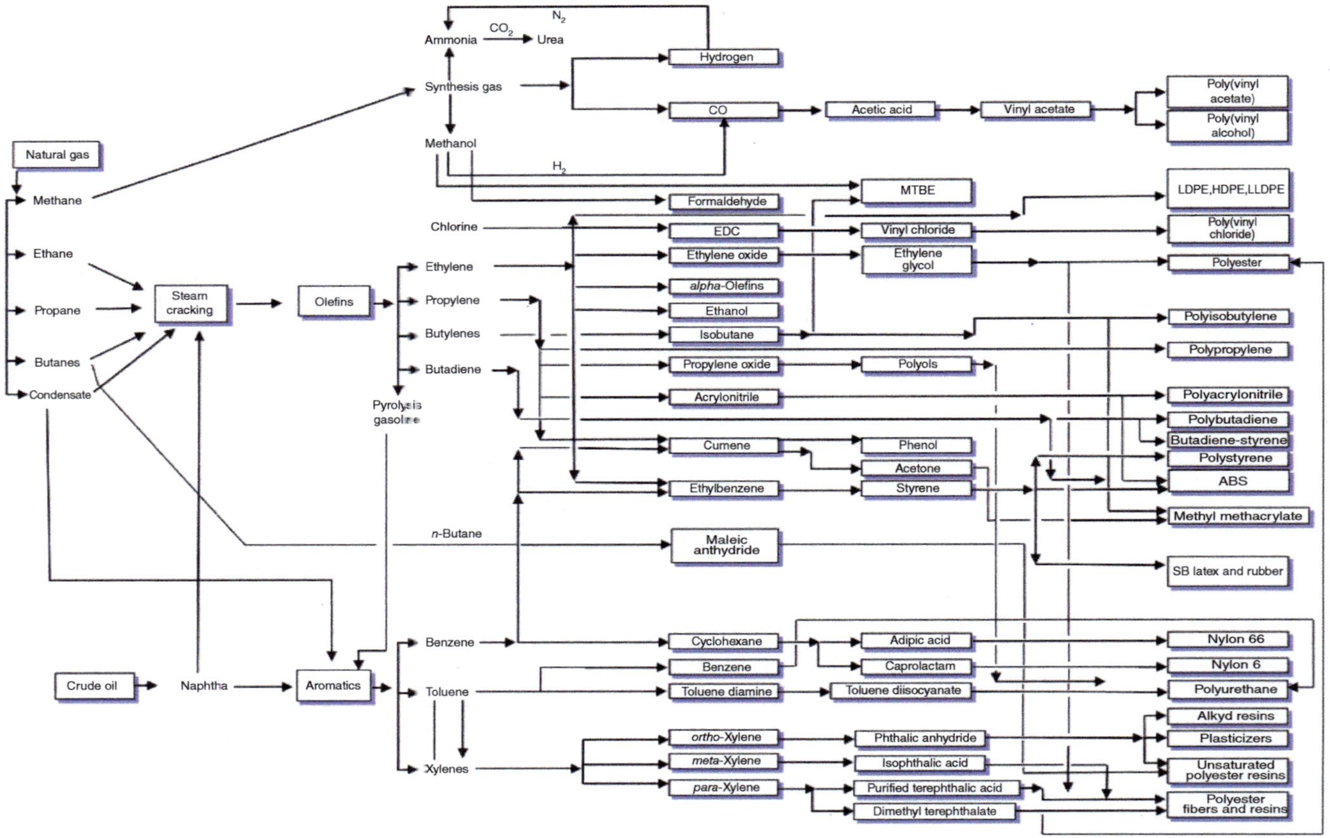

Figure 1.8 Illustrative example of products from building blocks.

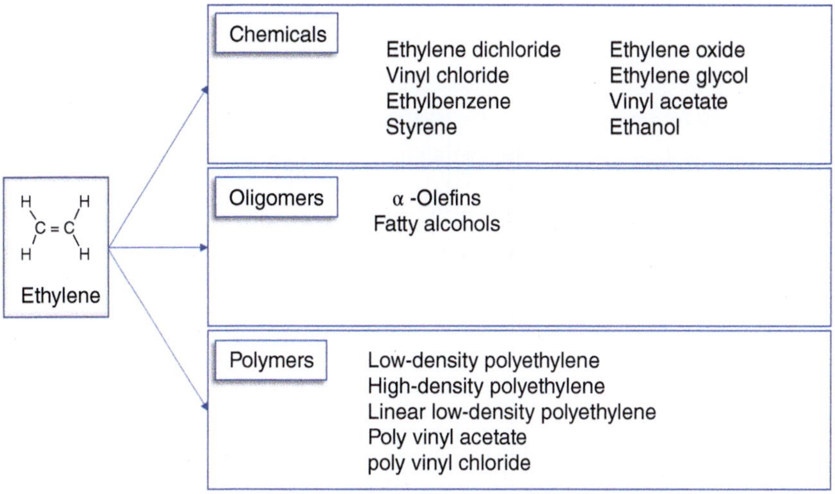

Figure 1.9 Examples of products from ethylene.

or shrouded in the complexity of the complicated mixtures they were contained in as described earlier. However, one of the first commercial complexes built for the large-scale production of a petrochemical was in the early 1900s in West Virginia. With its start in 1927, the Blaine Island plant of the Carbide and Carbon Chemicals Corporation (later known as Union Carbide and recently acquired by Dow Chemical) was the first truly integrated petrochemical plant. It produced predominantly ethylene through a process, which we now call steam cracking of hydrocarbons [13].

For any building block, intermediate or product to be useful, it must have an end use, which adds values for consumers. For example, the largest end use of ethylene (almost 70%) is in the manufacture of polyethylene. Polyethylene is broadly used in food packaging, in engineered polymers, and in many other applications. With such important and growing applications, it is no wonder that the production of ethylene has been growing at slightly higher than GDP (about 1.5× the global GDP growth rate on average from 1990 through 2017 [14]).

One aspect that makes the petrochemical building blocks particularly interesting, however, is the versatility to make multiple different products from the same molecule. This versatility is illustrated in Figure 1.8, which shows the major building blocks and only a few of the primary products, which can be produced from them. For example, the same ethylene molecule used to make polyethylene can also be used to make ethylene glycol, which is used as a coolant, as an ingredient in cosmetics and foodstuffs, and as a key component of polyester fibers and bottles.

Indeed, dozens of other petrochemicals, fine and specialty products can be derived from ethylene; the major products and further intermediates produced by ethylene are shown in Figure 1.9. The same is true for the other olefins, and for most of the other building blocks, making the evaluation petrochemical markets and economics a particularly rich and exciting field. We will discuss this in depth in Chapter 2.

References

1 Background. Naturalgas.org (archived from the original on 9 July 2014).
2 U.S. Energy Information Administration. Eia.gov (retrieved on 6 August 2013).
3 Beschloss, M. U.S. now World's leading natural gas producer. *Desert Sun* (retrieved on 4 November 2014).
4 Organic origins of petroleum. US Geological Survey (archived on 27 May 2010).
5 Speight, J.G. (1999). *The chemistry and technology of petroleum* (3rd ed., rev. and expanded ed.). New York: Marcel Dekker. pp. 215–216, 543. ISBN: 0824702174. OCLC 44958948.
6 Hyne, N.J. (2001). *Nontechnical Guide to Petroleum Geology, Exploration, Drilling, and Production*, 2e, 1–4. Tulsa, OK: Penn Well Corp. ISBN: 087814823X. OCLC 49853640.
7 American Chemistry Council, Economics & Statistics Department (July 2018). Plastics and Polymer Composites in Light Vehicles.
8 Faraday, M. (1825). On new compounds of carbon and hydrogen, and on certain other products obtained during the decomposition of oil by heat. *Philosophical Transactions of the Royal Society* 115: 440–466.
9 (2017). *CEH: Ethylene*, Chemical Economics Handbook Series. IHS Markit.
10 IHS Markit Note (11 April 2018). The tight link between methanol and olefins. https://cdn.ihs.com/www/blog/20180411-Chem-MTO-Infographic.pdf (accessed 03 September 2020).
11 IHS Markit: Methanol demand growth driven by methanol-to-olefins, China demand. *Hydrocarbon Process* (13 June 2017). https://www.hydrocarbonprocessing.com/news/2017/06/ihs-markit-methanol-demand-growth-driven-by-methanol-to-olefins-china-demand.
12 (2018). *CEH: Mixed Xylenes*, Chemical Economics Handbook Series. IHS Markit.
13 *Modern life made possible.* West Virginia Department of Commerce, Press Center. https://urethaneblog.typepad.com/my_weblog/2014/03/the-first-ethylene-cracker-in-the-us-was-in-wv.html.
14 Eskew, B. (2018). US petrochemicals: the growing importance of export markets. *IHS Markit Presentation at EIA Energy Conference* (04 June 2018). https://www.eia.gov/conference/2018/pdf/presentations/blake_eskew.pdf.

2

Petrochemical Markets

2.1 Introduction

Describing the web of production and uses of petrochemicals often evokes a number of different adjectives: complex, integrated, confusing are among the most common descriptors. In our experience, however, through the introduction and explanation of a few simple economic and technical concepts, the business end of petrochemical production and use can be quite simple and exciting. The key to unpacking this complexity is the concept of the core building blocks for modern petrochemicals, which we introduced in Chapter 1. As a reminder, these core building blocks are:

- Ethylene
- Propylene
- C4 Mono- and Di-Olefins
- Benzene
- Toluene
- Xylenes
- Synthesis Gas

We call them the building blocks of the modern petrochemical industry because, quite simply, nearly 90% of the volume of petrochemical products produced and traded are derived from these seven compounds or mixtures [1].

The building blocks are often referred to as primary or *commodity* chemicals. Simply put, a commodity chemical is often described as a chemical material made at large scale and which is used and traded globally. While it is true that the building blocks are, indeed, commodities, they are not the only commodity chemicals. So why does this definition matter? Because *commodities* have commonalities, which make them easy to *trade* – to sell between companies, and even between regions. For commodity chemicals, these commonalities are often chemical specifications – what concentration, what level of impurities, and what form (liquid, solid, solution, etc.) are the typical specifications used in the commodity chemical industry. With this set of properties to define the commodity, one company can buy the commodity from one or more different suppliers at different times and have confidence that they can use that commodity to make the end products they want to make.

Modern Petrochemical Technology: Methods, Manufacturing and Applications, First Edition.
Santi Kulprathipanja, James E. Rekoske, Daniel Wei, Robert V. Slone, Trung Pham, and Chunqing Liu.

In Chapter 1, we also briefly described how these building blocks are made. In addition to how they are made, how much and where (geographically) the building block is made is also important to understand. Collectively, this information makes up what can be called *sources of supply*, or just *source* or *supply* in short. Of course, there is no use making a chemical or building block if there is no use for it, or *demand* for the chemical. Sometimes this use is also called a sink, where the term "sink" represents a use or consumption of the chemical in question. The spatial and time balance and trade of these chemical commodities is done through *markets*. While often not physical locations like they historically have been, markets are the processes and institutions through which goods change hands. Matching sellers (supply) with buyers (demand) is the key function of the market. Often times, today, this is done electronically without the need for the buyer and seller to meet or, in some cases, even know each other. Through this process, the commodities, which have common specifications, can be traded between companies, regions, or even globally.

While not the intent of this chapter, it is important that the reader understand some basic fundamentals regarding the economic principle regarding the balance of supply and demand [2]. Briefly, the process by which supply and demand are balanced is through the market-clearing price [3]. Generally speaking, suppliers will want to supply more of a commodity if the price is high (subject to capacity and other constraints, of course). Conversely, the higher the price of a commodity, the less likely people who use the commodity (the demand side) will want to buy it and the more likely they will seek alternatives. This opposite response to the price of a commodity means that there will be a price at which supply and demand are perfectly balanced – the so-called market-clearing price. This is important because supply of and demand for chemical commodities: (i) follows these same principles; and, (ii) will fluctuate throughout time dependent upon many variables, including:

- raw material availability
- labor availability
- success or problems with production operations
- success or problems with operations, which consume the chemical
- logistics problems
- and many other factors

Figure 2.1 shows how this simple yet powerful economic theory works in an illustrative example.

Now that we have introduced the building blocks, the concept of commodity markets, and the principles of supply and demand, we can start to take a closer look at the market situation with respect to our major building blocks of the modern petrochemical industry.

2.2 The Market for Ethylene

One of the largest *volume* building blocks and commodity chemicals is also one of the smallest in molecular size – ethylene. At more than 162 million metric tons per

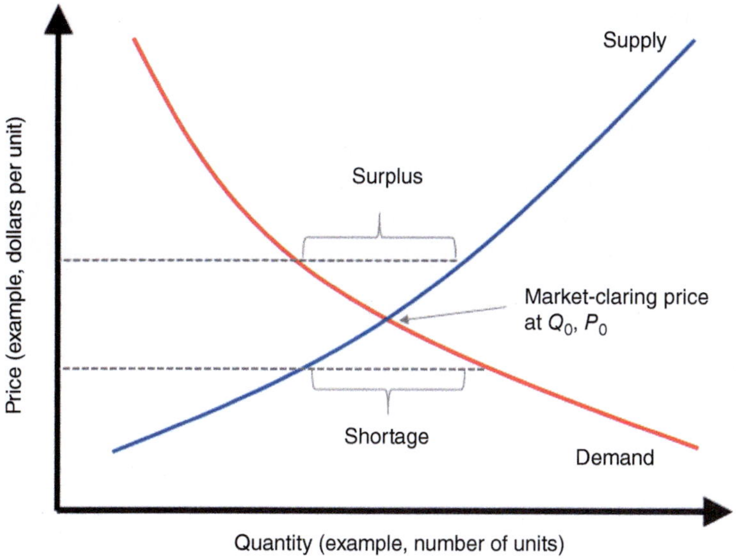

Figure 2.1 Illustration of supply, demand curves, and market-clearing price.

annum (estimated) consumption in 2019, ethylene is the largest of the hydrocarbon building blocks and second only to synthesis gas in terms of total annual consumption [1]. Ethylene is an exceptionally versatile molecule, being the starting point for hundreds, if not thousands, of final polymers and chemical compounds. Owing to the size and complexity of the markets and manufacturing processes for ethylene, production is concentrated in the hands of large, integrated refining and petrochemical companies. In 2019, it has been estimated that approximately 46% of the world's ethylene production came from assets of the top 10 producing firms [4].

Figure 2.2 shows a standard capacity-demand graphic for ethylene. The bar on the left-hand side illustrates the proportion of ethylene produced from the primary feedstocks. The vast majority (more than 95%) of the world's ethylene results from the thermal conversion of hydrocarbons through the steam-cracking process. We estimate about 40.4% of ethylene globally came from the cracking of naphtha, while 37.6% from that of ethane in 2019 [4]. The balance of the thermally derived ethylene came from LPG (Liquefied Petroleum Gas) (15.3%) and heavier hydrocarbon fractions (2.7%) [4]. Historically, naphtha has been the preferred feedstock for thermal cracking for several reasons. First, naphtha is readily available and transportable, often as the byproduct of producing heavy transportation fuels like diesel and jet fuel. Naphtha is too light to be converted into these heavier fuels and, instead, is converted into gasoline, which is typically a lower-volume demand transportation fuel outside of the United States. Second, until recently, the economic value of light naphtha was quite low. This changed with the rapid expansion of the polyester markets as light naphtha is the primary feedstock for the production of xylenes, most notably, *p*-xylene. Third, there are a number of very significant side-products formed in the thermal conversion of naphtha and heavier products

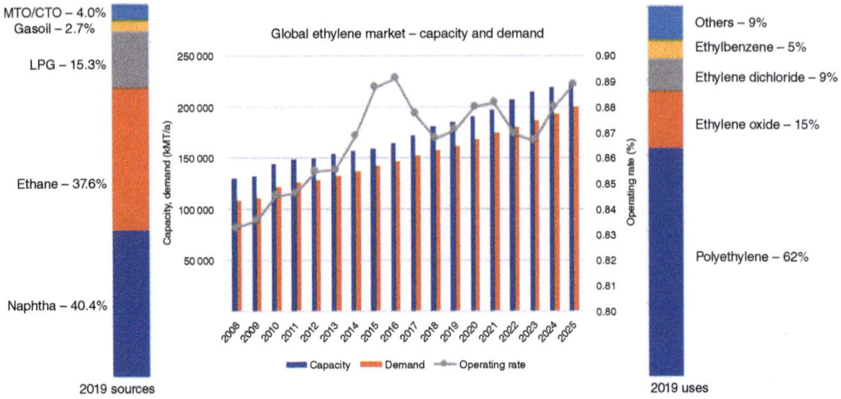

Figure 2.2 Global supply and demand dynamics for ethylene. Source: Based on data from IHS Markit and internal Honeywell UOP Estimates.

such as butadiene and benzene. Historically, these were a significant source of revenue for an olefins complex; recently, the co-product value of these chemicals has reduced, putting pressure on the economics of integrated naphtha crackers.

Ethane has always been used as a feedstock in places of abundance, such as the Middle East. It has recently returned to prominence in the United States with large ethane supplies becoming available as a result of the shale gas boom [5]. With the rapid addition of ethane-based ethylene in the United States, it is likely that ethane will equal or surpass naphtha as the predominant feedstock in the next five years [5, 6].

The remaining amount of ethylene (about 4% in 2019) is produced primarily from methanol or coal in the methanol/coal to chemicals process (M/CTO) [4]. While this technology is a small fraction of the total global ethylene production, it has been increasingly utilized in regions where coal and natural gas are more abundant than liquid hydrocarbons. Indeed, there has been a substantial growth in this nonconventional ethylene source in the past 10 years, driven by significant capacity additions in China. A significant co-product of the M/CTO process is propylene.

The middle portion of Figure 2.2 shows the capacity and demand balance (the bars) for ethylene globally. In 2019, the world's capacity to produce ethylene reached approximately 185.9 million metric tons per annum. This capacity has been increasing at average rate of about 3–4% annually for the past several years. Capacity addition is not smooth, however, as many years, even a decade, can be required to properly plan, permit, construct, and begin operation of a large ethylene complex. Indeed, the standard for a "world-scale" ethylene plant has increased to 1.0 million metric tons per annum for a naphtha-based plant or 1.5 million metric tons for an ethylene-based plant. About three to four of these plants are needed annually to keep up with the demand increases alone. When one factors in geographic growth due to abundant, cheap feedstocks (e.g. ethane in the United States) and higher demand rates due to concentrated economic growth (e.g. India and China), it is likely that between four and six plants will be added annually going forward. Clearly, both the production of ethylene and the construction of ethylene complexes are large and growing businesses.

As of 2019, the capacity to produce ethylene exceed the demand for ethylene by about 12–13%. This is evident in Figure 2.2 by the difference in the size of the bars representing capacity and demand. It is also shown in the line, which is the ratio of demand to capacity and is frequently termed the operating rate. This is the global operating rate needed for supply from the available capacity to meet the demand. As such, it does not represent the operating rate at any individual plant; it is simply a mathematical construct, which allows the balance of supply and demand. Despite being grounded only in this theoretical balance, it is a useful number to watch. Production plants of any kind, ethylene plants included, are rarely able to operate at full rates for 365 days a year – plant upsets, weather, equipment breakdowns, logistics challenges all conspire to reduce the theoretical limit to a practical limit. As such, when the global operating rate approaches a critical value, it is frequently a strong indication that more capacity – either through expanding existing facilities or building new ones – must occur for supply and demand to remain balanced.

In the chemical industry, the "bellwether" operating rate value is typically around 90–92%. When operate rates are expected to exceed this value, supply disruptions become more likely. As can be seen from Figure 2.2, the global operating rate for ethylene has been rising for most of the past decade as the demand has recovered from the global recession of 2008–2009, consuming some of the overcapacity, which resulted from a construction boom in the Middle East [7] that occurred around the same time. Our forward expectation is that the global operating rate will remain in a tight band between 87% and 90%, which is a historically high level for the operating rate. This is likely to mean regional supply pressures can occur more frequently due to plant impairments as has been experienced in recent years [8].

The right-hand side of the Figure 2.2 illustrates the uses of ethylene. Production of polyethylene in various grades dominates the demand for ethylene, with about 62% globally. A significant fraction of the polyethylene produced (over 50% by most estimates) is used in nondurable customer packaging, including food packaging and bottles; sadly, in the United States, less than 30% of this material was recycled in 2017 [9]. Single-use and nondurable uses of plastics were under significant scrutiny in 2019 [10], and the continued trajectory for polyethylene is unclear. However, the short-term offers few alternatives to the cost and convenience of polyethylene. It will be interesting to see how the environmental and the economic challenges around nondurable plastics are resolved in the coming years.

The production of ethylene for use in ethylene oxide manufacture follows at a distant 15%. Ethylene oxide has three predominant uses: as a freezing-point depressant and heat-transfer improver for mixtures in water in automotive antifreeze, in surface-active nonionic surfactants, and as a key raw material in the production of polyesters. The production of polyester represents the majority of the demand for ethylene oxide. Ethylene dichloride (9%) and ethylbenzene (5%) account for most of the remaining share of ethylene consumption, being used in the vinyls and styrenics plastics chains, respectively. The remaining demand (9%) is spread across a number of small products throughout the plastic and chemical value chains.

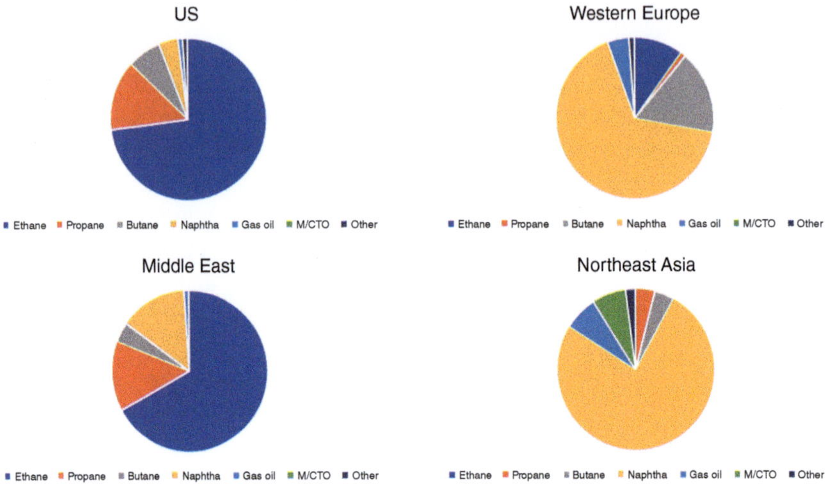

Figure 2.3 Regional variation in feedstock sources for ethylene. Source: Figure created using data adapted from IHS Markit [11].

Throughout this section, we have used global facts and figures to illustrate the features of the market for ethylene. It is important to note, however, that the regional markets for ethylene (and others of the building blocks) can be quite different from geography to geography. While it is beyond the scope of this book to describe all of these regional variations, we have constructed Figure 2.3 to illustrate the differences that can occur. Figure 2.3 compares the sources of ethylene in four large producing regions – the United States, Western Europe, the Middle East, and Northeast Asia – in 2016 [11]. The United States and the Middle East show large percentages of ethylene produced from ethane because of the availability of the feedstock. Western Europe and China show very low proportions from ethane, but quite high proportions of ethylene produced from naphtha. In addition, China shows about 7% of its ethylene produced from coal and methanol to olefins technology – a pathway not used in any of the other regions. These significant regional differences lead to a rich and varied global market.

2.3 The Market for Propylene

At almost 114 million metric tons estimated to be produced in 2019 [12], propylene is one of the top commodity chemicals produced as an intermediate for a large variety of chemicals compounds. Because of the existence of both aliphatic and olefinic bonds in the molecule, the chemical versatility of this molecule is substantial; indeed, propylene is the smallest such molecule with both key types of carbon–carbon bonds. The chemical reaction's versatility and ability to synthesize so many unique compounds stem directly from the presence of these highly varied bonds. Though not as concentrated as ethylene, more than 35% of the world's propylene production comes from the top 10 producing companies. Because of the

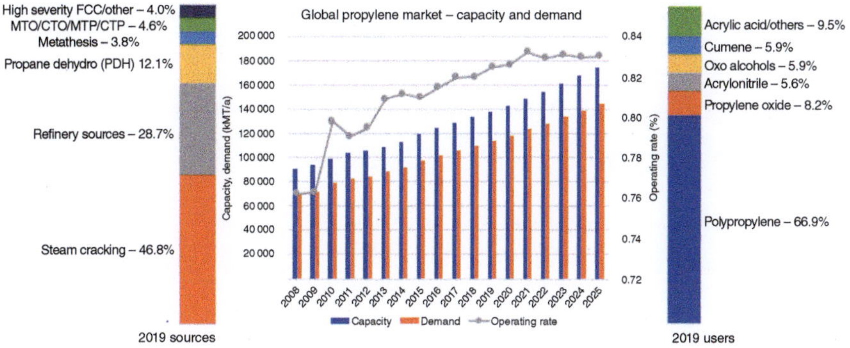

Figure 2.4 Global supply and demand dynamics for propylene. Source: Based on data from IHS Markit and internal Honeywell UOP Estimates.

complexity and size of propylene production and markets, these companies also tend to be the world's largest integrated oil refining and petrochemical companies.

Propylene is marketed typically in three major grades – minor, more specialized grades do exist – referred to as refinery, chemical, and polymer grade. They are listed in order of increasing purity. Refinery grade contain typically between 50% and 75% propylene, while chemical grade is typically 92–98% propylene. Polymer grade is always equal to or greater than 99.5% propylene, but there are multiple different impurity specifications (on methylacetylene and propadiene, for example), which alter the grades slightly. Our standard view for the capacity and demand balance for all propylene grades is shown in Figure 2.4. Starting from the left-hand bar, we can see that the sources of propylene are much more varied than the sources of ethylene. Only about 47% of the propylene produced in 2019 was through steam cracking, as described in Chapter 1 and in Section 2.2. This compares to more than 95% of the ethylene produced coming from the same process. Indeed, propylene is the primary "co-product" of the thermal conversion of hydrocarbons to make ethylene. However, as hydrocarbon feedstocks for the production of ethylene shift toward abundant ethane from heavier hydrocarbons, the amount of propylene produced as a co-product declines. This is illustrated in Table 2.1, which presents the relative production of key intermediates from the thermal conversion of different hydrocarbon species.

A significant portion of the world's sources of propylene comes from refineries. The so-called fluid catalytic–cracking process, or FCC unit, which focuses on the production of naphtha for motor fuels, also produces an off-gas byproduct, which contains a substantial quantity of propylene. From Figure 2.4, we can see that this source accounts for about 29% of the world's supply of propylene. Adding the production of propylene from FCC (byproduct of gasoline manufacture) and steam cracking (byproduct of ethylene manufacture) shows that slightly more than 75% of the world's propylene results in recovery of this value material as a byproduct of manufacturing routes primarily focused on other compounds (gasoline and ethylene). This is one manner in which the markets become complex and linked – for example, a producer who is focused on the ethylene market must also

Table 2.1 Relative amounts of byproducts, including propylene, from thermal cracking of various feedstocks.

Feedstock	Ethane	Propane	Butane	Naphtha
Feed amount	1243	2264	3361	3230
Products				
Ethylene	1000	1000	1000	1000
Propylene	20	347	376	516
Butylenes	—	64	126	320
Pyrolysis gasoline	—	177	274	730
Fuel gas	150	719	623	813

Source: Adapted from Nexant Inc. [5].

watch and make assessments of the propylene market to understand the feasibility of new capacity investment, or of increasing or decreasing operating rates.

The remainder of the source of propylene, slightly less than 25% of the world's supply, comes from processes, which are described as "on-purpose" propylene production. As the name implies, these processes have the production of propylene as their primary purpose. These processes have gained substantially in popularity in recent years, driven by a separation in growth rate between propylene and ethylene. The consumption of propylene has typically been growing at a slight multiple to the production of ethylene – for example, in recent years, propylene demand growth rates have been between 4.0% and 4.5%, while ethylene demand has been increasing between 3.0% and 3.5%. Additionally, gasoline demand has been declining, resulting in lower operating rates for FCC units and lower production of propylene. This has put pressure on propylene markets as recently as 2012–2014, resulting in the need for additional sources of propylene. These on-purpose methods are frequently seen as more expensive routes, but often simplify the business of propylene as they result in fewer (often times only one or two) products, which need to be marketed.

Most significant in the growth of these on-purpose methods has been propane dehydrogenation (also called PDH) technology, which now accounts for more than 10% of the world's production of propylene. This process starts with propane, an abundant and relatively cheap raw material, and makes only propylene as the product, thereby simplifying the business operations and diversifying both the raw materials and the production routes to propylene [13]. Significant new capacity for PDH units has been built in China using important propane from the United States and the Middle East. In addition to PDH, methanol to olefins (MTO) and coal to olefins (CTO) units produce propylene besides ethylene. This production technology has also seen a growth in adoption in China in recent years with its large reserves of coal [14].

The middle portion of Figure 2.4 shows the global capacity and demand balance (the bars) for propylene. The world's capacity to produce propylene has increased steadily in recent years between 4.0% and 4.5%, with capacity reaching approximately 138.1 million metric tons in 2019 [12]. As with ethylene, capacity addition is not perfectly smooth due to the size of the plants and the timing of construction decisions; however, the capacity curve is much smoother owing to the smaller capacity of typical propylene plants, particularly the on-purpose plants. As with ethylene, the difference in the size of the bars for the capacity to produce and the demand for propylene is accounted for by the operating rates of the production plants, shown in Figure 2.4 as the line. The operating rate for propylene has remained quite constant over the last 20+ years, from a low of about 76% during the global financial crisis of 2008–2009 to a high of about 83–85% from 2004 to 2007. Because of the demand growth and the relatively stable operating rate, the environment for new investment in propylene production has remained strong and the market has remained quite healthy over this same time.

Finally, the bar on the right-hand side of Figure 2.4 shows the approximate distribution of uses of propylene in 2019. Like with ethylene, the large majority of propylene produced (66.9%) gets consumed for the production of various grades of polymer, in this case, polypropylene. Other significant uses of propylene include the production of propylene oxide (8%), acrylonitrile (6%), and a variety of other chemical compounds, including cumene/phenol, oxo alcohols, acrylic acid, and superabsorbent polymers. While some polypropylene ends up in single-use plastics (bottle caps, closures, some food packaging), a larger percentage of polypropylene ends up in durable goods (carpets, other fibers, engineered plastics, etc.). This makes polypropylene likely to be less subjected to the scrutiny placed on single-use plastic materials than, for example, polyethylene.

The production of propylene has seen dramatic shifts in raw material uses and geographic distribution in recent years, and these shifts are likely to continue [12]. Most of these shifts have been caused by observable and predictable macroeconomic trends. As we have previously mentioned, the shift in feedstocks for the steam cracking of hydrocarbons from heavier materials (e.g. naphtha) to lighter hydrocarbons (e.g. ethane) has resulted in a lowering of the propylene produced as a byproduct of ethylene (see Table 2.1). Adding to this decline in traditional supply has been a decline in gasoline demand, which results in less propylene produced as a byproduct from the FCC unit. Occurring simultaneously has been the broad industrialization of China and the increased demand of the materials produced from propylene (e.g. polypropylene) to be available in China. The resulting increase in production of propylene in China has been enormous, as seen in Figure 2.5. Indeed, since 2013, China has overtaken North America as the region with the highest capacity to produce and, by 2021, will be nearly double the production capacity of the next largest region (see Table 2.2, [15]) – all occurring in a span of less than 10 years. Much of this capacity expansion has been enabled by the combination of increased propane supply from shale exploration in the United States as well as the recent boom in shipping availability [16]. It is these types of shifts that make the petrochemical markets dynamic, exciting, and, at times, frustrating.

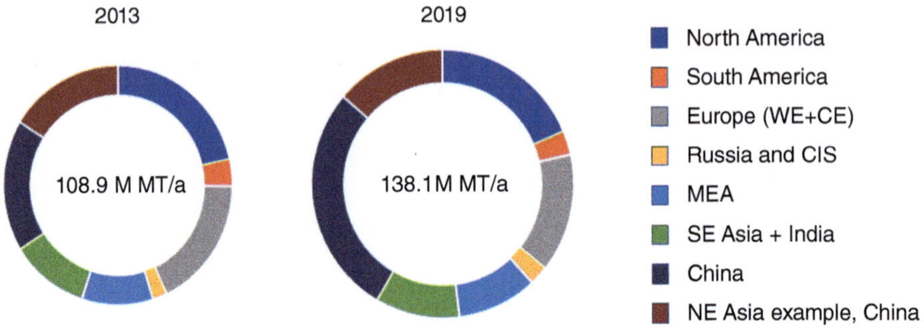

Figure 2.5 Regional distribution of propylene capacity – 2013 vs. 2019.

Table 2.2 Regional propylene capacity changes – 2013–2021.

	2013	2016	2019	2021
North America	23 500	24 100	25 600	26 100
South America	4 000	4 000	4 000	4 400
Europe	19 500	19 500	19 500	19 900
Russia and CIS	2 200	2 700	3 400	3 700
MEA	10 900	12 500	13 800	13 800
SE Asia and India	11 900	12 500	14 500	17 100
China	19 400	30 100	38 400	44 100
Other NE Asia	17 500	17 800	18 900	19 300

2.4 The Market for C4 Olefins and Diolefins

Like the markets for ethylene and propylene, which are large and have widely varied end products for which they are intermediates, the C4 olefins and dienes are produced in slightly smaller quantities but also have large and widely varying uses. However, there is one complication – while we group these C4 olefins and diolefins for convenience, there are really three separate compounds we need to consider: *n*-butenes, isobutene, and butadiene. While the end uses of the three separate compounds are quite different, the sources of these compounds are often the same or at least highly integrated as we will see in the discussion below. For simplicity, we will split the three classes of compounds into *two* distinct markets: butylenes, comprising *normal* and *iso* C4 olefins, and butadiene.

In 2019, the world's capacity to produce butylenes was approximately 80.1 million metric tons per annum, broken down into 46.3 million tons per year of *n*-butenes, and 33.8 million tons per year of *i*-butenes [17]. These butylenes are produced through three main refinery and chemical processes: FCC (as we saw with propylene in Section 2.3), steam cracking (as with both ethylene and propylene), and the direct conversion of isobutane to isobutene through the butane dehydrogenation

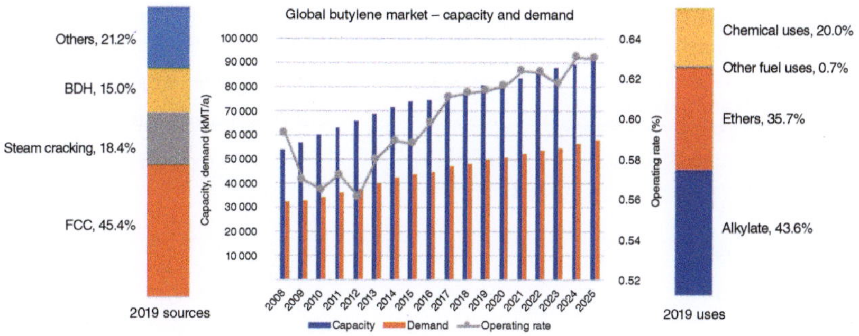

Figure 2.6 Global supply and demand dynamics for butylenes.

process (analogous to the PDH process for propylene production). In 2019, the capacity to produce butylenes was roughly estimated as 45.4% FCC, 18.4% steam cracking, 15% BDH (butane dehydrogenation), and 21.2% from a variety of other smaller chemical processes (as byproducts) [17]. This information on the capacity to produce butylenes is summarized in Figure 2.6.

The demand for butylenes is quite a bit smaller than the world's capacity to produce, with 2019 seeing a global butylene demand of approximately 49.0 million metric tons [17]. This demand is split nearly equally between normal (23.9 million) and isobutenes (25.1 million). Interestingly, we can also segment the butylene demand into demand for fuels (approximately 39.7 million metric tons in 2019) and demand for further petrochemical production (approximately 9.4 million metric tons). Butylenes are used in fuels in predominantly two different ways. The olefins can undergo alkylation or dimerization chemistry to produce a hydrocarbon product of longer chain length and a higher degree of branching. This results in a higher-octane value for the molecule, which is beneficial for gasoline internal combustion engines [18]. The second method involves reacting the olefins with an alcohol to make an ether, for example, methyl tertiary butyl ether (MTBE), ethyl tertiary butyl ether (ETBE), or tertiary amyl methyl ether (TAME). These ether fuel additives are prized for both their octane values, which are typically higher than pure hydrocarbons, but also because they add oxygen to the gasoline, which has been shown to improve combustion leading to lower levels of air pollutants [19].

In addition to fuel uses, the butylenes are used as a raw material for the production of a number of chemical compounds, including propylene (through mixed ethylene/butylene metathesis), 1-butene (for polymer additives), monomethyl acrylate (MMA), polyisobutylene (PIB), *sec*-butyl alcohol/methyl ethyl ketone (MEK), to name a few. Because the end use typically has a specific chemical requirement (for example, MTBE is made from isobutylene, not *n*-butylenes), the estimation of a capacity utilization is quite complex and depends strongly on the process and the ratio of *n*- and isobutenes produced by the process. As a result, we have shown both the fuel and chemical demand levels and the *n*- and isobutene demand breakdowns on the right-hand side of Figure 2.6 [17].

It is tempting to attempt to separate out the different butylene compounds and do a market balance on each butylene component. While this can be approximated at

Table 2.3 Butylenes produced from steam cracking and refinery sources (thousands of metric tons per annum) for selected years.

	Steam crackers			FCC		
Year	*n*	*i*	Ratio	*n*	*i*	Ratio
2000	2715	4124	0.658	11 056	7820	1.414
2005	3295	4935	0.668	9 875	8608	1.147
2010	2684	4825	0.556	8 051	8591	0.937
2015	2895	5376	0.539	12 845	9115	1.409
2020	4562	5745	0.794	14 746	9423	1.565

Source: Data and format adapted from IHS Markit [20].

the aggregate level, it is difficult to break down the specific sources of butylenes. For example, what is the number of metric tons per annum of *n*-butene produced from steam cracking? The answer to this, even at an approximate level, would require a very detailed analysis involving factors (at a minimum) such as the feedstock and capacity of all the steam-cracking units, and the extent to which the units recover the butylenes produced. This is well beyond the intention of this chapter. Instead, we can look at the aggregate production of butene isomers from the various sources and derive some conclusions.

Table 2.3 shows the reported annual production of the two primary categories of butylenes, *n*-butene and isobutene, for selected years (2000, 2005, 2010, 2015, and an estimate for 2020) from both steam-cracking operations and from refineries [20]. A couple of interesting trends can be observed. First, the amounts of C4 olefins declined from 2000 to 2010 and have since begun to rebound. This is true for both FCC and steam-cracking sources. The decline in C4 olefin production in steam cracking corresponds to the period of time of a large shift in feedstock economics favoring lighter feeds like ethane and propane. During this period, despite increasing supply of ethylene from steam cracking, the amount of C4 olefins produced declined because the feedstocks got much lighter on average. However, once all the units, which can change to lighter feedstocks, have done so, then additions of ethylene production also mean additions of C4 olefin production, and this is seen in the later years. For the FCC units, the production numbers were quite stable during the period of 2000–2010 and then began to increase substantially from 2010 to 2020. The increase since 2010 corresponds to a period of time when alkylate for fuels became increasingly valuable from refineries due to global clean fuel standards [21] resulting in more alkylate (and therefore C4 olefin) production.

Our second observation is that the ratio of normal to iso olefins resulting from the steam-cracking operation is much more consistent over the years shown than the ratio from the FCC unit. Steam cracking is a thermal process and there are limited choices to be able to adjust the ratio of normal to iso, and largest among this is the feedstock. As a result, the ratio remains much more stable as not all plants can adjust feedstock readily. As discussed above, the shift to lighter feedstocks is consistent with the slight increase in normal to iso ratio toward the end of the period shown. The ratio of normal to iso olefin produced from the FCC process from 2000 to 2020 varies widely from a low of 0.937 to a high of 1.565. There are likely multiple reasons for this observation, but one involves changes to the FCC process. The FCC process uses a catalyst, one which has been engineered over the last several decades to produce more isobutane and isobutene as the branched isomers are more valuable to the refinery for alkylation. This is most likely the reason for the significant shift from 2000 to 2010.

It is important to note that there is currently substantial overcapacity for the production of butylenes, but it is not homogeneously distributed between *n*-butenes and isobutene. The demand for *n*-butenes corresponds to a utilization of capacity of slightly higher than 50%, while the same estimate for isobutene shows a utilization of close to 75%. Despite isobutene being the less thermodynamically stable isomer, there is a stronger demand growth for isobutene due to its significant use across many compounds – alkylate, fuel additives, and chemical uses – than there is for *n*-butenes.

Although produced in much lower volume than butylenes, butadiene is a much more valuable compound used exclusively in chemical production. In 2019, it is estimated that global demand for butadiene will reach almost 13 million metric tons against a capacity of approximately 13.4 million metric tons. The production of butadiene is quite tight – meaning that producing assets operate a quite high utilization level. Indeed, the utilization of butadiene production assets runs typically in the range of 90–95% (see Figure 2.7), which is quite high for petrochemical production assets. There are several reasons for this. First, butadiene is produced almost exclusively as a byproduct of the production of ethylene in crackers; butadiene is only economically recovered from heavy feedstocks used in crackers (e.g. naphtha and heavier feeds). As a result of being a byproduct and being a sought-after material, the production of butadiene is quite lucrative, resulting in asset owners wanting to operate the assets as much as possible. Second, the crackers, which provide the source, also have typically a very high utilization ratio because these are very capital-intensive operations and return on investment requires high utilization.

While the only appreciable source of butadiene currently available is as a byproduct from naphtha and heavier feed crackers, there are multiple uses of butadiene and multiple competitive options, depending on the end market. Butadiene demand growth has been rather unsteady over the past decade with historical year-over-year growth varying from 1% to almost 5% from 2008 to 2018. Key drivers for this are the aforementioned tightness of the supply. Consumers view this tightness as risk and therefore look for alternative raw materials, which can be substituted, and this substitution creates a rather "lumpy" demand profile. Tight supply also increases the price at which people sell butadiene in the market, which opens up opportunities for competition as well. In addition, one of the key substitute products (natural rubber, see below) is an agricultural product promoted by several governments in Southeast Asia in recent years as a means for rural economy improvement. In Figure 2.7, the forward growth curve for butadiene has been estimated at about 4%

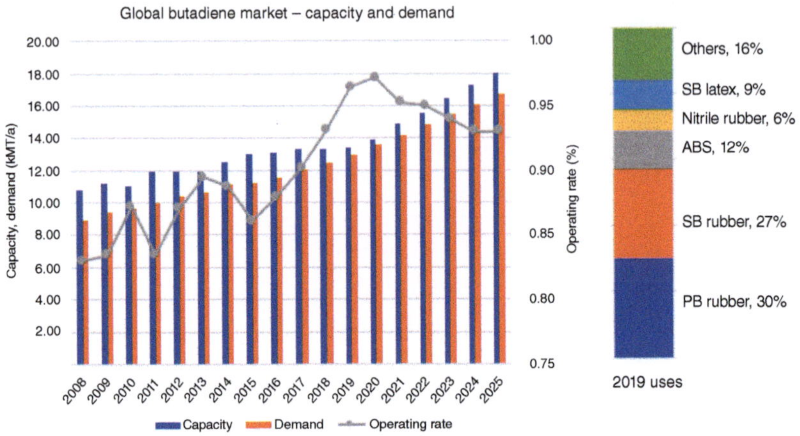

Figure 2.7 Global supply and demand dynamics for butadiene.

growth year-over-year, but this will likely vary substantially for the same economic and competitive reasons mentioned above.

Butadiene is used as raw material for a number of engineered polymers and elastomers. Figure 2.7 also shows the global demand by use of butadiene. Note that there have been substantial shifts over relatively short periods of time in the uses. Butadiene rubber and styrene-butadiene rubber (SBR) remain the largest uses, supporting 30% and 27% of the global demand, respectively, in 2019. However, over the last 10 years, the consumption toward butadiene rubber has overtaken SBR because of increased uses of butadiene rubber in high-performance tires for the automotive market [22]. Within tires, butadiene rubber is always used in conjunction with SBR and often with natural rubber, an agricultural product from the rubber tree [23]. Indeed, adjustments to the produces' formulae for production of tires often change as the relative cost of synthetic polymers and natural rubber changes. Other significant uses for butadiene rubber include footwear, wire and cable insulation, and conveyor belts [22]. Besides SBR and butadiene rubber, butadiene is key raw material for the production of a co-polymer known as acrylonitrile-butadiene-styrene, or ABS. This is rapidly growing engineered polymer used for high-performance applications such as household appliances, automotive parts and interiors, and other demanding applications.

The recovery of butylenes and butadiene from refineries and steam crackers is an elegant, complicated, and highly integrated process. Figure 2.8 shows a simplified generic flow scheme for the recovery processes resulting in butadiene, isobutylene, and *n*-butylenes and a few critical products of the C4's as well. Crude unsaturated C4 streams from steam crackers and refinery sources are shown on the left along with isobutane, which can be converted into isobutylene through a dehydrogenation process similar to that described early for on-purpose propylene production. In general, only the unsaturated C4 streams from steam crackers will have a high enough concentration of butadiene to warrant recovery, as shown in Figure 2.8. When butadiene recovery is warranted, it is passed to a butadiene extraction unit,

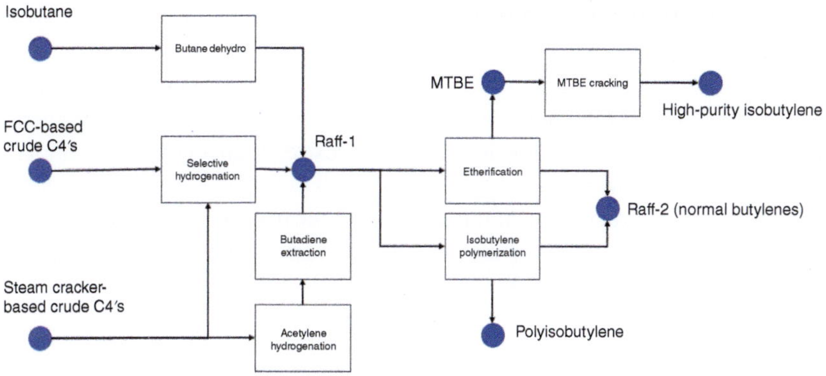

Figure 2.8 Generic recovery scheme for C4 unsaturates.

which uses solvent extraction technology to selectively absorb the butadiene, and then strip the butadiene from the solvent using heat [24]. Licensors of this process technology include BASF, McDermott, Nippon-Zeon, LyondellBasell, and others [25]. If butadiene is not desired, or not present in high enough concentration to be economically recovered, the unsaturated C4 stream is typically passed through a selective hydrogenation unit, which will convert the diene and acetylenes into mono-olefins. The effluent from the selective hydrogenation units will now look very much like the effluent from the butadiene extraction units. The industry has termed this stream "Raffinate-1" or "Raff-1." This Raff-1 stream will be a combination of isobutylene, normal butylenes, and butanes (saturated C4 compounds). It is quite different from the effluent of an isobutane dehydrogenation unit, which is predominantly isobutylene and butanes, but for the purposes of Figure 2.8, the streams can be considered together from this point forward.

The Raff-1 stream then will generally pass to an isobutylene utilization or recovery section. Isobutylene can be used to make MTBE (by condensation with methanol), which is a high-octane gasoline additive used in many countries, or to make PIB. If ultra-high purity isobutylene is desired, one can make MTBE first to recover the isobutylene from the Raff-1 mixture, and then crack the MTBE back into isobutylene and methanol. This reaction is equilibrium limited, but by separation of the isobutylene you can recycle the MTBE to extinction and have high recovery of very high isobutylene. This is frequently practiced to derive high-quality butyl rubber. Following the recovery or consumption of the isobutylene, the industry has termed this effluent stream as "Raffinate-2" or "Raff-2." This stream is now predominantly normal butylenes contained in saturated butanes and can be used further processed to make pure normal butylenes, alkylate for motor fuel applications, solvents, or other products [17]. There are a number of different options and processes, which can utilize these unsaturated C4 streams, and Figure 2.8 should be considered only one example configuration. The versatility of these unsaturated C4 streams results in a wide number of important specialty compounds with critical uses in our economy.

2.5 The Market for Aromatic Compounds

Our next "building block" is actually a number of building blocks, which encompasses carbon numbers ranging from 6 to 8 but have an important functional group, which distinguished them: all of these building blocks contain an aromatic ring. An aromatic ring, also known as an arene compound, is an organic compound, which has a planar "cyclic" structure of atoms [26]. For the purposes of our building blocks, we can further limit this to compounds containing only carbon and hydrogen, and further limit it to four distinct chemical compounds and their isomers:

- benzene, the simplest hydrocarbon aromatic, with molecular formula C_6H_6, resulting in the atoms being arranged in a planar structure;
- toluene, having formula C_7H_8, where one of the hydrogens in benzene has been replaced with a CH_3 methyl group;
- xylene, having the formula C_8H_{10}, where two of the hydrogens of benzene have been replaced with CH_3 methyl groups. There are three isomers defined by the positional relationship of the methyl groups – *ortho*-xylene, *meta*-xylene, and *para*-xylene, and
- ethylbenzene, closely related to xylene and having the same formula, but where the benzene ring is substituted one time by a C_2H_5 ethyl group instead of two methyl groups.

Each of these four is characterized as containing a ring composed of six carbon atoms in a planar structure. The first three in the list above are the primary aromatic building blocks, which we will consider in this section. We will first consider these compounds separately, and in Chapter 5 we discuss their interconversion through a process known as transalkylation.

2.5.1 The Market for Benzene

Benzene was originally a byproduct of the coking process for the steel industry [27] and is still today produced from this source. However, the market for benzene has grown large enough to support its own processes for its production, which has decoupled benzene market dynamics from those of the steel industry to a large extent. In 2019, the world's capacity to produce benzene exceed 65.7 million metric tons per annum [28], which comes from five key sources. Reformate, which accounted for about 39% of the world's benzene capacity in 2019, can be described as the benzene that is extracted from the products of naphtha reforming. Naphtha reforming is a key unit operation in the manufacture of both transportation fuels and aromatic petrochemicals like benzene, toluene, and xylenes [28]. The second-largest source of benzene capacity is pyrolysis gasoline (32%), or pygas for short, which is a byproduct of steam cracking of hydrocarbons to produce ethylene. A large amount of benzene (16%) can be produced in processes, which can be generically described as transalkylation, which is defined here as the interchange of alkyl groups among the aromatic rings. This is a key process in the efficient manufacturing of *p*-xylene [29], where as much of 50% of the *p*-xylene in a complex can be produced through the

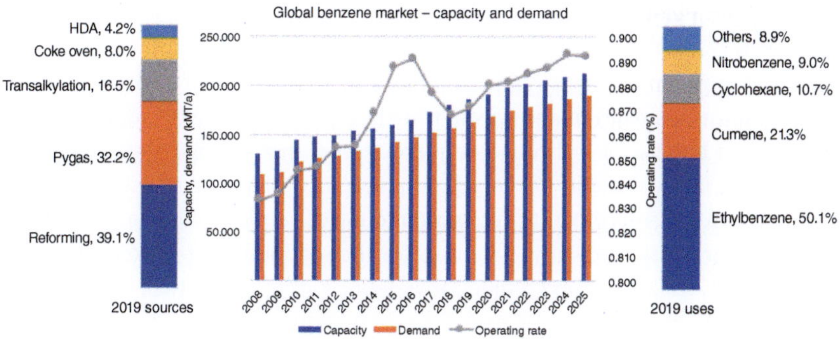

Figure 2.9 Global supply and demand dynamics for benzene.

transalkylation process. Extraction of benzene from liquids resulting from the coking process of coal used for steel manufacturing remains a significant contributor to the capacity to produce benzene (8%). The remainder of benzene production capacity is attributed to a process called hydrodealkylation (HDA), accounting for about 5% of global capacity. In the HDA process, toluene or (rarely) other alkyl-substituted aromatic compounds are converted – usually in a thermal process – into methane and benzene. This capacity usually runs intermittently as the economics of the conversion of toluene into benzene are frequently not favorable.

The fraction of global capacity production given in the paragraph above represents the total capacity amounting to a possible production of approximately 65.55 million metric tons of benzene in 2019. In reality, only about 49 million metric tons of benzene were produced. Figure 2.9 shows the sources of supply of benzene in 2019, and the inset table in Figure 2.9 shows the approximate operating rate for the five different key sources of supply (e.g. the actual production divided by the capacity to produce) in 2019. As mentioned above, the lowest operating rate is observed for the swing capacity, HDA-based benzene, at about 31%. The highest estimated operating rate in 2019 is observed for the production of benzene as a byproduct of *p*-xylene production through the transalkylation process (almost 89%). This is because, as a byproduct, production economics are less dependent on benzene alone. Indeed, the amount of benzene produced by a *p*-xylene production facility has decreased dramatically over the years driven by economics strongly favoring *p*-xylene over benzene.

The total production of benzene in 2019 was approximately 48.97 million metric tons. The production of benzene has been increasing at an average rate close to 2% for nearly a decade and this growth rate is expected to continue to the next several years [28]. Benzene is used for a wide variety of end products, including a number of monomers, which are essential for production of polymers such as styrenics, phenolics, and nylon. Almost 50% of the world's demand for benzene is driven by the need to produce ethylbenzene (through benzene alkylation with ethylene [28]); ethylbenzene is a critical precursor for the production of styrene, which, in turn, is dehydrogenated into styrene – the starting monomer for polystyrene and related styrenic materials. The next largest demand for benzene is through the production of cumene (about 21%), which is similarly produced by the alkylation of benzene with propylene [28]. Cumene is further converted into phenol and acetone through an oxidation process [30]. Phenol is a critical monomer for the production of the phenolic polymer chain, which includes phenolic resins (through condensation with formaldehyde), nylon through hydrogenation to cyclohexanone, polycarbonates, and bisphenol-A through condensation with acetone. All these plastics compounds are in wide use in household products as well as engineered materials. Cyclohexane is a key intermediate in the production of adipic acid and caprolactam, both of which are critical starting materials for the production of various grades of nylon fibers [28]. In 2019, it is estimated that about 11% of the world's benzene will be hydrogenated to produce cyclohexane for these nylon polymers. Together, these three value chains (styrenics, phenolics, and nylons) represent about 82% of the demand for benzene production. The remainder of the benzene produced is used in the synthesis of nitrobenzene (10%, intermediate in polyurethane), chlorobenzene (2%, dyes and herbicides), maleic anhydride (2%, coatings and surfactants), and other key intermediates for the vast petrochemical industry. The market for benzene is described in Figure 2.9 using our common format.

2.5.2 The Market for Toluene

The market dynamics for toluene are very interesting because while toulene is recovered in quite large quantities, it is produced in much larger quantities, which are not isolated but are instead sold or further processed. Toluene production is driven by demand for three very large commodities – gasoline, ethylene, and *p*-xylene – and is isolated and produced as a byproduct of making these three large commodities. As a result, the "capacity" to produce toluene is somewhat difficult to assess – many millions of metric tons are produced in naphtha reforming for gasoline, for example, and are never isolated and produced; this toluene, however, could be quite easily isolated for production and sale through fractionation, extraction, or some combination of both. In our analysis of capacity, we only account for facilities with demonstrated capability and practice of producing toluene. Toluene has a number of uses – ranging from high-value uses in chemicals to modest-value use as a gasoline-blending component. Indeed, much of the toluene, which is isolated, actually ends up being blended back into the gasoline market where local needs for higher octane exist [31].

Despite the complexities, the capacity to recover toluene can be estimated based on the process units, which exist at refinery and petrochemical locations around the world. Based on that analysis [31], it is estimated that the capacity to isolate toluene in 2019 is approximately 43.5 million metric tons per annum (see Figure 2.10). The capacity to produce will likely increase at a compound annual growth rate of about 3–4% over the next 5–10 years as new capacity comes onstream. Very little in the way of capacity additions are likely in the United States and Western Europe, while the majority of new capacity investment will be in China, Southeast Asia, and the Middle East as these are the regions where more toluene is needed for production of chemicals as well as blending into gasoline. The large majority of toluene is produced in naphtha reforming, about 77% in 2019, with naphtha crackers for the production

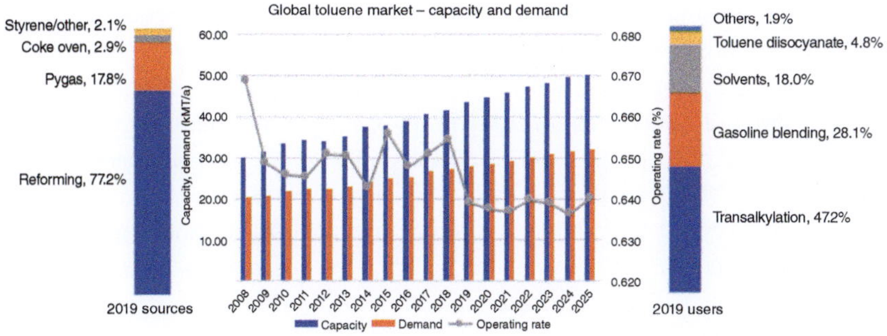

Figure 2.10 Global supply and demand dynamics for toluene.

of ethylene being the second largest source at about 18% [31]. The remainder of the produced toluene comes from minor processes like extraction from coking ovens and coal-based chemical processes.

The demand for toluene was approximately 27.8 million metric tons in 2019, growing at a rate between 2% and 3% annually [31]. This growth rate is expected to continue for the next several years as *p*-xylene demand continues to increase at about 4.5–5.5% growth per year. While toluene and the transalkylation reaction are a significant source of xylenes for the production of *p*-xylene (see Chapter 5), most of the new complexes being developed globally are moving toward net-zero isolation of toluene [32], resulting in a slower growth rate for toluene demand. Also, large-scale blending of toluene for gasoline property (vapor pressure and octane) improvement is already practiced, and this has two significant consequences. First, the demand growth for gasoline blending will be muted as most regions are now using toluene, leaving growth equivalent of the overall gasoline demand growth only. Second, this growth in gasoline blending has set a floor or minimum price for toluene in the market as producers can simply sell toluene for gasoline blending. This has closed off a traditional market of *p*-xylene production from toluene and mixed xylenes. Complexes used to be able to make acceptable margins by purchasing toluene and mixed xylenes on the market and converting them into *p*-xylene. In the past 10 years, this has become a very difficult and low-margin business.

Toluene is isolated and produced for three basic reasons: the production of benzene and xylenes (BX; usually related to *p*-xylene production); the production of chemicals from toluene; and, gasoline blending. The largest demand of these three is for the production of benzene and xylenes (BX), with demand reaching about 13.1 million metric tons in 2019 [31]. The next largest demand is for blending into gasoline, which reached approximately 7.8 million metric tons in 2019 [31], though demand growth will slow in the coming years as indicated above. The final category of toluene demand, chemicals production from toluene, includes a diverse set of compounds, including toluene diisocyanate, benzoic acid, and common and specialty solvents. This category accounted for about 7.0 million metric tons of toluene demand in 2019 [31]. Over the next decade, it is likely that the highest demand growth will remain aligned with the demand for *p*-xylene, while growth of toluene-derived chemicals will start to approach a similar growth rate.

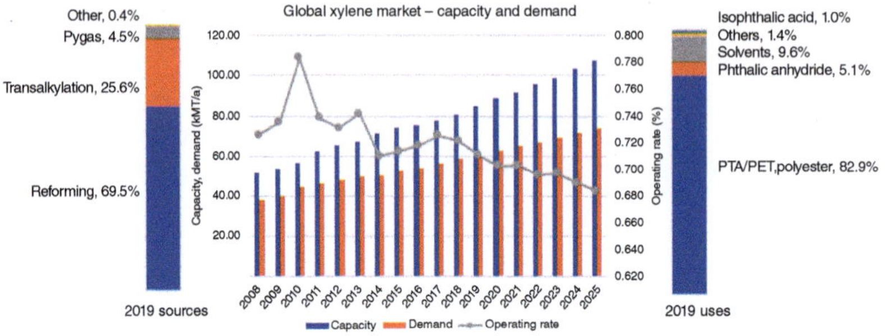

Figure 2.11 Global supply and demand dynamics for xylenes.

2.5.3 The Market for Xylene

Mixed xylenes are a key intermediate in the production of several polyester compounds and have grown to become a very significant part of the petrochemicals businesses. Because of the rapidly growing demand for polyethylene terephthalate, in particular, the market for mixed xylenes has exploded by nearly a factor of 3 since 2000 [33]. All three of the xylene molecules have end uses in plastic materials, but more than 83% of the mixed xylenes end up as *p*-xylene [33]. However, mixed xylenes typically contain about 23% *p*-xylene at equilibrium, and more than 50% *m*-xylene – for which the smallest market exists. As a result, much of the product of mixed xylenes involves extracting xylene isomer of interest (usually *p*-xylene) and then re-equilibrating the *p*-xylene depleted stream to approach the equilibrium distribution again [34]. This isomerization-separation loop is the basis for a significant growth in the polyester industry and has become quite energy and material efficient over the last three decades of research and development [34].

The world's capacity to produce mixed xylenes is estimated to be approximately 84.7 million metric tons in 2019 and has been growing at a compound annual growth rate of about 4.0–4.5% for several decades (see Figure 2.11). As consumer use of polyester materials, PET (polyethylene terephthalate) bottles, and other plastics involving polyesters has grown, the capacity to produce mixed xylenes – the raw material source for these polyester materials – has continued to keep pace. However, the majority of the new capacity to produce mixed xylenes has and will continue to be added in Asia – specifically, China, Korea, India, and, to a lesser extent, Southeast Asia. Indeed, since 2007, more than 94% of the capacity added to produce mixed xylenes has been in Asia [29]. The underlying trend in the polyester chain has been one of growth of production in China, and the development of a supply chain of *p*-xylene and mixed xylene to serve that market development has been a highly regional story. Indeed, while China had a capacity to produce mixed xylenes of around 5 million metric tons in 2006, capacity has grown to nearly 22 million metric tons in 2019 [33]. That being said, there have been some developments outside of Asia, particularly in the Middle East, which will continue as resource owners look to increase the value they gain from bringing those raw materials to market.

Mixed xylenes come from a variety of sources, the most abundant of which are naphtha reforming, pyrolysis gasoline from naphtha crackers, transalkylation processes, and other sources, including coal-derived chemicals [33]. Almost 70% of the world's supply of mixed xylenes comes from naphtha reforming, with much of this being on-purpose or petrochemical-oriented naphtha reforming. Some continue to be extracted from motor fuel reformers, though this is becoming less and less attractive due to the demands on octane in motor fuels. Only about 4–5% of the mixed xylenes are supplied from naphtha crackers, and this percentage is likely to decline over the next decade [33]. The increased demand for xylenes, and the shift to lighter (ethane) feedstocks in crackers, virtually assures the importance of pyrolysis gasoline will decline over the next decade. It will, however, remain very relevant from specific geographic regions. About 25–30% of the xylenes are supplied from the various transalkylation processes, which involve moving alkyl groups, specifically methyl groups, across the aromatic rings. This will remain roughly steady, and will grow with the naphtha-reforming capacity; in a modern xylene complex today, it is very rare to build reformers without a transalkylation unit.

As mentioned in an earlier paragraph of this section, the great majority of mixed xylenes produced (83%) are used to make *p*-xylene, which, in turn, is used to make purified terephthalic acid and, ultimately, polyester products (fibers, films, and bottles). Slightly more than 5% of the demand for mixed xylenes is for the production of *o*-xylene, a key raw material in the manufacture of phthalic anhydride (PA). PA is uses in many PVC (polyvinyl chloride)-based products as a plasticizer. About 1% of the demand for mixed xylenes is driven by *m*-xylene, despite the fact that this is the isomer available in highest quantity when the xylenes are at equilibrium. *m*-Xylene is used to add ductility and impact resistance to plastic bottles [33]. All of this means the demand for mixed xylenes has been growing 3.5–4.5% per year and will continue to do so as long as polyester demand growth continues. In 2019, the demand for mixed xylenes was approximately 60.2 million metric tons per annum [33], with about 60% of this mixed xylene being consumed in Northeast Asia.

2.6 The Market for Synthesis Gas

Synthesis gas is a widely used fuel-gas mixture, which is primarily composed of carbon monoxide and hydrogen [35, 36] and has primary uses in power generation, synthesis of liquid fuels, use as a gaseous fuel, separated and used as industrial gases, and for the synthesis of chemicals. Our primary focus in this section is on the use of synthesis gas for chemical synthesis, which is slightly larger than 50% of the overall synthesis gas market in 2019. While there are many different chemicals that are derived from synthesis gas, there are two chemicals that make up the vast majority of the synthesis gas chemicals industry – methanol and ammonia. We focus our discussion on these two chemicals in Sections 2.6.1 and 2.6.2.

2.6.1 The Market for Ammonia

Nitrogen is an essential element for all life as we know it – both plants and animals – and is used as a fundamental building block or catalyst by most biological organisms, providing fuel for growth. In essence, ammonia therefore fuels human, animal, and plant growth on the plant to a large extent. With the population of the planet expected to reach 9.8 billion people by 2050, it is estimated that the food supply will need to increase by 70% from 2016 levels to provide the appropriate nutrition [37]. An essential path to closing this gap is agricultural productivity, which is enabled by higher levels of fertilizer and soil quality, which plays directly to the increased demand for ammonia. While nitrogen is captured through air separation, natural gas or coal are the primary sources of hydrogen for the synthesis of ammonia. The emergence of a coal-to-chemicals industry in China and favorable economics in North America driven by cheap natural gas created a strong environment for ammonia growth beginning in the mid-2000s. The last decade has been one of strong increase in capacity and demand, driven by these strong economics.

The capacity to produce ammonia had been quite stagnant for some years, and then underwent a renaissance in the mid-2000s. In 2004, for example, capacity to produce stood at about 156.8 million metric tons per annum; by the end of 2019, that number reached nearly 255 million metric tons of capacity [37, 38]. Capacity has increased at a growth rate of about 3.0–3.5% over this period of time, and is expected to increase by the same or higher growth rate for the next 5–10 years. Figure 2.12 shows the typical capacity and demand balance for ammonia. Natural gas is used as the hydrogen source for about 70% of the world's capacity of ammonia [36, 37]. As a consequence, ammonia synthesis had broadly moved to areas with cheap natural gas such as North Africa and the Middle East. The shale gas revolution in North America has changed this landscape, with US-based shale gas being among the lowest-cost natural gas in the world. Construction of a number of ammonia and ammonia-based chemicals plants has been started since 2008 and more are planned. In spite of this increase in capacity in the United States, China has maintained its position with about one-third of the capacity to produce ammonia from the late-2000s through to today. Going forward, it is expected that the proportion of capacity in China will drop, driven by investment in the Middle East and North

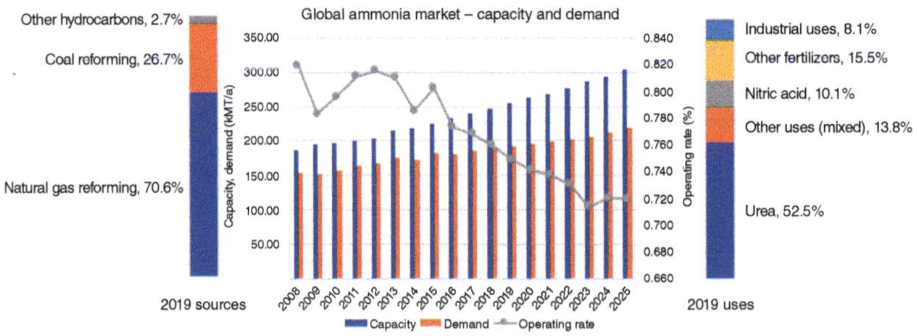

Figure 2.12 Global supply and demand dynamics for ammonia.

America. Most of China's ammonia capacity is based on coal as the hydrogen source, and this capacity is coming under more and more environmental scrutiny.

The demand for ammonia has increased steadily across the last two decades at about 2.0–2.5% with demand reaching 190.9 million metric tons per annum in 2019 [37, 38]. About one-third of the demand is driven by China, consistent with its capacity – ammonia and ammonia-derived chemicals is one area where China has achieved a strong level of self-sufficiency. At least 81% of the ammonia demand is driven by fertilizers, which should be connected to population growth and increased nutritional needs [37]. Slightly more than 50% of the ammonia produced ends up as urea products for use directly or indirectly as fertilizers [36, 37]. After urea, there are a number of smaller ammonia-based products like phosphate, nitrate, sulfate, etc., which share fertilizer and other markets as end-use applications. There are dozens of other applications for ammonia, from refrigerants to cleaners to fermentation feed, but the consumption for these uses is minor [39].

2.6.2 The Market for Methanol

Methanol has moved from a relatively modest market to a large, diverse, rapidly growing chemical market over the last decade or so. With a production capacity of approximately 28.9 million metric tons in 1995, methanol grew modestly to about 39.5 million metric tons in 2004 for a compound annual growth rate of about 3.5% [40]. In 2019, the capacity to produce methanol had increased to more than 124.1 million metric tons per annum, an annual compound growth rate of greater than 10% since 2004 [41]. The methanol industry is one of the most fragmented of our building blocks, meaning that only 28% of the world's capacity is controlled by the top 10 producers [40]. Geographically, almost 54% of the capacity to produce methanol is within Northeast Asia, which, in this case, is almost exclusively China. This boom in capacity addition has been in response to a number of new uses for methanol, but has also resulted in a lower operating rate compared to the historical norms (see Figure 2.13).

The story around methanol demand over the past two decades is a very complicated one with many different drivers as new uses and applications for methanol have developed. In the late 1990s, the major drivers for methanol consumption

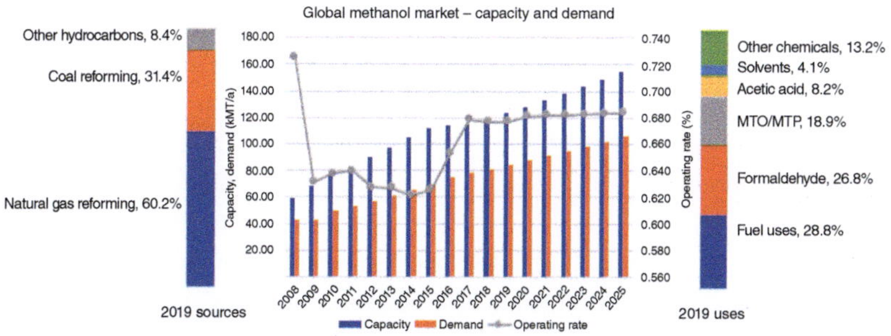

Figure 2.13 Global supply and demand dynamics for methanol.

were formaldehyde and a fuel oxygenate additive called MTBE at about 7.7 million metric tons and 5.9 million metric tons, respectively. Formaldehyde is used to produce resin products for the wood products and building products industry [40]. Since the late 1990s, despite a ban on MTBE use in the United States, the fuel usage of methanol has grown dramatically reaching 25.7 million metric tons in 2019 [42]. This comes in the form or MTBE, another ether called tertiary amyl methyl ether (TAME), dimethyl ether, and direct fuel blending; in 2019, direct fuel blending into gasoline had reached almost 7 million metric tons, from 0 in the late 1990s [41]. In addition, a new market for methanol – converting methanol into ethylene and propylene – has emerged in China, consuming over 13 million metric tons of methanol in 2019 [41]. In addition, formaldehyde use has continued to grow, reaching 22.6 million metric tons per annum in 2019. Considering all these factors, and the ups and downs of various different end uses, the demand for methanol is estimated to be just slightly above 84 million metric tons in 2019 [41]. Since 2008, the demand has nearly doubled in 11 years for a 6.4% annual growth rate.

Looking at the split of demand by use in 2019, we find that about 27% of the demand is driven by formaldehyde and building products end uses, which is the largest single chemical use. However, more than 28% of the methanol produced ends up in the fuels sector through MTBE and other fuel ethers, direct gasoline blending, and the biodiesel industry. The fastest growing use, and third largest, is the MTO application (about 19%), where methanol is converted to ethylene and propylene (see Chapter 3). The remainder of the methanol is used in acetic acid (8%), solvents (4.1%), and many other small chemical compounds [42]. This is summarized in the capacity-demand figure for methanol in Figure 2.13.

References

1 IHS Markit (15 August 2019). CEH Petrochemical Industry Overview.

2 The authors recommend Pindyck, R.S. and Rubinfeld, D.L. (2005). *Microeconomics*, 6e. Pearson-Prentice Hall for a complete description.

3 Pindyck, R.S. and Rubinfeld, D.L. (2005). *Microeconomics*, 6e, 24. Pearson-Prentice Hall.

4 (2017). *CEH Ethylene*. IHS Markit; internal estimates from Honeywell UOP.

5 Nexant Inc. (2014). The global petrochemical industry. Presentation, Amsterdam (7 April 2014).

6 (2017). *CEH Ethylene*, 18. IHS Markit; internal Honeywell UOP estimates.

7 (2017). *CEH Ethylene*, 98. IHS Markit (Middle East Capacity for Ethylene graphic).

8 Baumgarten, S. (17 October 2016). BASF confirms second death in Germany explosion. https://www.icis.com/explore/resources/news/2016/10/17/10045568/basf-confirms-second-death-in-germany-explosion/ (accessed 04 September 2020).

9 Toto, D. (17 December 2018). Plastic bottle recycling rate declines in 2017. https://www.recyclingtoday.com/article/2017-plastic-bottle-recycling-rate/ (accessed 04 September 2020).

10 Loepp, D. (30 May 2019). Under pressure, plastics find place in the global circular economy. https://www.plasticsnews.com/blog/under-pressure-plastics-find-place-global-circular-economy (accessed 04 September 2020).

11 *CEH Ethylene*. IHS Markit.

12 (2017). *CEH Propylene*. IHS Markit and internal Honeywell UOP estimates.

13 Myers, R.A. (2018). *Handbook of Petrochemical Product Processes*, 2e. McGraw-Hill, Section 10.3.

14 EChemI Website (12 July 2018). 2018 China newly added methanol and MTO/CTO capacity statistics. http://www.echemi.com/cms/13515.html (captured 08 November 2019).

15 (2017). *CEH Propylene*. IHS Markit.

16 Rekoske, J.E. (2016). Technoeconomic impacts of abundant natural gas liquids on the chemical industry. *Presentation at the 11th Natural Gas Conversion Symposium*, Tromso, Norway (6 June 2016). Natural Gas Conversion Board: Washington, DC.

17 (2017). *CEH Butylenes*. IHS Markit and internal Honeywell UOP estimates.

18 U.S. Department of Energy, Energy Information Agency (13 February 2013). Alkylation is an important source of octane in gasoline. https://www.eia.gov/todayinenergy/detail.php?id=9971 (captured 08 November 2019).

19 U.S. Department of Energy, Energy Information Agency Document. MTBE, oxygenates and motor gasoline. https://www.eia.gov/outlooks/steo/special/pdf/mtbe.pdf (captured 08 November 2019).

20 *CEH Butylenes*. IHS Markit.

21 Kranz, K., duPont STRATCO Alylation Training. (September 2008). Intro to alkylation chemistry. https://silo.tips/download/intro-to-alkylation-chemistry.

22 Encyclopedia Britannica. Butadiene rubber. Online Edition. https://www.britannica.com/technology/butadiene-rubber (captured on 23 October 2019).

23 Sethurai, M.R. and Mathew, N. (eds.) (1992). *Developments in Crop Science: Volume 23, Natural Rubber*. Elsevier Science.

24 White, W.C. (2007). *Chemico-Biological Interactions* 166 (1–3): 10–14.

25 Nizamoff, A.J. (2013). On-Purpose Butadiene. Nexant ChemSystems PERP Report.

26 Carey, F.A. and Sundberg, R.J. (1990). *Advanced Organic Chemistry Part A: Structure and Mechanisms*, 499. New York: Plenum Press.

27 Spitz, P.H. (1988). *Petrochemicals: The Rise of an Industry*, 42. New York: Wiley.

28 (2017). *Benzene World Analysis*. IHS Markit.

29 (2018). *Mixed Xylene. World Economics Handbook*, 15. IHS Markit.

30 Matar, S. and Hatch, L.F. (1994). *Chemistry of Petrochemical Processes*, 265. London: Gulf Publishing.

31 (2018). *CEH Toluene*. IHS Markit and internal Honeywell UOP estimates.

32 Internal Honeywell UOP analysis.

33 (2018). *Mixed Xylene. World Economics Handbook*. IHS Markit, and internal Honeywell UOP estimates.

34 Myers, R.A. (2018). *Handbook of Petrochemical Product Processes*, 2e. McGraw-Hill, Section 13.3.

35 Chauvel, A. and Lefebvre, G. (1989). Hydrogen, synthesis gases and their derivatives. *Petrochmical Processes, Part 1: Synthesis Gas Derivatives and Major Hyxrocarbons*, 27. Paris: Editions Technip.

36 Speight, J.G. (2019). *Natural Gas, A Basic Handbook*, 2e. London: Gulf Professional.

37 (2017). *CEH Ammonia*. IHS Markit.

38 Internal Honeywell UOP estimates.

39 Appl, M. (1999). *Ammonia: Principle and Industrial Practice*. Wiley-VCH Verlag GmbH.

40 (2017). *CEH Methanol*. IHS Markit.

41 (2017). *CEH Methanol*. IHS Markit and internal Honeywell UOP estimates.

42 Olah, G.A., Goeppert, A., and Prakash, G.K.S. (2018). *Beyond Oil and Gas: The Methanol Economy*, 3e. Wiley-VCH.

Section II

Olefins and Synthesis Gas

3

Olefins and Synthesis Gas Production Technologies

3.1 Introduction

Olefins are key building blocks of chemical industry. Light olefins, ethylene and propylene, are the most important olefins used in the manufacture of a large number of chemicals in commodities and specialties such as polyethylene, ethylene dichloride, ethylene oxide, polypropylene, propylene oxide, and acrylonitrile. Butylene is also in high demand for further processing to produce methyl-*tert*-butyl ether (MTBE) and alkylate for gasoline blending. Heavy normal-olefins (C_{10}–C_{14}), derived from straight run kerosene, are the main feedstock component of linear alkylbenzene (LAB), the most common raw material in the manufacture of biodegradable household detergents. In this chapter, the manufacturing technologies having commercial significance for producing these olefins will be discussed, including steam cracking, paraffin (light paraffin and heavy normal-paraffin) dehydrogenation, and methanol to olefins. There are other processes in a refinery, e.g. fluid catalytic cracking (FCC), that produce light olefins as byproducts. These technologies will not be discussed in this chapter.

Synthetic gas (syngas), a mixture of carbon monoxide and hydrogen, is used in large volumes for the production a wide variety of commodity chemicals such as hydrogen, methanol, ammonia, olefins, and MTBE. Hence, it represents a suitable alternative to produce base chemicals from resources other than crude oil. Syngas manufacturing technologies discussed in this chapter include conversion of natural gas to syngas and coal gasification to syngas.

3.2 Steam Cracking

In the steam-cracking process, a hydrocarbon feedstock is mixed with steam and cracked at elevated temperatures in a tubular reactor. It can accept a variety of hydrocarbons, ranging from gases (ethane, propane, and butane) to liquids (naphtha and gas oils), as shown in Figure 3.1. The reactor effluent gases are first liquefied and then separated by distillation columns as ethylene, propylene, a crude C_4–fraction, pyrolysis gasoline, and heavies.

Modern Petrochemical Technology: Methods, Manufacturing and Applications, First Edition.
Santi Kulprathipanja, James E. Rekoske, Daniel Wei, Robert V. Slone, Trung Pham, and Chunqing Liu.

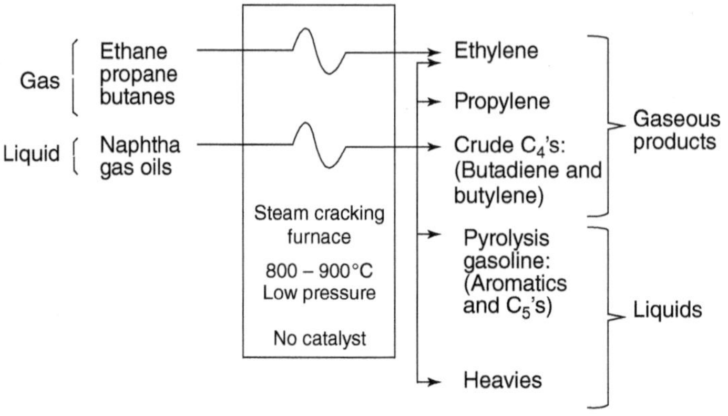

Figure 3.1 Steam-cracking technology.

There are three steps in the steam-cracking process: pyrolysis and cooling, compression and acid gas removal, and cryogenic cooling and product separation.

Pyrolysis and cooling: The feedstock is initially mixed with a dilution steam and heated to 815–900 °C at 1.7–2.4 bars. The resulting products are immediately cooled to 340–510 °C and further cooled to 37–43 °C in a water quenching tower.

Compression and acid gas removal: The cracked gas stream is compressed from atmospheric pressure to 35–38 bars. During the compression, acid gases (H_2S and CO_2) are removed by scrubbing the stream. The resulting stream is dried over a solid adsorbent.

Cryogenic cooling and product separation: The dried gas is cooled to between −90 and −130 °C in a refrigeration train. The resultant gas, consisting primarily of hydrogen, methane, and carbon monoxide, is purified to recover hydrogen; the remainder is burned as plant fuel. Low-temperature fractionation separates the dried products. Ethane and propane are generally recycled, while hydrogenation of acetylene, propyne, and propadiene yields ethylene and propylene.

For the conventional endothermic cracking process, ethylene yield is improved by raising the cracking temperature and reducing residence time. The operating conditions are constrained by the metallurgy of the furnace tubes and coking tendency in the cracking coils. The formation of coke deposits in cracking furnace coils remains one of major obstacles in steam-cracking operations. Coke formation and subsequent accumulation in furnace components result in significant energy and maintenance costs. Typically, a naphtha pyrolysis furnace is decoked every 15–40 days. Maximum cycle time is 60–100 days.

Ethylene fractionation separates ethylene as a high-purity product (>99 wt%) from ethane, which is combined with propane and recycled for cracking. Propylene is separated from propane in the propylene fractionation section as a chemical-grade product (typically 93–95 wt% min) or more frequently as a polymer-grade propylene (>99.5 wt%).

A mixed C_4 stream is produced from the cracking of naphtha and heavy feedstocks providing an important source for butadiene and butylene. More than 95% of world

Table 3.1 Typical products from cracking various feedstocks in a 450 KMTA ethylene plant (KMTA).

Products	Ethane	Propane	*N*-Butane	Naphtha	Vac gas oil
H_2-Rich gas	30	20	15	15	306
CH_4-Rich gas	40	300	250	220	200
Ethylene	450	450	450	450	450
Propylene	10	290	190	260	260
Butadiene	10	30	40	80	80
Butylenes/butanes	5	20	80	130	80
Pyrolysis gasoline + BTX + others	20	140	160	990	1200
Fuel oil	0	10	30	300	610
Ethylene yield (wt%)	81	43	40	37	24
Total	575	1260	1215	2445	3186

supply of finished butadiene is extracted from the mixed C_4 produced in the steam crackers. The production of mixed C_4 stream is strongly dependent on cracking severity and feedstock specification. Typically the amount of mixed C4 produced per ton of ethylene will decrease when running the cracker at higher severity.

The selection of feedstock for steam cracking is considered as one of the most important issues for a petrochemical complex. Steam crackers are designed for feed flexibility so that they can process a predefined range of feedstocks. Product dependence on feedstock is shown in Table 3.1.

An ethane cracker is preferred when the feedstock is available in sufficient amounts. As a result of shale gas development in the United States producing high abundance of ethane, ethane cracking has become more attractive due to its low cost, but byproduct credits from naphtha-based crackers in Europe and Asia could be enough to keep them profitable maintaining a large share of the olefin market [1].

In a naphtha-based cracker, the feedstock contains normal-paraffins, isoparaffins, naphthenes, and aromatics. Ethylene and other olefins are formed primarily from paraffins and naphthenes in the feedstock. Normal-paraffins are the preferred components for high ethylene and propylene yields. Isoparaffins produce much smaller yields of ethylene and propylene than normal-paraffins.

An option to maximize the profitability from a naphtha-based cracker is to integrate it with a refiner by using UOP MaxEne™. The MaxEne process can simultaneously increase the yield of high value products from a naphtha-based cracker and catalytic reforming unit. This process is designed to extract normal-paraffins from full-range naphtha, consisting of molecules in the C_6 to C_{10} range. The separation uses a simulated countercurrent contacting scheme, in which normal-paraffins are first adsorbed in a series of fixed beds of adsorbent and then desorbed using a desorbent. The extract stream with high-purity normal-paraffins makes a superior

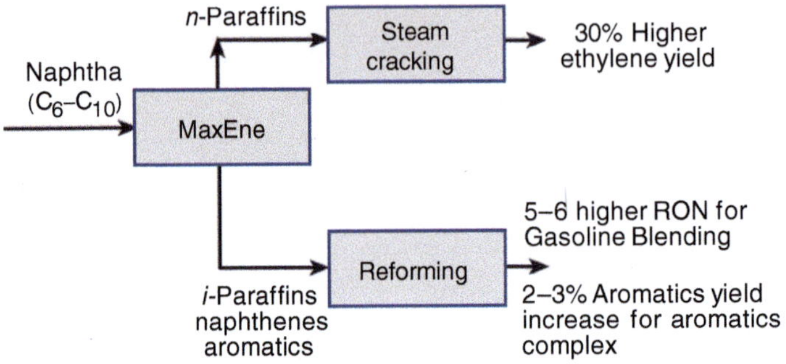

Figure 3.2 UOP MaxEne process.

steam cracker feed, while the raffinate stream enriched with naphthene/aromatic is optimal as a reforming feed (Figure 3.2). The process enables:

- 30% ethylene yield increase from existing naphtha cracker with no loss in propylene,
- 5–6 RON increase of reformate from reforming unit to gasoline pool, and
- 2–3% aromatics yield increase for reformate from reforming unit to aromatics complex.

3.3 Light Paraffin Dehydrogenation

3.3.1 Process and Catalyst

The majority of propylene and butylene are produced as byproducts of petroleum refineries (FCC) and steam crackers. Thus, most of propylene and butylene are byproducts of other products, specifically gasoline and ethylene. However, when production capacity is not coupled with demand for those byproducts, a supply/demand imbalance can occur.

The paraffin dehydrogenation process provides petrochemical producers with a catalytic, on-purpose means for making propylene or butylene from their corresponding paraffin, independent of demand for gasoline and ethylene.

Paraffin dehydrogenation for the production of light olefins has been in use since the late 1930s. Catalytic dehydrogenation of butanes over a chromium catalyst was practiced to produce butylenes, which were then dimerized to octenes and hydrogenated to octanes to yield high-octane aviation fuel. Dehydrogenation of butanes over a chromium catalyst was first developed and commercialized at Leuna in Germany and was also independently developed by then Universal Oil Products (UOP) in United States, together with ICI in United Kingdom.

In the dehydrogenation process using chromium catalysts, the catalyst is contained in a fixed-bed reactor. A significant amount of coke is deposited on the catalyst during the dehydrogenation; therefore, a number of reactors are used in parallel – some for dehydrogenation while the rest are being purged or regenerated.

In the late 1980s, the application of chromium catalysts with cyclical operation was extended by Houdry to the dehydrogenation of propane to propylene and isobutane to isobutylene. The new process application was called Catofin™ [2].

In 1959, a fluidized-bed dehydrogenation technology with chromium catalyst was developed and commercialized in the former Soviet Union. This reactor design avoids the cyclical operation by circulating the catalyst from the reactor to the regenerator. Circulating regenerated catalyst is used to provide the heat of reaction in the riser, while spent catalyst is reheated by carbon burn in the regenerator. The technology was commercialized by Yarsintez [3] and further developed and commercialized by Snamprogetti [4, 5].

The Philips STAR process also features regeneration of the catalyst on a cyclic basis using a fixed-bed fired-tube reactor. Steam is used to lower the partial pressure of the hydrocarbons and to slow down the deposition of coke on the catalyst providing for extended cycle times. For continuous operation, various furnace modules (banks of tubes) can be operated in parallel with several modules operating in the process mode, while one is in the regeneration mode.

In the early 1970s, UOP introduced continuous catalyst regeneration (CCR™) technology that enabled noble metal catalysts to remain at their most desirable activity for several years without having to shut down the reactor for catalyst regeneration. The combination of noble metal catalysts operating at high severity in conjunction with CCR technology made it possible to design, build, and economically operate large catalytic dehydrogenation units that can produce light olefins, in particular, propylene and butylene, at high selectivities. This technology is known as the UOP Oleflex™ process.

The UOP Oleflex process can dehydrogenate propane, isobutane, normal-butane, or isopentane feedstocks separately or as a mixture of spanning two consecutive carbon numbers. The UOP Oleflex process and uses of the Oleflex process to produce propylene, MTBE, and alkylate are discussed in Sections 3.3.2 and 3.3.3 of this chapter.

As shown in Figure 3.3, the UOP Oleflex process is separated into three different sections:

- Reactor section,
- Product recovery section, and
- Catalyst regeneration section.

3.3.1.1 Reactor Section

Hydrocarbon feed is mixed with hydrogen-rich recycle gas. This combined feed is heated to the desired reactor inlet temperature and converted at high olefin selectivity in the reactors.

The reactor section consists of several radial-flow reactors, charge and interstage heaters, and a reactor feed-effluent heat exchange. The diagram shows a unit with four reactors, which would be typical for a unit processing propane feed. Three reactors are used for butane dehydrogenation. Three reactors are also used for blends of propane and butane feed, supplying heat through interstage heaters, placed between

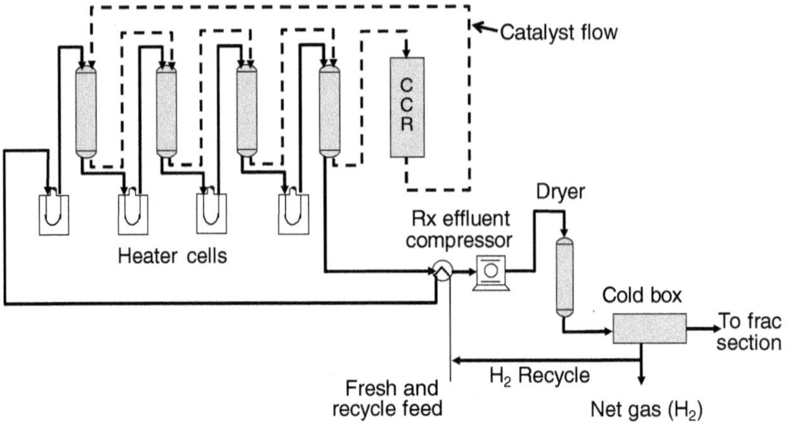

Figure 3.3 UOP Oleflex process flow.

the reactors. The catalyst is collected on the bottom of the last reactor, transported pneumatically to the regenerator. The effluent leaves the last reactor, exchanges heat with the combined feed, and is sent to the product recovery section.

3.3.1.2 Product Recovery Section

The reactor effluent is cooled, compressed, dried, and sent to a cryogenic separation system. The dryers serve two functions: (i) to remove trace amounts of water formed from the catalyst regeneration, and (ii) to remove hydrogen sulfide. The treated effluent is partially condensed in the cold separation system and directed to a separator.

Two products come from the Oleflex product recovery section: separator gas and separator liquid. The gas from the cold high-pressure separator is expanded and divided into two streams: recycle gas and net gas. The net gas is recovered at 85–93 mol% hydrogen purity. The impurities in the hydrogen product consist primarily of methane and ethane. The separator liquid, which consists primarily of the olefin product and unconverted paraffin, is sent downstream for processing.

3.3.1.3 Catalyst Regeneration Section

The regeneration section, shown in Figure 3.4, is similar to the CCR unit used in the UOP Platforming™ process. The CCR perform four functions:

- Burning the coke off the catalyst,
- Redistributing the platinum,
- Removing excess moisture, and
- Reducing the catalyst prior to returning to the reactors.

The slowly moving bed of catalyst circulates in a loop through the reactors and the regenerator. The cycle time around the loop can be adjusted within broad limits but is typically anywhere from 5 to 10 days. The catalyst activity can be restored without interrupting the catalytic dehydrogenation process in the reactor and recovery sections.

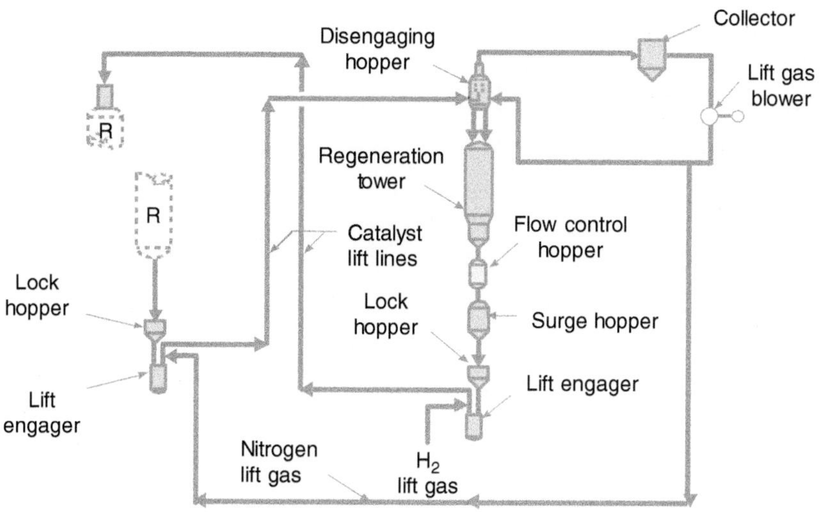

Figure 3.4 Oleflex regeneration section.

The Oleflex process for light paraffin dehydrogenation uses a platinum catalyst supported on alumina base. The key role of the dehydrogenation catalysts is to accelerate the main reaction while controlling other reactions. The alumina-supported platinum catalysts are highly active but are not selective to dehydrogenation. In addition, the catalyst rapidly deactivates due to fouling by heavy carbonaceous materials. Therefore, the properties of platinum and the alumina support need to be modified to suppress the formation of byproducts and to increase catalyst stability. Arsenic, tin, germanium, lead, and bismuth are among metals reported as platinum activity modifiers [6]. Platinum is a highly active catalytic element and is not required in large quantities to catalyze the reactions when it is dispersed on a high surface area support. The high dispersion is also necessary to achieve high selectivity to dehydrogenation relative to undesirable side reactions, such as cracking. A typical high surface area alumina supports employed having acidic sites that accelerate skeletal isomerization, cracking, oligomerization, and polymerization of olefinic material, and enhance "coke" formation. Alkali or alkaline earth metals assist in the control of the acidity [6]. Modified catalysts possess high activity and high selectivity to mono-olefins. The major byproducts are controlled kinetically.

3.3.2 UOP C_3 Oleflex Complex

The main reaction and side reactions for propane dehydrogenation are shown below.

Main reaction:

$$C_3H_8 \leftrightarrow C_3H_6 + H_2$$

Side reactions:

$$C_3H_8 \leftrightarrow C_3H_4 + 2\,H_2$$

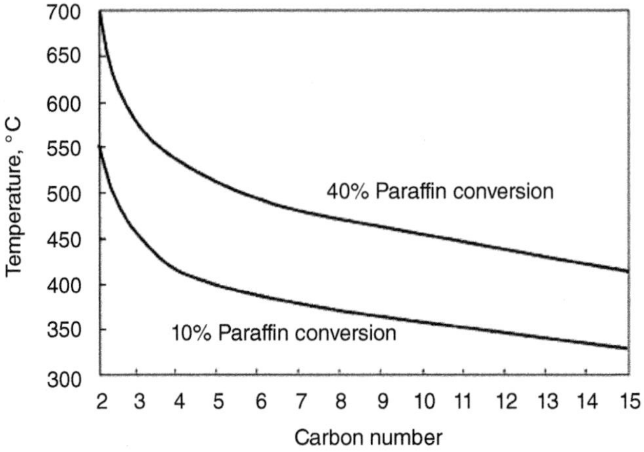

Figure 3.5 Temperatures required to achieve 10% and 40% conversion of C_2–C_{15} normal-paraffins at 1 bar. Source: Bhasin et al. [6]. © 2001, Elsevier.

$$C_3H_8 \rightarrow C_2H_4 + CH_4$$

$$C_2H_4 + H_2 \leftrightarrow C_2H_6$$

The main propane dehydrogenation is operated at equilibrium favored by high temperature and low pressure. Figure 3.5 [6] shows the temperatures required for the dehydrogenation of light paraffins are much higher than those for heavy paraffins. For 40% conversion of propane requires a temperature of 600 °C. It is advisable to increase the conversion per pass as much as possible so that the cost of recycling the unconverted reactant and that of separating the reactant and product is reduced. However, high temperatures also enhance side reactions, such as coke formation, causing fast catalyst deactivation and thermal cracking reactions that reduce the selectivity to the desired product. Thus, all factors need to take into consideration to optimize the process economics.

Oleflex process units typically operate in conjunction with fractionators and other process units within a production complex. In a propylene complex (Figure 3.6), a preheated propane-rich liquefied petroleum gas (LPG) feedstock is introduced to a depropanizer to reject butanes and heavier hydrocarbons. The depropanizer overhead is then directed to the Oleflex unit. The once-through propane conversion closely approaches the equilibrium value defined by the Oleflex process conditions, to selectively produce propylene and hydrogen with minimized side products. Two product streams are created within the C_3 Oleflex unit: a hydrogen-rich vapor product and a liquid product rich in propane and propylene. The net hydrogen can be exported directly, upgraded to PSA hydrogen, or used as fuel within the plant if hydrogen is not in demand in the vicinity of the C_3 Oleflex complex.

Trace levels of methyl acetylene and propadiene are removed from the Oleflex liquid product by selective hydrogenation. The SHP process, consisting of a single liquid-phase reactor, selectively saturates diolefins and acetylenes to mono-olefins without saturating propylene.

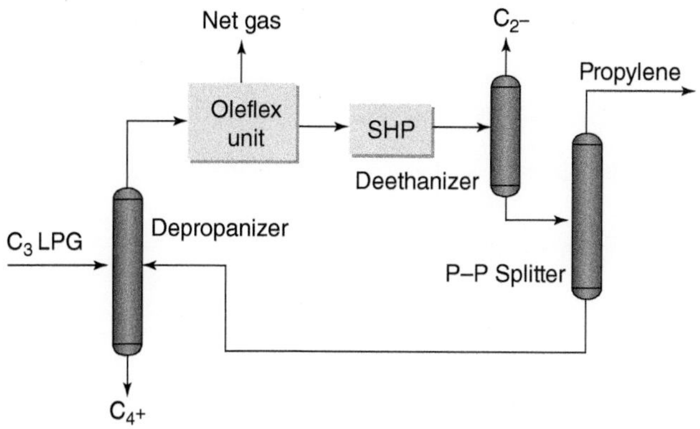

Figure 3.6 UOP C_3 Oleflex complex for propylene production.

Ethane and lighter material enter the propylene complex in the feed and are also created by nonselective reactions within the Oleflex unit. These light ends are rejected from the complex by a deethanizer column. The deethanizer bottoms are then directed to a propane–propylene (P–P) splitter. The splitter produces high-purity propylene as the overhead product. Typical propylene purity ranges from 99.5 and 99.8 wt%. Unconverted propane from the Oleflex unit concentrates in the splitter bottoms and is returned to the depropanizer for recycle to the Oleflex unit.

UOP's latest Oleflex complex design with new generation of catalyst, DeH-26, consumes less than 1.12 MT of propane per MT of propylene produced [7] (Figure 3.7).

Investment in a C_3 Oleflex complex can be an excellent feedstock and end-product diversification play to increase the operating flexibility of the refinery. The propane feedstock from natural gas liquid (NGL) plants can be supplemented with propane produced within the refinery and sent to the C_3 Oleflex complex. The C_3 Oleflex complex can also be used to upgrade lower-purity refinery-grade propylene (RGP) streams produced by the FCC unit to chemical-grade propylene (CGP) or polymer-grade propylene (PGP) in the propylene complex P–P splitter to realize an additional product value uplift. Additionally, the hydrogen produced within the

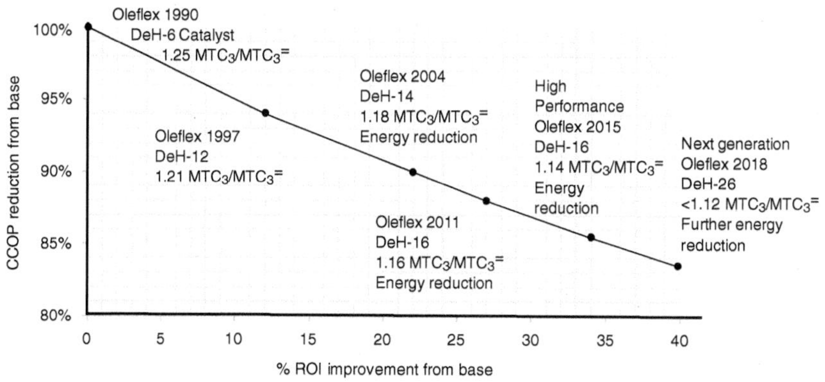

Figure 3.7 UOP Oleflex technology evolution.

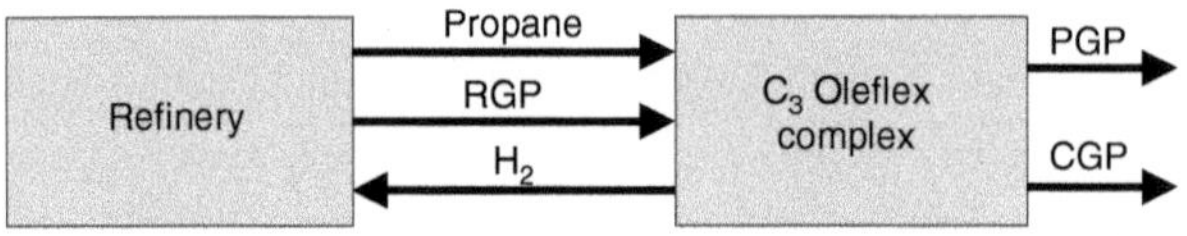

Figure 3.8 UOP C_3 Oleflex complex integration with an existing refinery.

C_3 Oleflex complex can be used for hydroprocessing needs within the refinery or exported for sale (Figure 3.8).

3.3.3 UOP Iso-C_4 Oleflex Complex

A typical iso-C_4 Oleflex complex configuration is shown in Figure 3.9 for the production of methyl tertiary butyl ether (MTBE) from butane and methanol. MTBE usage grew in the early 1980s in response to octane demand resulting initially from phase-out of lead from gasoline and later from rising demand for premium gasoline. Ethanol can be substituted for methanol to make ethyl tertiary butyl ether (ETBE) with the same process configuration. Furthermore, isopentane may be used in addition to or instead of field butane to make tertiary amyl methyl ether (TAME) or tertiary amyl ethyl ether (TAEE).
Three primary catalytic processes are used in an MTBE complex:

- Paraffin isomerization to convert normal butane into isobutane,
- Dehydrogenation to convert isobutane into isobutylene, and
- Etherification to react isobutylene with methanol to make MTBE.

Field butane, a mixture of normal-butane and isobutane obtained from natural gas condensate, is fed to a deisobutanizer (DIB) column. The DIB column prepares an isobutane overhead product, rejects any pentane or heavier material in the DIB bottoms, and makes a normal-butane side cut for feed to the butane isomerization unit.

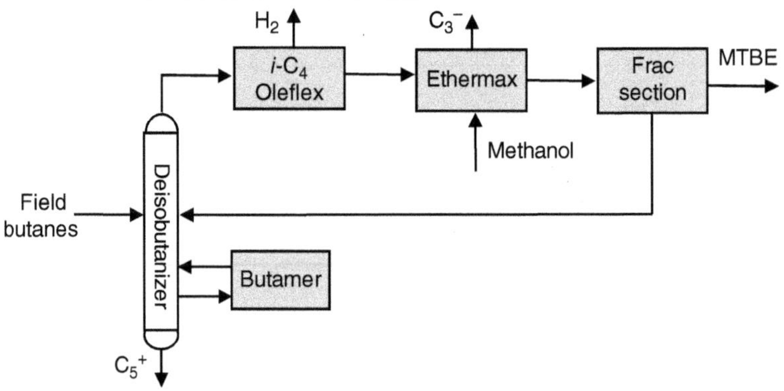

Figure 3.9 UOP iso-C_4 Oleflex complex.

The DIB overhead is directed to the iso-C_4 Oleflex unit, in which isobutane is selectively converted to isobutylene and hydrogen. The main reaction and side reactions are shown below:

Main reaction:

$$i\text{-}C_4H_{10} \leftrightarrow i\text{-}C_4H_8 + H_2$$

Side reactions:

$$i\text{-}C_4H_{10} \leftrightarrow n\text{-}C_4H_{10}$$

$$n\text{-}C_4H_{10} \rightarrow n\text{-}C_4H_8 + H_2$$

$$n\text{-}C_4H_8 \leftrightarrow C_4H_6 + H_2$$

Cracking of butanes to lighter materials also occurs as side reactions. The main isobutane dehydrogenation is operated at equilibrium favored by high temperature and low pressure.

Two product streams are created within the iso-C_4 Oleflex unit: a hydrogen-rich vapor and a liquid product rich in isobutane and isobutylene. The iso-C_4 Oleflex liquid product is sent to an etherification unit, where methanol reacts with isobutylene to make MTBE. Raffinate from the etherification unit is depropanized to remove propane and lighter materials. The depropanizer bottoms are then dried, saturated, and returned to the DIB column (Figure 3.9).

Although MTBE is a good gasoline-blending component, it has properties that cause problems. MTBE is more soluble in water than most other components of gasoline. If the gasoline containing MTBE is spilled on the ground or leaks out of underground storage tanks, MTBE could contaminate groundwater. Even fairly small amount of MTBE in water can give it an unpleasant taste and odor, making the water undrinkable. MTBE also does not break down (biodegrade) easily, so it is harder to clean up once contamination occurs. In the late 1990s, many community drinking water suppliers in the U.S. were found to have detectable levels of MTBE. As a result, MTBE has been phased out for gasoline blending in the United States in late 1990s. U.S. refineries were then faced with the challenge of replacing the lost volume and octane value of MTBE in the gasoline pool. Isooctane was identified as a cost-effective alternative to MTBE. It utilized the same isobutylene feeds used in MTBE production and offered excellent gasoline-blending value. Furthermore, isooctane production can be achieved in a low-cost revamp of existing MTBE plant to the UOP InAlk™ process unit.

As shown in Figure 3.10, the Oleflex dehydrogenation unit can be easily integrated with downstream dimerization of isobutylene followed by hydrogenation (UOP InAlk process). The hydrogen from the Oleflex can be used to hydrogenate isooctene to isooctane, a high-octane gasoline-blending component. The UOP process for this combination is the UOP InAlk complex.

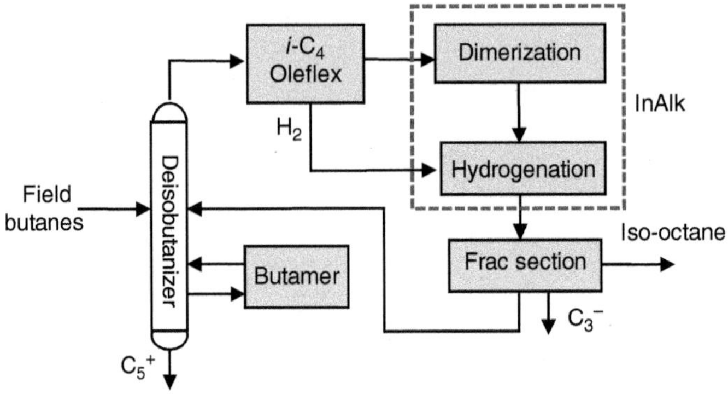

Figure 3.10 UOP InAlk complex.

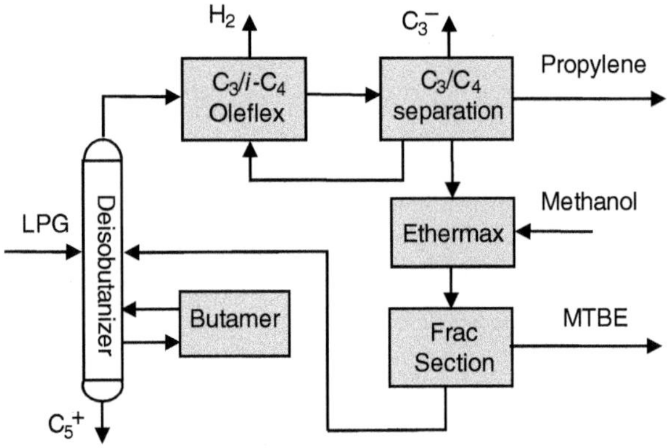

Figure 3.11 UOP C_3/C_4 Oleflex complex.

3.3.4 UOP C_3/C_4 Oleflex Complex

Mixed dehydrogenation of propane and isobutane within the same reactor system has also been of interest in the marketplace. UOP has developed and commercialized co-feeding propane and isobutane to a dehydrogenation unit for co-production of propylene and MTBE (Figure 3.11).

3.3.5 UOP Mixed-C_4 Oleflex Complex

Demand for alkylate, nonaromatic source of octane has increased with more stringent fuel specifications. Cheap, excess butane from shale gas (approximately 60% normal-butane and 40% isobutane) provides an upgrading opportunity to meet the need for high octane alkylate. Refiners are interested in butane dehydrogenation to feed sulfuric acid alkylation. By leveraging its existing isobutane Oleflex experience, UOP developed mixed-C_4 Oleflex technology to provide suitable feedstock for sulfuric acid alkylation. This was accomplished by maximizing the conversion while managing the increased rate of heavies' formation associated with normal-butane

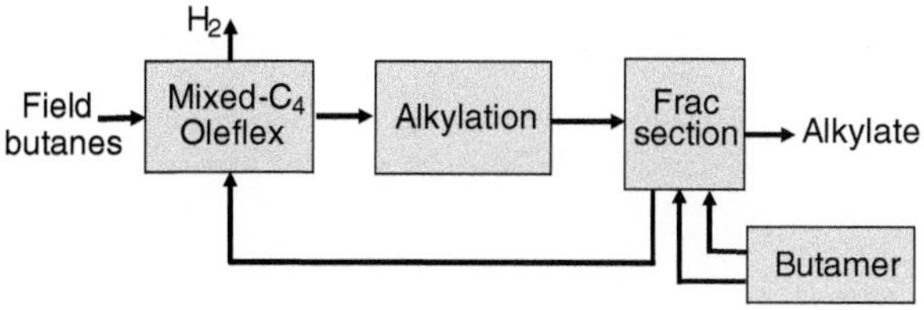

Figure 3.12 UOP mixed-C_4 Oleflex complex.

dehydrogenation. UOP mixed-C_4 Oleflex complex (Figure 3.12) has the following features:

- Capability to process any mixture of butanes, including 100% normal-butane,
- Field butanes fed directly to the mixed-butane Oleflex – elimination of the need for deisobutanizer, and
- Alkylate octane increased by 1–2 numbers ((R+M)/2) using the feed from mixed-butane Oleflex relative to the use of 100% isobutylene feed.

3.4 Heavy Normal-paraffin Dehydrogenation

Normal-paraffins are linear hydrocarbons of C_{10}–C_{14} chain lengths that are usually separated from the kerosene of crude oil by close-cut distillation, followed by isolation from branched-chain hydrocarbons and aromatics by selective adsorption with molecular sieves. The normal-paraffins are dehydrogenated to produce corresponding normal-olefins as a raw material for producing LAB.

In 1960s, UOP started development work on heterogeneous platinum catalysts supported on an alumina base for the dehydrogenation of heavy normal-paraffins. The resulting successful process, known as the UOP Pacol™ process, was first commercialized in 1968. The advent of the UOP Pacol process marked a substantial transformation in the detergent industry and contributed to the widespread use of LAB on an economical, cost-effective basis. More than 40 Pacol units have been built or are under design or construction; practically all new LAB capacity built on a worldwide basis over the last two decades makes use of UOP's Pacol catalytic dehydrogenation process.

In the Pacol process (Figure 3.13) [8], the normal-paraffins are dehydrogenated in a vapor-phase reaction to corresponding mono-olefins the presence of hydrogen over a selective platinum dehydrogenation catalyst. An adiabatic radial-flow reactor with feed preheat is normally used to compensate for the endothermic temperature drop and to minimize pressure drop within an efficient reactor volume. Relatively high space velocities are used so that only a modest amount of catalyst is required. Hydrogen and some byproduct light ends are separated from the dehydrogenation reactor effluent, and a part of this hydrogen gas is recycled back to the dehydrogenation reactor to minimize coking and enhance catalyst stability. The separator liquid is an equilibrium mixture of linear of olefins and unconverted normal-paraffins, which are charged to a reactor for the selective conversion of diolefins to mono-olefins.

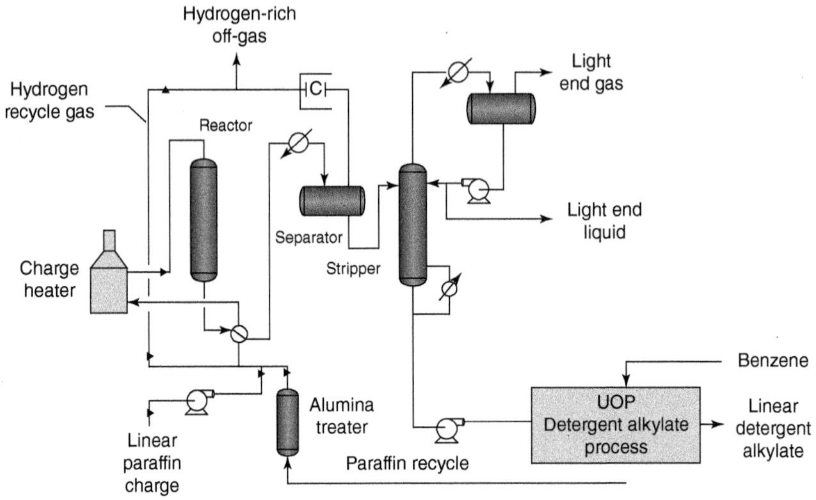

Figure 3.13 UOP pacol dehydrogenation process.

The equilibrium conversion for the dehydrogenation reaction is determined by temperature, pressure, and hydrogen partial pressure. As expected, the equilibrium conversion increases with temperature and decreases with pressure and with increasing hydrogen-to-hydrocarbon ratio. Kinetically, the overall conversion depends on space velocity; excessively high space velocities do not allow for sufficient conversions, and space velocities that are too low lead to lower selectivities because of the onset of side and competitive reactions.

Since first commercial operation of the Pacol process in 1968, several advances with modified platinum catalyst have been made showing significantly lower coking and doubling the catalyst stability. For longer carbon chain paraffins, diffusion also becomes a critical parameter, and thus large-pore low-density support is needed to access all platinum sites. Further catalyst performance improvement has been realized by introducing finely dispersed thin platinum layer over an inert core [8].

Figure 3.14 illustrates the flow scheme of an integrated complex featuring the UOP Unionfining™, Molex™, Pacol, DeFine™, PEP™, detergent alkylation, and Detal™ processes.

There are two major sections in a LAB complex: production of normal-paraffins and production of LAB.

3.4.1 Production of Normal-paraffins

Kerosene prefractionation is often used to tailor the kerosene feed to the desired carbon range. Kerosene is stripped of light ends and heavier components so that the heart cut, containing the desired normal-paraffins for the production of LAB of a certain range of molecular weight, is produced.

The distillate Unionfining process hydrotreats kerosene at sufficient severity to remove sulfur, nitrogen, olefins, and oxygenate compounds, which might otherwise poison the Molex adsorbent.

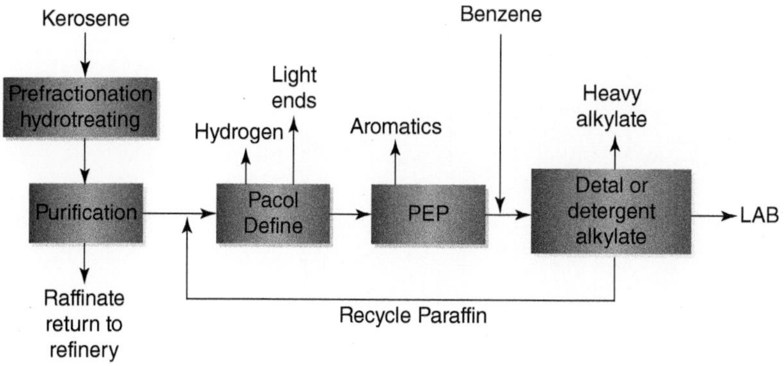

Figure 3.14 Integrated LAB complex.

The Molex process is a liquid-state separation of normal-paraffins from branched and cyclic components using Sorbex™ technology. The simulated moving-bed adsorptive separation results from using a proprietary multi-port rotary valve. The extract stream is a high-purity normal-paraffins stream. The raffinate stream, consisting mainly of iso- or cyclic-kerosene range components, is often blended into jet fuel.

3.4.2 Production of LAB

In the Pacol process, the normal-paraffins are dehydrogenated in a vapor phase reaction to corresponding mono-olefins over a highly selective and active catalyst.

The DeFine process is a liquid-phase, selective hydrogenation of Diolefins in the Pacol reactor effluent to corresponding mono-olefins over a catalyst bed.

The PEP process allows the selective removal of aromatics in the feed to the Detal or detergent alkylation unit.

Detergent alkylation is a process in which benzene is alkylated with mono-olefins produced in the Pacol unit to LAB using HF acid as a catalyst. In 1995, the first commercial Detal process unit using a solid-bed catalyst alkylation process was commissioned. This technology abolished the use of liquid acid in the plant, reducing capital investments, maintenance costs, and waste treatment.

The LAB technology is the most economical technology available today, and more than 70% of the world's LAB is produced using UOP technologies.

3.5 Methanol to Olefins

Conversion of methanol to hydrocarbons was first developed by Mobil to produce gasoline (MTG) in 1975 using ZSM-5 catalysts [9]. Selectivity of methanol to ethylene and propylene over ZSM-5 was generally low, with selectivities favoring heavier more highly branched hydrocarbons and aromatics.

$$2\ CH_3OH \leftrightarrow CH_3\text{-}OH\text{-}CH_3 \rightarrow C_2 - C_5\ \text{Olefins} \rightarrow \begin{array}{l}\text{Iso-parafins}\\ \text{Aromatics}\\ C_6{}^{+}\text{Olefins}\end{array}$$

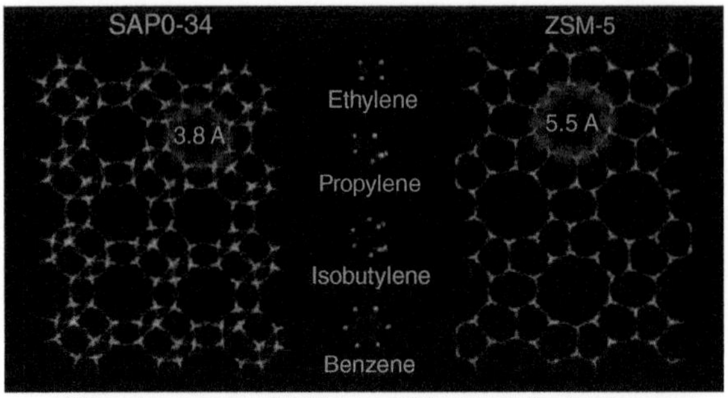

Figure 3.15 Framework of SAPO-34 and ZSM-5. Source: Chen et al. [13]. © 2005, Elsevier.

Characteristics of ZSM (medium-pore zeolite) for conversion of methanol are summarized below:

- Major olefin product is propylene,
- Significant C_5^+/aromatics byproduct formation, and
- Slow deactivation by coking.

In the 1980s, a group of scientists at Union Carbide (the group later became part of UOP LLC) discovered SAPO-34, a silicon-, aluminum-, and phosphorous-based molecular sieve, which is an excellent catalyst for conversion of methanol to ethylene and propylene [10–12].

As shown in Figure 3.15, SAPO-34 has a smaller pore size (about 3.8 Å) compared to that of ZSM-5 (about 5.5 Å) along with the small sizes of certain organic molecules are keys to the MTO process [13].

The cavity dimensions and pore opening of SAPO-34 influence the diffusion of heavy and/or branched hydrocarbons, and this leads to high selectivity to the desired small linear olefins. Another key feature of the SAPO-34 molecular sieve is its optimized acidity relative to aluminosilicate-based zeolitic materials. The mild acid function on SAPO-34 affects conversion of olefins to paraffins by hydride transfer reaction, leading to lower paraffinic byproduct formation. A further advantage of SAPO-34 is that the majority of the C_4–C_6 fraction product is olefinic. As discussed later in this section, these olefinic compounds make the heavy byproduct suitable for upgrading by olefin cracking.

For methanol conversion, SAPO-34 exhibited the following features:

- Major olefin products are ethylene and propylene,
- Fast deactivation by aromatic coke, and
- SAPO molecular sieves more stable than corresponding zeolite structure.

Based on discoveries of SAPO-34, UOP developed a process based on SAPO-34 [14, 15]. Working with Norsk Hydro in Norway, UOP successfully developed UOP/Hydro MTO process. In development of the process, UOP built on its in-house FCC experience for fluidized reactor and regenerator.

Unrelated to MTO at the time, Atofina was at work in the 1990s developing olefin-cracking technology. Shortly after, in 2000, Atofina (which later became part of Total Petrochemicals and nowadays of Total Refining and Chemicals) and

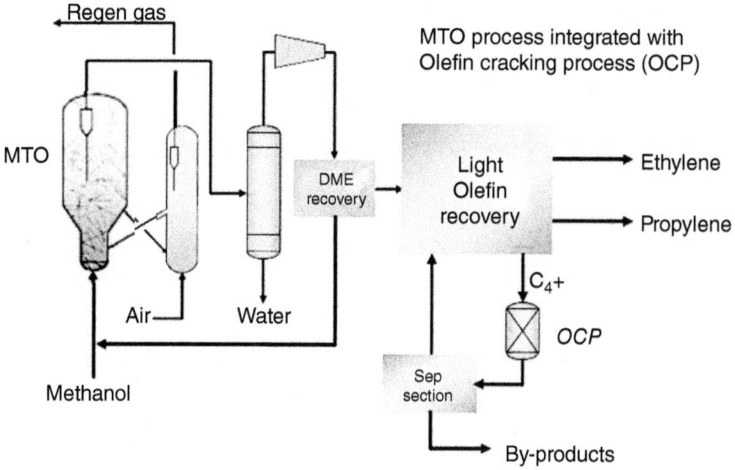

Figure 3.16 UOP advanced MTO process.

UOP formed a joint alliance to further develop olefin-cracking technology. This collaboration led to development of the Total Petrochemicals/UOP olefin-cracking process (OCP).

The OCP has been integrated with UOP/Hydro MTO process – this combination of processes is the basis for advanced MTO.

A major milestone for MTO commercialization was the start-up of the semicommercial unit in 2009, fully integrated MTO demonstration unit in Belgium, which successfully demonstrated the performance of the integrated UOP/Hydro MTO Process with the OCP.

The UOP/Hydro MTO process utilizes a fluidized reactor and regenerator system to convert methanol to olefins using the SAPO-34 catalyst. The UOP/Hydro MTO process can be operated on "crude" or undistilled methanol as well as with pure (grade AA) methanol. The choice of feedstock quality generally depends on project-specific situations because there can be advantages in either case. Figure 3.16 illustrates a simple flow diagram for the UOP advanced MTO process.

The methanol feed is preheated and then introduced into the reactor. The conversion of methanol to olefins requires a selective catalyst that operates at moderate-to-high temperatures. The reaction is exothermic, so heat can be recovered from the reaction. Carbon or coke accumulates on the catalyst and requires removal to maintain catalyst activity. The coke is removed by combustion with air in a catalyst regenerator system. A fluidized-bed reactor and regenerator system is ideally suited for the MTO process because it allows for heat removal and CCM. The reactor operates in the vapor phase at temperatures between 350 and 550 °C and pressures between 2 and 4 bars. A slipstream of catalyst is circulated to the fluidized-bed regenerator to maintain high activity. The operation of the reactor system is characterized as stable steady-state.

The reactor effluent is cooled and quenched to separate water from the product gas stream. The product gas is compressed and then unconverted oxygenates are

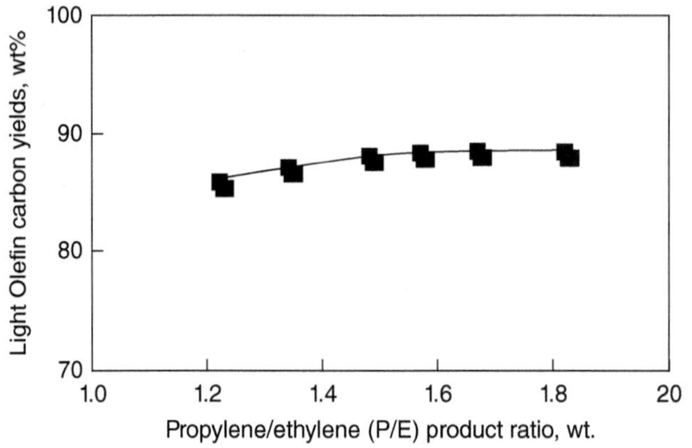

Figure 3.17 Propylene-to-ethylene ratio in UOP advanced MTO process.

recovered and returned to the reactor. The reactor provides very high conversion so there is no need for a large recycle stream. After the oxygenate recovery section, the effluent is further processed in the fractionation and purification section to remove contaminants and separate the key products from the byproduct components. Ethylene and propylene are produced as polymer-grade products and sent to storage. The C_4–C_6 fraction is sent to the OCP reactor where it is selectively converted to light olefins, majority of which is propylene. Typically, the propylene-to-ethylene ratio in OCP reactor effluent is about 4. The OCP product is depropanized, the C_3 and lighter fraction is sent to MTO product recovery section, and residual C_4 plus fraction is taken as byproduct fuel.

As shown in Figure 3.17, the advanced MTO process, which is the integrated MTO-OCP processes, can produce propylene to ethylene product ratio between 1.2 and 1.8 [16] to help meet the growing demand for propylene and additional flexibility is achievable with the technologies, if desired.

3.6 Syngas Production

Syngas is a gas mixture consisting primarily of hydrogen, carbon monoxide, and very often some amount of carbon dioxide. Syngas can be produced from many sources, including natural gas, coal, biomass, or virtually any hydrocarbon feedstock, by reaction with steam (steam reforming), carbon dioxide (dry reforming), or oxygen (partial oxidation). Historically, natural gas has been the primary feedstock for syngas production, but over the last 10 years or so, coal has been used as the feedstock for making syngas in China since an abundance of source of natural gas is not available in the country.

Syngas production technologies are the focus of the sequent sections, including removal of impurities in raw natural gas, conversion of natural to syngas, and coal gasification to syngas.

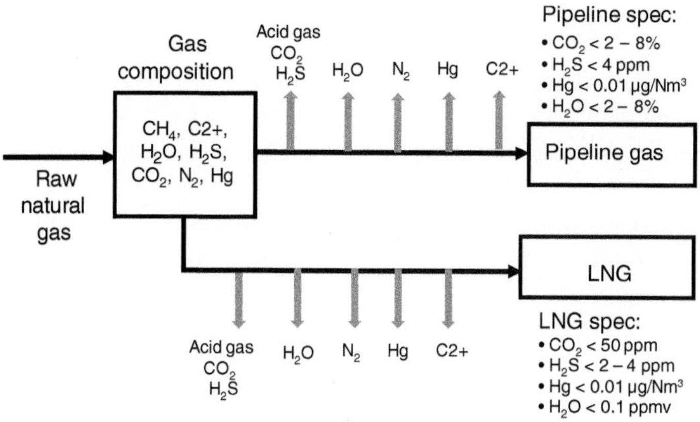

Figure 3.18 Requirements for removal of natural gas impurities.

3.6.1 Removal of Impurities in Raw Natural Gas

Raw natural gas typically consists of methane, heavier gaseous hydrocarbons, acid gases (CO_2, H_2S), water, and mercury. Acid gases in combination with water are highly corrosive and can rapidly destroy pipeline and equipment. Furthermore, H_2S is hazardous for human beings and CO_2 reduces natural gas stream heating value. Thus, the concentration of impurities in the raw natural gas must be reduced to meet the quality specification for pipeline transmission or producing liquefied natural gas (LNG) (Figure 3.18).

The technologies that are commonly used to remove acid gases include chemical and physical adsorption processes, cryogenic condensation, and membranes.

Chemical adsorption: The amine-based process is commonly used for chemical adsorption of acid gases. Figure 3.19 shows the UOP Amine Guard™ FS process [17]. CO_2 and H_2S react chemically with the amine solvent (USARSOL™) and form dissolved chemical compounds in the amine adsorber. The solvent is regenerated in the amine stripper by application of heat. For natural gas treating, the Amine Guard is capable of selective removal of H_2S and partial removal of CO_2 for pipeline specification. Another commonly used process for acid gas removal from natural gas is UOP Benfield™ process [18]. It is a thermally regenerated cyclical solvent process using an activated, inhibited hot potassium carbonate solution to remove CO_2 and H_2S (Figure 3.20).

Physical adsorption: Physical adsorption processes are used when the feed gas is characterized by high partial pressures of acid gases and low temperatures. In a physical adsorption process, the solvent interacts only physically with the dissolved gas with thermodynamic properties such that the relative adsorption of CO_2 and H_2S is more favored over the other components of the gas mixture. CO_2 and $H2_S$ are then removed by the strippers. Figure 3.21 shows the UOP Selexol™ technology for H_2S removal and CO_2 capture [19].

Membranes: Membranes are thin, semipermeable barriers that selectively separate certain components. They are characterized by permeability or capacity (flux) and selectivity. Feed gas enters along the side of the membrane and as the feed gas travels the membrane, CO_2, H_2S, and H_2O permeate through the membrane and are collected in the permeate tube. The driving force for transport is the high-feed and low-permeate pressures. The gas on the feed side does not permeate leaves through the side of the element opposite the feed position. In high H_2S applications, most polymers have been shown to have very little resistance to H_2S plasticization ("softening" of the membrane) and as a consequence can only be used at very low H_2S partial pressure limits. UOP's cellulose acetate members in contract have been shown to be applicable at extreme H_2S partial pressure conditions [20]. Figures 3.22 and 3.23 show one-stage and two-stage UOP Separex™ processes, respectively. The simplest membrane-processing scheme is the one-stage process. A feed gas is separated into a permeate stream rich in acid gas and hydrocarbon-rich residual stream. In high H_2S-removal applications, hydrocarbons permeating with the acid gases may be rejected and therefore are not lost to future gas sales. If re-injection is not a problem, then multistage system can be used to achieve higher hydrocarbon recoveries.

Mercury removal: The primary reason for removing mercury from natural gas is to protect downstream equipment and catalysts. UOP HgSIV™ regenerative system is designed to dry the natural gas and remove mercury (Figure 3.24) [21]. Since the sorption sites for mercury removal are separate from additive to the dehydration sites, mercury removal is accomplished by replacing a portion of the dehydration-grade molecular sieve with HgSIV adsorbents. The dryer bed size does not have to be increased to remove both water and mercury. Mercury is removed from HgSIV adsorbents at conventional dryer regeneration temperatures. Properly regenerated HgSIV adsorbents can be disposed of as nonhazardous waste. Mercury can be removed from the regeneration gas by condensation when the mercury level in the feed gas is high. Treatment of the HgSIV spent regeneration gas is usually not needed for low-to-moderate mercury feed gas levels.

3.6.2 Conversion of Natural Gas to Syngas

The catalyst systems used for syngas production from natural gas and subsequent processes are susceptible to deactivation by sulfur. To protect the catalysts, there is a natural gas purification section to convert sulfur species to hydrogen sulfide, followed by a zinc oxide–based adsorption process to remove the hydrogen sulfide [22].

$$\text{Hydrodesulfurization:} \qquad \text{R}-\text{S}+\text{H}_2 \rightarrow \text{R}-\text{H}+\text{H}_2\text{S}$$

$$\text{H}_2\text{S Adsorption:} \qquad \text{ZnO(s)}+\text{H}_2\text{S(g)} \rightarrow \text{ZnS(s)}+\text{H}_2\text{O(g)}$$

There are three types of process widely used by industry for conversion of natural gas to syngas: steam reforming, partial oxidation, and autothermal reforming.

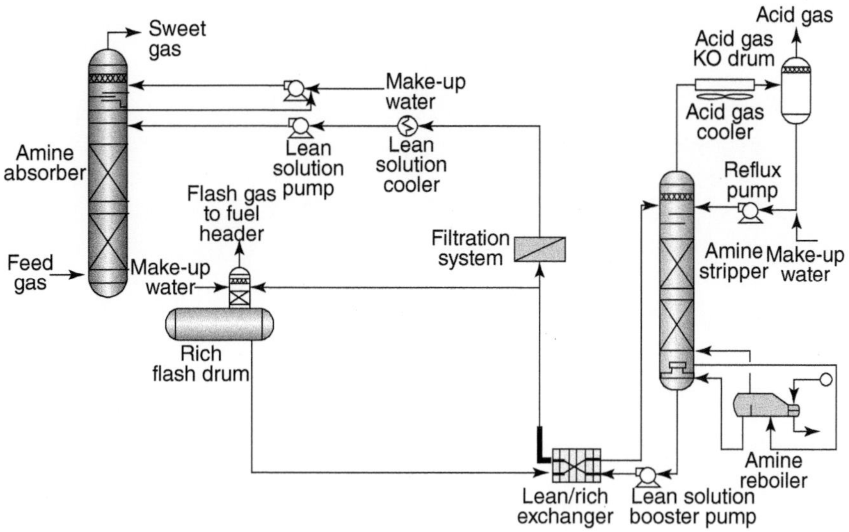

Figure 3.19 UOP amine guard FS process (1-stage) for acid gas removal. Source: Based on "UOP Amine GuardTM FS Technology for Acid Gas Removal", https://honeywelluop.chinacloudsites.cn/wp-content/uploads/2011/02/UOP-Amine-Guard-Technology-for-Acid-Gas-Removal-tech-presentation.pdf [17].

3.6.2.1 Steam Reforming

Steam reforming of natural gas is one of the most common processes for syngas production. In this catalytic process, methane reacts with steam at 800–900 °C and moderate pressure (around 40 bars) in presence of a catalyst to produce syngas [23]. The H_2/CO ratio produced in the reforming reaction is close to 3.

$$CH_4 + H_2O \rightarrow 3H_2 + CO$$

The steam reforming of methane is an endothermic reaction. The reaction heat is provided by sending the feed stock consisting of a mixture of desulfurized natural gas and steam (typically up to three steam to carbon ratio) to a tubular reactor. The reformer tubes are packed with catalyst and heated externally by burners to achieve high temperatures needed for considerable syngas yields, making steam reforming a major energy consumer. Several different transition metals can be used as catalyst in the reforming process, but nickel is the most widely used because it is usually the cheapest and leads to satisfactory results [24].

Because this process of steam reforming leads to highest H_2/CO ratio, it is considered ideal for hydrogen production.

3.6.2.2 Partial Oxidation

The partial oxidation of methane is a process whereby methane reacts directly with oxygen or air in the presence of a catalyst, and product of this reaction is syngas with a good H2/CO ratio. The partial oxidation reaction is shown below.

$$CH_4 + ½\,O2 \rightarrow CO + 2H_2$$

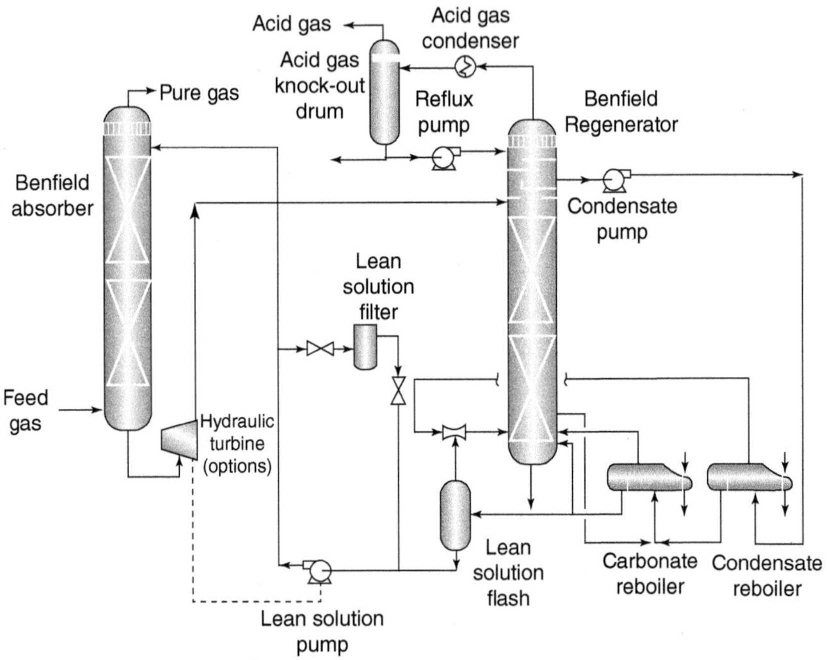

Figure 3.20 UOP benfield process for acid gas removal.

In a noncatalytic partial oxidation, the production of syngas is carried out at temperatures of 1500–1200 °C without a catalyst. A noncatalytic partial oxidation process was developed by Texaco and Shell, which results in high syngas yields at high temperature and pressures [25]. The problems related to the homogeneous process are high temperatures, long residence time as well as excessive coke formation, which reduce the controllability of the reaction [26]. The use of a catalyst in the production of syngas lowers the reaction temperature to around 800–900 °C [27]. Methane is converted with oxygen (or air) over a noble metal (Pt, Rh, Ir, Pd) or a non-noble metal (Ni, Co) catalyst to syngas in a single-step process. The catalytic partial oxidation can be used only if the sulfur content of natural gas is below 60 ppm [26]. Higher sulfur content would poison the catalyst and therefore noncatalytic partial oxidation should be used for such fuels. Many studies have been conducted for the catalytic partial oxidation on reaction mechanism, reactor configuration, reactor simulations, and catalyst systems [26].

3.6.2.3 Autothermal Reforming

The autothermal reforming of methane is a combination of both steam reforming and partial oxidation. Thus, in the steam reforming, there is contact with a flow of gaseous oxygen in the present of a catalyst. The autothermal reforming of methane process was designed to save energy, because the thermal energy required is generated by the partial oxidation of methane [28]. The H_2/CO ratio of the syngas obtained in the autothermal reforming is a function of the gaseous reactant fractions introduced in the process input. Typically, the H_2/CO ratio ranges between 1 and 2 [29].

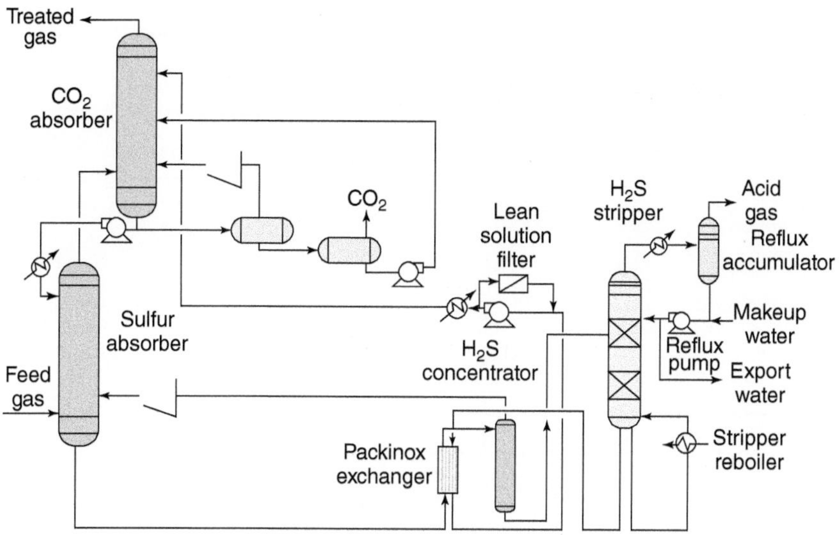

Figure 3.21 UOP selexol technology for H_2S removal and CO_2 capture. Source: Based on "UOP SelexolTM Technology Applications for CO_2 Capture", https://www.uwyo.edu/eori/_files/co2conference09/mike%20clark%20revised%20selexol_wyoming_conference.pdf [19].

Figure 3.22 One-stage flow scheme.

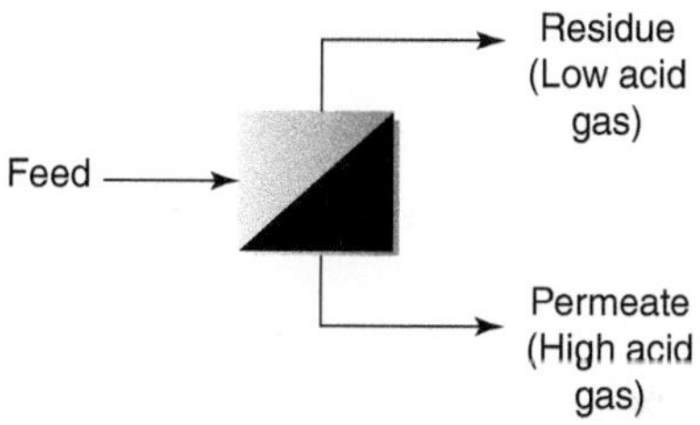

Figure 3.23 Two-stage flow scheme.

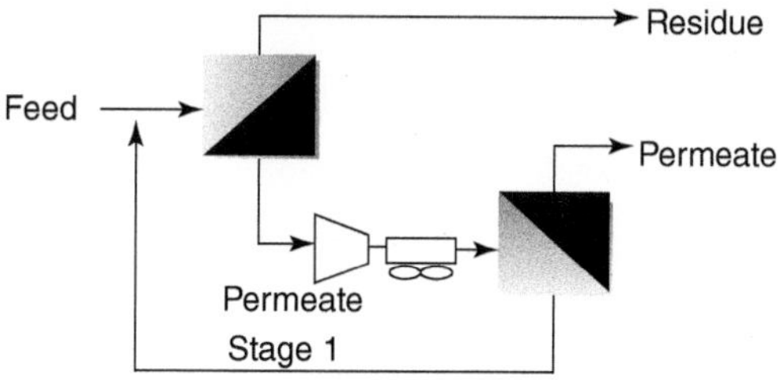

The autothermal reforming is also utilized as a "secondary reformer" (for lowering the CH_4 residue), and it is placed after a primary steam reformer in syngas plants integrated with ammonia synthesis reactors [30].

3.6.3 Coal Gasification to Syngas

Coal gasification is an old process for converting coal partially or completely to combustible gases. After purification, these gases can be used as fuels or as raw materials

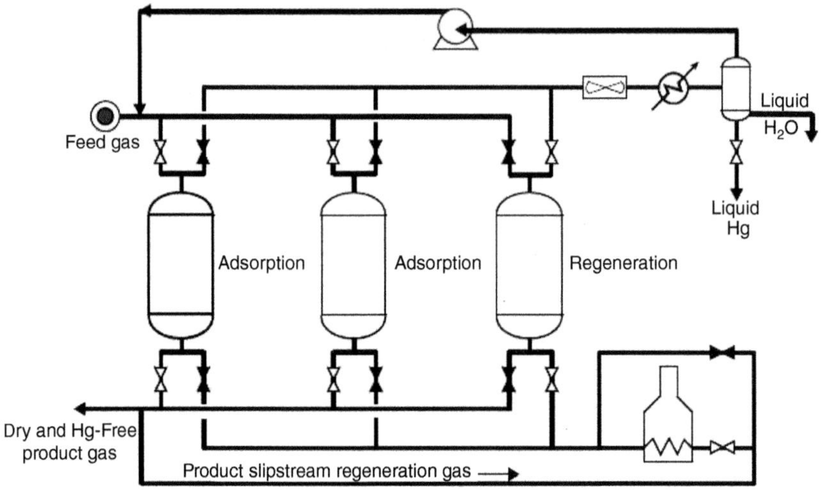

Figure 3.24 UOP mercury removal and recovery system. Source: Based on Corvini et al. [21].

Table 3.2 Major chemical reactions in goal gasification.

Gasification with oxygen	$\mathbf{C + \frac{1}{2}\,O_2 \rightarrow CO}$
Combustion with oxygen	$C + O_2 \rightarrow CO_2$
Gasification with carbon dioxide	$C + CO_2 \rightarrow 2CO$
Gasification with steam	$C + H_2O \rightarrow CO + H_2$
Gasification with hydrogen	$C + 2H_2 \rightarrow CH_4$
Water gas shift	$CO + H_2O \rightarrow CO_2 + H_2$
Methanation	$CO + 3H_2 \rightarrow CH_4 + H_2O$

for chemical manufacturing. In recent years, production of chemicals by coal gasification has had a significant growth led by China. China does not have sufficient sources of crude oil and natural gas and coal is the dominant energy source, accounting for about 64% of China's total primary energy consumption [31]. China has constructed coal gasification plants in a major way to produce syngas for manufacturing of a variety of downstream chemicals.

Coal gasification is a process that converts coal into a gaseous mixture composed primarily of carbon monoxide, carbon dioxide, hydrogen, and methane at high temperature in the presence of oxygen and steam. The chemistry of gasification is quite complex. Some of the major chemical reactions are shown in Table 3.2.

As the coal is exposed to rising temperature in the gasifier, devolatilization and breaking of weaker chemical bonds occurs, releasing volatile gases such as tar vapors, methane, and hydrogen. The volatile products and some of the carbon react with limited oxygen to form carbon dioxide and carbon monoxide, releasing heat needed for the subsequent gasification reactions. The remaining carbon reacts with carbon dioxide, steam, and hydrogen to produce carbon monoxide, hydrogen, and methane. Water gas shift and methanation are reversible reactions taking place simultaneously based on the gasifier conditions.

There are three major types of gasifiers: moving-bed (also called fixed-bed), fluidized-bed, and entrained-flow. Each type is defined by how the reactor brings the coal in contact with the reacting gas and the temperature profile inside the gasifier.

Moving-bed reactor: In a moving-bed reactor, large particles of coal move slowly down through the gasifier while reacting with gases moving up (Figure 3.25) [32] at moderate pressures (25–30 bars). Reactions within the gasifier occur in different "zones." In the "drying zone" at top of the gasifier, the entering coal is heated and dried, while cooling the product gas before it leaves the reactor. The coal is further heated and devolatized by the high-temperature gas as it descends through the "carbonization zone." In the next zone, the "gasification zone," the devolatized coal is gasified by reaction with steam and carbon dioxide. Near the bottom of the gasifier, in the "combustion zone," which operates at the highest temperature, oxygen reacts with the remaining char. Due to the countercurrent flow and the inherent recuperation of the sensible energy in the gas through devolatilization and coal drying, this type of gasifier gives the highest cold gas efficiency of any of the gasifiers.

Fluidized-bed reactor: Fluidized-bed reactors employ back-mixing and efficiently mix feed coal particles with coal particles already undergoing gasification in the reactor vessel. Coal of small particles (<6 mm) is supplied through the side of the reactor, and oxidant and steam are supplied near the bottom (Figure 3.26) [32]. Due to the thorough mixing within the gasifier, a constant temperature is sustained in the reactor bed. The gasifiers normally operate at moderately high temperature to achieve an acceptable carbon conversion (e.g. 90–95%) and to decompose most of the tar, oils, phenols, and other liquid byproducts. However, the operating temperatures are usually less than the ash fusion temperature so as to avoid clinker formation and the possibility of de-fluidization of the bed. This, in turn, means that fluidized-bed gasifiers are best suited to relatively reactive coals, low-rank coals, and other fuels such as biomass [32].

Entrained-flow reactor: In entrained-flow systems, fine coal feed and oxidant (air or oxygen) and/or steam are fed co-currently to the gasifier (Figure 3.27) [32]. Entrained-flow gasifiers operate at high temperature and pressure with turbulent flow, which causes rapid feed conversion and allows high throughput. The gasification reactions occur at a very high rate (typical residence time is on the order of few seconds), with high carbon conversion efficiencies (98–99.5%). Entrained-flow gasifiers have the ability to handle practically any coal feedstock and produce a clean, tar-free syngas. Given the high operating temperatures, this type of gasifier melts the coal ash into vitreous inert slag. Different systems may use different coal feed systems (dry or water slurry) and heat recovery systems.

A comparison of moving-bed, fluidized-bed, and entrained-flow gasifiers is shown in Table 3.3 [32].

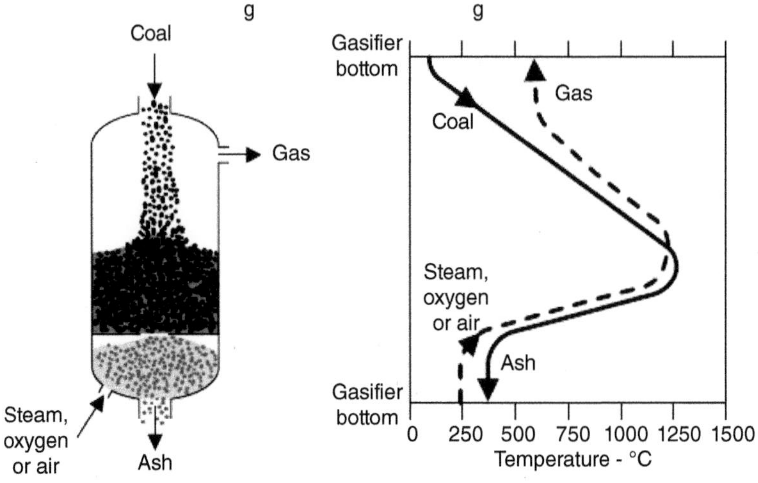

Figure 3.25 Moving-bed gasifier.

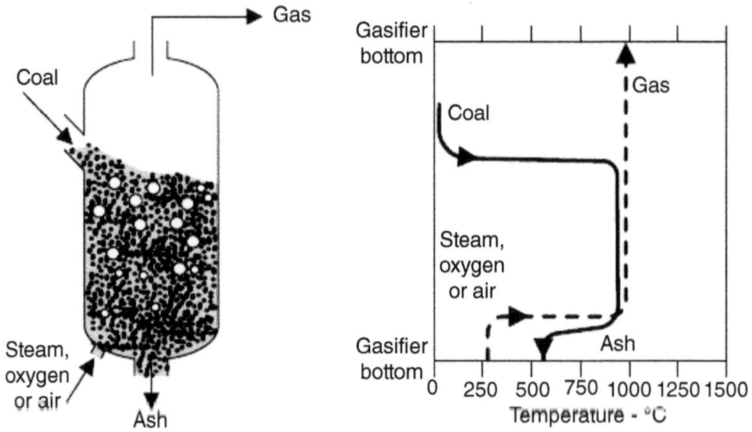

Figure 3.26 Fluidized-bed gasifier. Source: Based on U.S. Department of Energy, Gasification Technology [32].

Raw syngas should be purified and the H_2/CO ratio should be adjusted before it is used as feed for downstream processing. Desired H_2/CO values for various syngas applications are shown below:

- *H_2 production*: Pure H_2
- *Ammonia*: $H_2/CO = 3$
- *Methanol*: $H_2/CO = 2$

To meet the downstream process requirements, the syngas is passed through a multistage, fixed-bed reactor containing shift catalyst to convert carbon monoxide and water into additional H_2 and carbon dioxide according to the following reaction

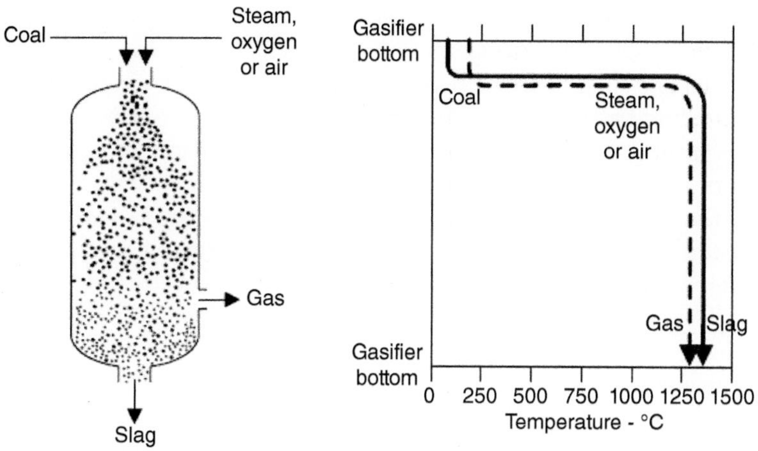

Figure 3.27 Entrained-flow gasifier. Source: Based on U.S. Department of Energy, Gasification Technology [32].

Table 3.3 Comparison of gasifier characteristics.

	Moving bed		**Fluidized bed**		**Entrained flow**
Ash condition	**Dry**	**Slagging**	**Dry**	**Agglomerate**	**Slagging**
Coal feed	~2 in.	~2 in.	~1/4 in.	~1/4 in.	~100 Mesh
Fines	Limited	Better than dry ash	Good	Better	Unlimited
Coal rank	Low	High	Low	Any	Any
Gas temperature (°F)	800–1200	800–1200	1700–1900	1700–1900	>2300
Oxidant required	Low	Low	Moderate	Moderate	High
Steam required	High	Low	Moderate	Moderate	Low
Issues	Fines and hydrocarbon liquids		Carbon conversion		Raw gas cooling

Source: Based on U.S. Department of Energy, Gasification Technology [32].

known as the gas water shift reaction (GWSR):

$$CO + H_2O \leftrightarrow CO_2 + H_2$$

GWSR is a moderately exothermic reversible reaction. Therefore, with increasing temperature, the reaction rate increases, but the conversion of reactants to products becomes less favorable.

To take advantage of both the thermodynamics and kinetics of the reaction, the industrial-scale GWSR is conducted in multiple adiabatic stages consisting of a high-temperature shift followed by a low-temperature shift with intersystem cooling [33]. The initial high-temperature shift at 310–450 °C results in incomplete conversion of carbon monoxide and a 2–4% carbon monoxide exit composition. To shift the equilibrium toward hydrogen production, a subsequent low temperature shift reactor at 200–250 °C is employed to produce a carbon monoxide exist composition of less than 1%. Due to the different reaction conditions, different catalysts must be employed at each stage to ensure optimal activity. The commercial high-temperature shift catalyst is the iron oxide-chromium oxide catalyst and the low-temperature shift catalyst is a copper-based catalyst. The copper-based catalyst can be easily poisoned by sulfur compounds, whereas the iron-based catalyst is sulfur tolerant. If sulfur is present in the feed, a guard bed should be used to remove the sulfur compounds to protect the copper-based catalyst [33].

References

1 Foster, J. (2013). *Can Shale Gale Save the Naphtha Crackers?* London: Platts.Platts special report: petrochemicals
2 Graig, R.G. and Spence, D.C. (1986). Catalytic dehydrogenation of liquefied petroleum gas by the Houdry Catofin and Catadiene processes. In: *Handbook of Petroleum Refining Processes* (ed. R.A. Meyers). New York (Chapter 4.1: McGraw-Hill.
3 Sanfilippo, D., Buonomo, F., Fusco, G., and Miracca, I. (1998). Paraffins activation through fluidized bed dehydrogenation: the answer to light olefins demand increase. *Studies in Surface Science and Catalysis* 119: 919–924.
4 Iezzi, R. Bartolini, A. (1997). Process for dehydrogenating light paraffins in a fluidized bed reactor. US Patent 5,633,421.
5 Luckenbach, E.C. Zenz, F.A. Papa, Giovanni Bertolini, A. (1997). Fluidized bed reactor and process for performing reactions therein. US Patent 5,656,243.
6 Bhasin, M.M., McCain, J.H., Vora, B.V. et al. (2001). Dehydrogenation and oxydehydrogenation of paraffins to olefins. *Applied Catalysis A: General* 221: 397–419.
7 Shakur, M. (2018). New generation of Oleflex™ technology. *Presented in UOP International Technology Conference*, Beijing, China.
8 Vora, B.V. (2012). Development of dehydrogenation catalysts and processes. *Topics in Catalysis* 55: 1297–1308.
9 Chang, Silvestri, Smith, R.L. Aromatization reactions. US Patents 3,894,103; 3,928,483.
10 Kaiser, S.W. Production of light olefins. US Patent 4,499,327; 1985.
11 Kaiser, S.W. and Arabian, J. (1985). *The Sciences and Engineering* 10: 361.
12 Lewis, J.M.O. (1998). Methanol to olefins process using silicoalumino-phosphate molecular sieve catalysts. *Catalysis* (ed. J.W. Ward). Amsterdam: Elsevier.
13 Chen, J.Q., Bozzano, A., Glover, B. et al. (2005). Recent advancements in ethylene and propylene production using the UOP/hydro MTO process. *Catalysis Today* 106: 103–107.

14 Kaiser, S.W. Production of light olefins. US Patents 4,499,327; 4,677,242; 1987.

15 Vora, B.V., Marker, T.L., Barger, P.T. et al. (1997). Economic route for natural gas conversion to ethylene and propylene. *Studies in Surface Science and Catalysis* 107: 87–98.

16 Gregor, J.H. (2012). Maximize profitability and olefin production with UOP's advanced MTO technology. *HIS word Methanol Conference*, Madrid, Spain (27–29 November 2012).

17 "UOP Amine Guard™ FS Technology for Acid Gas Removal"; https://www.uop.com/tech-library.

18 Echt, W.; (2013). "Technologies for efficient purification of natural and synthesis gases"; www.uop.com (accessed 4 September 2020).

19 "UOP Selexol™ Technology for acid gas removal"; https://www.uop.com/tech-library.

20 Cnop, T. Dortmundt, D. Schott, M.; Continued development of gas separation membranes for highly sour service. https://www.uop.com/tech-library.

21 Corvini, G. Stiltner, J. Clark, K.; Mercury removal from natural gas and liquid streams. https://docplayer.net/20958934-Mercury-removal-from-natural-gas-liquid-streams-john-markovs-uop-des-plaines-illinois-jack-corvini-uop-houston-texas.html.

22 Liu, K., Song, C., and Subramani, V. (2010). *Hydrogen and Syngas Production and Purification Technologies*. Hoboken, NJ: Wiley.

23 Saumel, P. (2003). GTL technology challenges and opportunities. *Catalysis Bulletin of the Catalysis Society of India* 2: 82–99.

24 Neiva, L.S. and Gama, L. (2010). A study on the characteristics of the reforming of methane: a review. *Brazilian Journal of Petroleum and Gas* 4 (3): 119–127.

25 Pena, M., Gomez, J., and Fierro, J.L. (1996). New catalytic routes for syngas and hydrogen production. *Applied Catalysis A: General* 144: 7–57.

26 Saleh Al-Sayari, A. (2015). Recent development in the partial oxidation of methane to syngas. *The Open Catalysis Journal* 6: 17–28.

27 Liu, J.A.; (2006). Kinetics, catalysts and mechanism of methane steam reforming. Master Thesis of Science in Chemical Engineering, Worcester Polytechnic Institute.

28 Ayabe, S., Omoto, H., Utaka, T. et al. (2003). Catalytic autothermal reforming of methane and propane over supported metal catalysts. *Applied Catalysis A: General* 241: 261–269.

29 Palm, C., Cremer, P., Peters, R., and Stolten, D. (2002). Small-scale testing of a previous metal catalyst in the autothermal reforming of various hydrocarbon feeds. *Journal of Power Sources* 106: 231–237.

30 Ghoneim, S.A., El-Salamony, R.A., and El-Temtamy, S.A. (2016). Review on innovative catalytic reforming of natural gas to syngas. *World Journal of Engineering and Technology* 5: 116–139.

31 National Bureau of Statistics (2016, 29 February). *National Economy and Social Development Statistic Bulletin 2015 (Internet)*. Beijing: National Bureau of Statistics of the People's Republic of China http://www.stats.gov.cn/english/PressRelease/201602/t20160229_1324019.html.

32 Overview of DOE's gasification program; U.S. Department of Energy, Gasification Technology. Available online: http://www.netl.gov/technologies/coalpower/gasification/pubs/pdf/DOE%20Gasification%20Program%Overview%202009%2009-03%20vls.pdf.

33 Byron, S.R.J., Loganthan, M., and Shantha, M.S. (2010). A review of the water gas shift reaction. *International Journal of Chemical Reactor Engineering* 8: 1–32. http://www.netl.doe.gov/technologies/coalpower/gasification/pubs/pdf/DOE%20Gasification%20Program%20Overview%202009%2009-03%20v1s.pdf (accessed on 20 September 2009)

4

Uses of Olefins and Synthesis Gas

4.1 Introduction

As discussed in Chapter 3, ethylene and propylene are the main petrochemical building blocks. Ethylene is used as a feedstock to manufacture polyethylene, ethylene chloride, ethylene oxide (EO), and other petrochemicals (Figure 4.1). These materials are very useful for the packaging, plastic processing, construction, houseware, and textile industries. Most propylene is used to make polypropylene, but it is also a feedstock to produce propylene oxide, acrylic acid, cumene, and many other chemical derivatives (Figure 4.2). Not only plastic processing, the packaging industry, and the furnishing section, but also the automotive industry are users of propylene and its derivatives.

In addition to ethylene and propylene, other olefins also find uses in fuel and petrochemical chemical applications. As discussed in Chapter 3, the leading applications of butylene are for production of gasoline-blending components (alkylate, MTBE, and ETBE) while heavy normal-olefins are primarily used for producing linear alkylbenzene (LAB), a dominant precursor of biodegradable detergents. Syngas, on the other hand, represents a suitable alternative, from resources other than crude oil, to produce chemicals with commercial significance such as hydrogen, ammonia, and methanol. In this chapter, the uses of olefins (ethylene and propylene) and syngas as raw materials to produce downstream products will be discussed.

4.2 Uses of Ethylene

4.2.1 Ethylene to Polyethylene

Polyethylene is the most widely used plastic, made by repeating $-CH_2-$ unit. There are three major types of polyethylene produced with different methods: low-density polyethylene (LDPE) referred to as high-pressure polyethylene, high-density polyethylene (HDPE) known as low-pressure polyethylene, and linear low-density polyethylene (LLDPE).

Modern Petrochemical Technology: Methods, Manufacturing and Applications, First Edition.
Santi Kulprathipanja, James E. Rekoske, Daniel Wei, Robert V. Slone, Trung Pham, and Chunqing Liu.

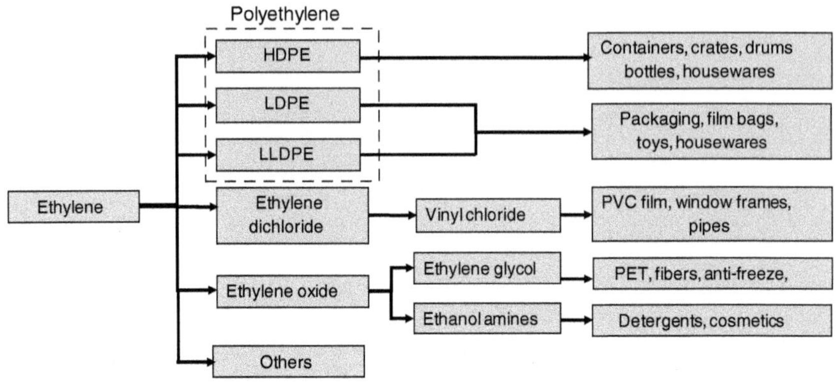

Figure 4.1 Derivatives of ethylene.

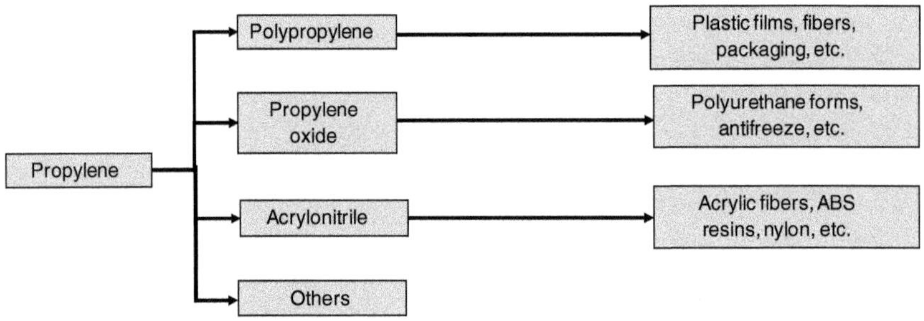

Figure 4.2 Derivatives of propylene.

The high-pressure method produces LDPE in tubular or autoclave reactors [1, 2] at a pressure between 1000 and 3000 bars and a temperature of 200–300 °C with oxygen or peroxide as initiator for the free radical polymerization. Unreacted ethylene at reactor outlet is removed and recycled. The resulting LDPE has a density range of 0.91–0.93 g/cc.

There are three processes developed for low-pressure ethylene polymerization: solution process, slurry process, and gas-phase process. Most of HDPE is made by the slurry process [3] with a co-monomer (butene-1, hexene-1, or octene-1), in the presence of a chromium on silica or Ziegler–Natta catalyst, at a pressure 15–35 bars and a temperature 80–110 °C. The HPPE has a density range of 0.94–0.96 g/cc.

LLDPE is mostly produced by low-pressure gas-phase and solution processes by polymerization of ethylene with alpha olefins (butene-1, henxene-1, octene-1) with Ziegle-Natta catalyst. LLDPE has a density of 0.91–0.94 g/cc.

The density difference is a result of variation in molecular structures. LPPE has low density due to its extensive brancking structure. It is softest and least crystalline of the polythylenes. HDPE is a denser material, more ridged, and less permeable since its polymer chains include less bracking showing high crystallinity. LLDPE has a basic linear structure with short branchings. When compared with LDPE, the

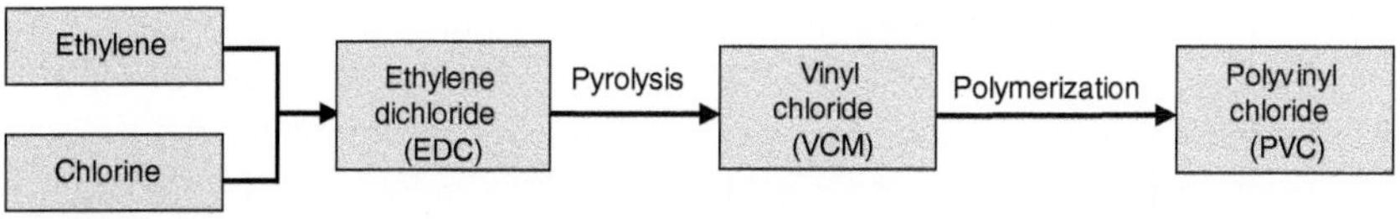

Figure 4.3 Major steps for manufacturing of PVC.

short chains in LLDPE are able to move against on another upon elongation without entangling together, resulting in improved strength and impact properties.

4.2.2 Ethylene to Polyvinyl Chloride

Polyvinyl chloride (PVC) is a commonly used plastic favored in construction, piping, and many other industries due to the properties of the polymer, including light weight, chemical resistance, and versatility. Manufacturing of PVC involves the steps shown in Figure 4.3.

Vinyl chloride (VCM) production involves the following reactions:

Direct chlorination $\quad CH_2 = CH_2 + Cl_2 \rightarrow ClCH_2 - CH_2Cl$

Oxychlorination $\quad CH_2 = CH_2 + 2\,HCl + \frac{1}{2}\,O2 \rightarrow ClCH_2 - CH_2Cl + H_2O$

Pyrolysis $\quad 2ClCH_2 - CH_2Cl \rightarrow 2CH_2 = CHCl + 2HCl$

Overall $\quad 2CH_2 = CH_2 + Cl_2 + \frac{1}{2}\,O_2 \rightarrow 2\,CH_2 = CHCl + H_2O$

Ethylene dichloride (EDC) is produced by direct chlorination or oxychlorination using hydrogen chloride as the chlorination agent. The direct chlorination process involves a liquid- or vapor-phase reaction of ethylene and chlorine using an $FeCl_3$ catalyst. This exothermic reaction is controlled by mass transfer and can be operated at 50–120 °C to minimize the undesired by-product formation. Oxychlorination is used in vinyl chloride production because it consumes the hydrochloric acid (HCl), a major byproduct of vinyl chloride production. The purified EDC is sent to a thermal cracker (pyrolyzer), which outputs VCM and HCl.

PVC is produced by polymerization of VCM as shown in Figure 4.4 [4].

There are three types of PVC manufacturing processes: suspension polymerization, emulsion polymerization, and bulk polymerization. PVC made from the suspension polymerization is by far the most common. The suspension polymerization affords particles with average diameters of 100–180 μm. VCM and water are introduced into the reactor along with a polymerization initiator and other additives. The

Figure 4.4 Polymerization of VCM to PVC. Source: Chanda and Roy [4].

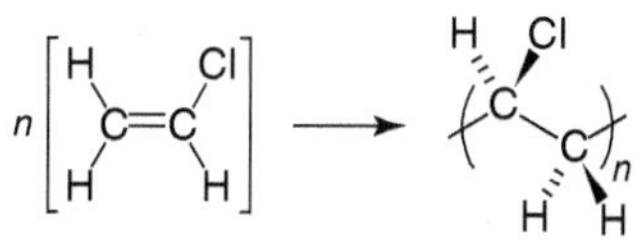

contents of the reaction vessel are pressurized and continually mixed to maintain the suspension and ensure a uniform particle size of the PVC resin. The reaction is exothermic and thus requires cooling. As the volume is reduced during the reaction (PVC is denser than VCM), water is continually added to the mixture to maintain the suspension [5].

4.2.3 Ethylene to Monoethylene Glycol

Monoethylene glycol (MEG) is mainly used for two purposes: (i) for antifreeze formulations, and (ii) as a raw material in the manufacture of polyethylene terephthalate (PET), which is subsequently used in the production of polyester fibers, resins, films, and other consumables. MEG is conventionally produced through the hydrolysis of EO, which is obtained via ethylene oxidation.

Nearly all the EO production capacity is based on direct oxidation using a silver catalyst.

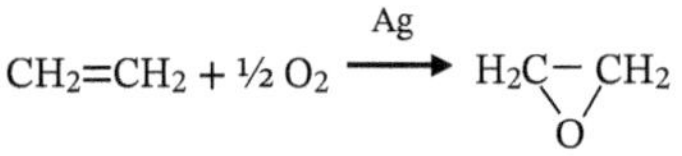

Air or high-purity oxygen is mixed with recycled gas and fresh ethylene [6]. Vapor-phase oxidation inhibitors such as EDC or vinyl chloride or other halogenated compounds are added in the inlet of the reactor to retard carbon dioxide formation [7]. The gas is fed to s multi-tube reactor in which the reactor temperature is controlled at 220–300 °C and pressure is set at 10–30 bars. The oxygen-based EO process is preferred over the air route in large plants due to higher yields and less catalyst required [7].

EO is further processed to produce MEG by a liquid-phase, noncatalytic hydrolysis reaction operated at an elevated temperature.

This conventional process was started in 1937 by Union Carbide [6]. In this reaction system, MEG inevitably reacts with remaining EO generating diethylene glycol and triethylene glycol.

Selectivity to the different glycols is controlled by varying the ratio of water to EO with large excess of water promoting MEG selectivity. Removing the excess water is energy intensive and requires capital investment in evaporators. These factors limit the amount of excess water, which can be used for control of the selectivity to MEG.

To maximize the MEG selectivity, many different systems have been studied. Mitsubishi and Shell developed a two-step process [8] called OMEGA process.

$$C_2H_4O + CO_2 \rightarrow C_3H_4O_3$$

$$C_3H_4O_3 + H_2O \rightarrow HOC_2H_4OH + CO_2$$

In the first step, EO is reacted with CO_2 to form ethylene carbonate. In the second step, the ethylene carbonate is catalytically hydrolyzed to MEG and CO_2 released in this second reaction step is recycled back to the first reaction step.

4.3 Uses of Propylene

4.3.1 Propylene to Polypropylene

Polypropylene (PP) is a thermoplastic polymer used in a wide variety of applications. It belongs to the group of polyolefins and is partially crystalline and nonpolar. Its properties are similar to polyethylene, but it is slightly harder and more heat resistant. It is a white, mechanically rugged material and has a high chemical resistance [9]. Polypropylene is the second-most widely produced commodity plastic (after polyethylene) and it is often used in plastic films, fibers, and packaging.

Polypropylene is produced by the chain-growth polymerization of monomer propylene.

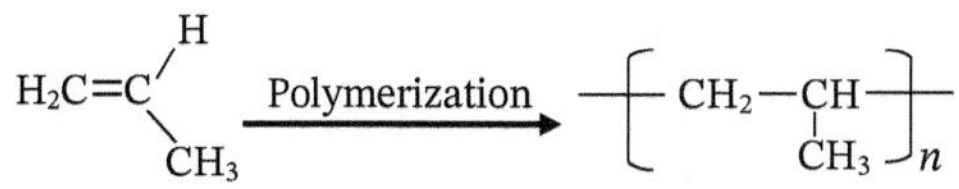

Polypropylene can be categorized as atactic polypropylene, syndiotactic polypropylene, and isotactic polypropylene. In case of atactic polypropylene, the methyl group ($-CH_3$) is randomly aligned, alternating for syndiotactic polypropylene, and evenly for isotactic polypropylene (Figure 4.5). This has an impact on the crystallinity and thermal properties.

The term tacticity describes how the methyl group is oriented in the polymer chain, usually indicated in percent, using the isotactic index. As the methyl group is in isotactic propylene consistently located at the same side, the higher the isotacticity, the greater the crystallinity, and thus also the softening point, rigidity, and hardness [10]. Only the isotactic form is of commercial interest as a thermoplastic. Atactic polypropylene, due to its lack of crystallinity, is too soft and rubber-like to be of wide commercial use as a thermoplastic, while syndiotactic polypropylene is more difficult to prepare using metallocene catalysts.

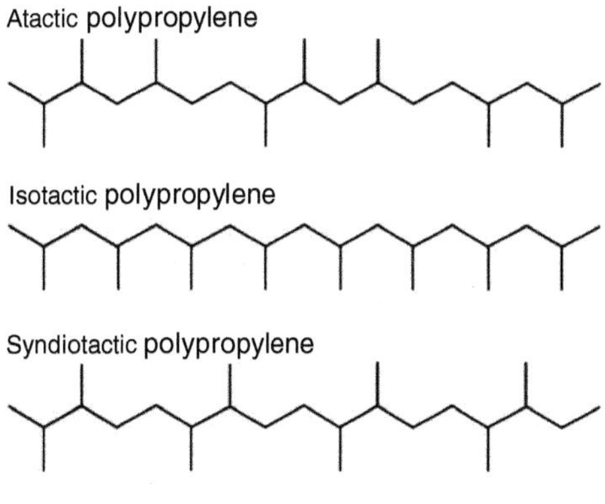

Figure 4.5 Molecular structures of polypropylene.

There are three different types of polypropylene: homopolymer, random copolymer, and impact copolymer. Homopolymer contains only propylene monomers having high stiffness, low density, good chemical resistance, and relatively high-temperature resistance. However, homopolymer has poor impact resistance, especially at low temperature. Polypropylene-containing ethylene as a comonomer in the polypropylene chains at levels in the range of 1–8%, and this is referred to as a random copolymer. Random copolymers have higher impact resistance, higher clarity, and more flexibility than homopolymers. Impact copolymer has an ethylene content of 45–46% giving better impact strength at lower temperatures, as well as higher melting points, than random copolymers.

The industrial production process can be grouped into gas-phase polymerization, bulk polymerization, and slurry polymerization. All state-of-the-art processes use either gas-phase or bulk reactor systems [11].

Gas-phase polymerization: The gas-phase polymerization is carried out in a fluidized-bed reactor, propylene is passed over a bed containing the heterogeneous (solid) catalyst, and the formed polymer is separated as a fine powder. Unreacted gas is recycled and fed back into the reactor.

Bulk polymerization: In bulk polymerization, propylene monomer is fed into the reactor loop in liquid phase. Insoluble polymer formed on the suspended catalysts precipitates out to form a slurry. The formed polymer is withdrawn and any unreacted monomer is flashed off.

Slurry polymerization: In the slurry polymerization, typically C_4–C_6 alkanes are utilized as inert diluent to suspend the growing polymer particles. Propylene is introduced into the mixture as a gas.

Ziegler–Natta catalysts are used in the polymerization process. These catalysts are generally produced by interaction of titanium and magnesium chlorides and aluminum alky compounds. New metallocene catalysts open the possibility to synthesize the polypropylene with tighter control of polymer molecular weight, molecular weight distribution, and crystalline structure as well as long-chain branched or blocky copolymers with good properties.

4.3.2 Propylene to Propylene Oxide

Propylene oxide production is the second-largest propylene outlet, after polypropylene, accounting for about 7–8% of the global propylene consumption. About 60–70% of propylene oxide is converted to polyether polyols. These polyols are building blocks in the production of polyurethane plastics. About 20% of propylene oxide is hydrolyzed into propylene glycol for further processing to make unsaturated polyester resins or for antifreeze and solvent applications. Other products derived from propylene oxide are polypropylene glycol, propylene glycol ethers, and propylene carbonate.

There are two major processes for the manufacture of propylene oxide: propylene chlorohydrin process and propylene oxidation process using peroxides.

The chlorohydrination process consists of formation of propylene chlorohydrin by the reaction between hypochlorous acid and propylene. The propylene chlorohydrin is epoxidized to propylene oxide by a 10% solution of lime or NaOH.

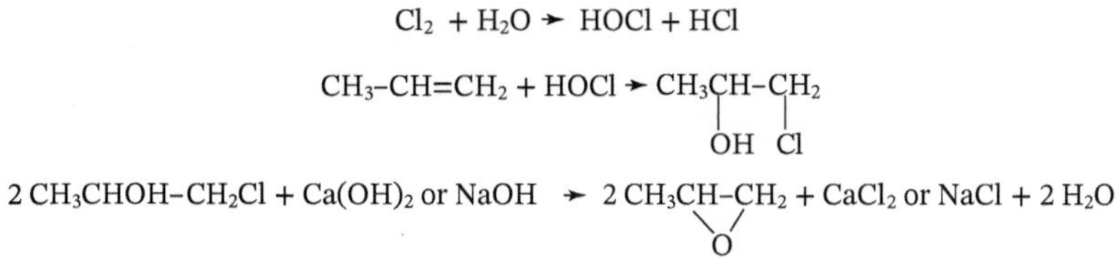

Byproducts formed during the reactions are 1,2-dichloropropane and chlorinated di-isopropyl ether.

Some of the disadvantages of the chlorohydrination process include use of costly chlorine, production of weak calcium chloride byproduct, and corrosion problem due to chlorine handling.

The other route to produce propylene oxide involves oxidation of propylene with an organic peroxide.

$$CH_3CH\text{-}CH_2 + RO_2H \rightarrow CH_3CHCH_2O + ROH$$

The process is practiced with four hydroperoxides [12]: *tert*-butyl hydroperoxide, ethylbenzene hydroperoxide, cumene hydroperoxide, and hydrogen peroxide. One of the main benefits of the process using hydrogen peroxide to produce propylene oxide (HPPO process) is that it eliminates the need for additional infrastructure or marketing of coproducts that are produced by using other types of hydroperoxides.

4.3.3 Propylene to Acrylonitrile

Acrylonitrile is a commodity petrochemical produced primarily from propylene. The primary use of acrylonitrile is as the raw material for the manufacture of acrylic and modacrylic fibers. Other major uses include the production of plastics acrylonitrile-butadiene-styrene (ABS) and styrene-acrylonitrile (SAN), nitrile rubber, adiponitrile, and acrylamide.

Acrylonitrile is produced by catalytic ammoxidation of propylene, also known as the SOHIO process.

$$2\,CH_3 - CH = CH_2 + 2\,NH_3 + 3\,O_2 \rightarrow 2\,CH_2 = CH\text{-}C{\equiv}N + 6\,H_2O$$

In the SOHIO process, propylene, ammonia, and air are passed through a fluidized-bed reactor containing catalyst at 400–510 °C and 1.5–3 bars. Acetonitrile and hydrogen cyanide are byproducts that are recovered for sale [13].

4.4 Uses of Syngas

4.4.1 Syngas to Hydrogen

Growth in the hydrogen market is driven primarily by regulations pertaining to desulfurization of fuel used in transportation and recent interest in the so-called hydrogen economy putting the hydrogen as energy carrier. Use of syngas from steam reforming of hydrocarbons is the most common way to produce hydrogen for refineries because of its reliable technology and its low production cost [14, 15].

As discussed in Chapter 3, the gas produced by the steam reforming reaction first passes through a shift reactor system, where carbon monoxide converts to carbon dioxide and hydrogen. Leaving the shift reactor, the gas contains high concentration of hydrogen, with carbon dioxide, carbon monoxide, and some unconverted methane (plus nitrogen if this was present in feed to the steam reformer). The pressure swing adsorption (PSA) system adsorbs the carbon monoxide, carbon dioxide, and other impurities in a fixed-bed adsorber at feed (high) pressure. The impurities desorb from the bed upon "swinging" the adsorber from the feed to the tail gas (low) pressure, and by using a high-purity hydrogen purge. The adsorbent does not adsorb the hydrogen. Apart from the pure hydrogen product (up to 99.999 vol%), the PSA system produces a low-pressure gas stream, the tail gas. It contains all of the impurities present in the feed gas and some of the hydrogen used for regeneration of the adsorbent. It is used as fuel gas for the steam reformer furnace to provide the heat input. Figure 4.6 shows a simplified diagram of UOP Polybed™ PSA system for purifying the hydrogen to meet downstream process requirements.

To purify the hydrogen in syngas for ammonia production, the syngas from the shift reactor is first washed with a solvent such as monoethanolamide or hot potassium to remove most of the carbon dioxide. Residual carbon dioxide is removed by catalytic methanation, which is conducted over a nickel catalyst. Exit gas from the methanator is cooled and compressed for ammonia synthesis.

4.4.2 Syngas to Ammonia

The manufacture of ammonia is crucial for the world's agricultural industry. The ammonia produced is used by far for manufacture of fertilizers, including urea and ammonium salts (ammonium nitrate, ammonium phosphates, calcium ammonium nitrate).

Hydrogen in syngas is purified, as discussed in Section 4.4.1, and then reacted with nitrogen (usually from air) at a hydrogen to nitrogen ratio of 3, over a catalyst,

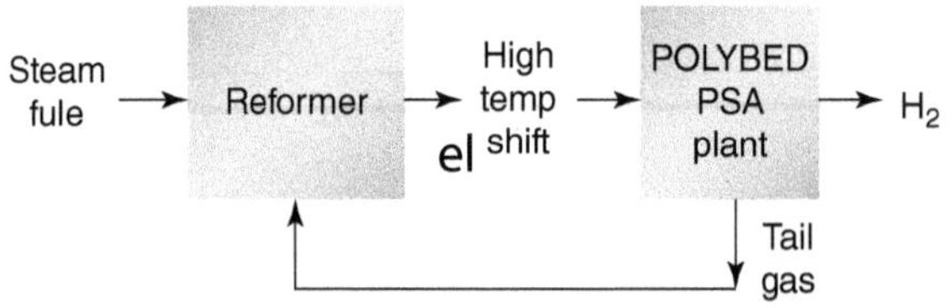

Figure 4.6 Integration of steam reforming with PSA system.

to produce ammonia.

$$N_2 + 3\,H_2 \leftrightarrow 2\,NH_3$$

A wide range of reaction conditions are used to produce ammonia, temperature between 400 and 500 °C and pressure between 150 and 250 bars. As the reaction is exothermic, cool reactants (nitrogen and hydrogen) are added to maintain stable temperatures of the reactors. On each pass, only about 15% conversion occurs, but any unreacted gases are recycled, and eventually an overall conversion of 97% can be achieved [16].

Improvements have been made in ammonia synthesis catalysts. Traditionally, ammonia catalysts have been ion-based materials with promoters. Recent research has shown a ruthenium-based catalyst is much more active to enable the process to take place at lower pressures.

4.4.3 Syngas to Methanol

The natural gas–based technology for making syngas gives higher carbon efficiency and lower production cost. For methanol production, the ideal syngas molar ratio is just over 2 : 1 (H_2:CO). The steam-reforming process produces a hydrogen-rich syngas stream where excess hydrogen is typically purged from the system in the methanol synthesis loop and is usually used as a fuel in the reformer.

The syngas from coal has a low H_2:CO ratio for methanol synthesis, requiring a water shift plant to increase the H_2:CO ratio, which has a negative impact on the overall carbon efficiency.

Methanol is produced via the following exothermic reactions:

$$CO + 2H_2 \rightarrow CH_3OH$$

$$CO_2 + 3H_2 \rightarrow CH_3OH + H_2O$$

The methanol synthesis reactions are carried out in a water-cooled reactor at pressure of 50–100 bars and temperature of 220–260 °C in presence of a copper-based catalyst.

References

1 Schuster, C.E. (2005). ExxonMobil high-pressure process technology for LDPE. *Handbook of Petrochemicals Production Processes* (ed. R.A. Meyers), pp. 14.45–14.58. NY: McGraw-Hill.

2 Mirra, M. (2005). Polimeri Europa polyethylene high-pressure technologies. *Handbook of Petrochemicals Production Processes* (ed. R.A. Meryers), pp. 14.59–14.70. NY: McGraw-Hill.

3 Smith, M. (2005). Chevron Phillips slurry-loop-reactor process for polymerizing linear polyethylene. *Handbook of Petrochemicals Production Processes* (ed. R.A. Meryers), pp. 14.31–14.44. NY: McGraw-Hill.

4 Chanda, M. and Roy, S.K. (2006). *Plastics Technology Handbook*, 1–6. CRC Press.

5 Allsopp, M.W. and Vianello, G. (2012). Poly(vinyl chloride). In: *Ullmann's Encyclopedia of Industrial Chemistry* (eds. Barbara Elvers, Stephen Hawkins, and William Russey), pp. 717–742. Weinheim: Wiley-VCH.

6 McKetta, J.J. (1984). *Encyclopedia of Chemical Processing and Design*, vol. 20. New York: Marcel Dekker Inc.

7 Othmer, K. (2004). *Encyclopedia of Chemical Technology*, 4e, vol. 9. New York: Marcel Dekker Inc.

8 Shell Magazine 2006. *Spring*; https://www.shell.com/business-customers/catalysts-technologies/licensed-technologies/petrochemicals/ethylene-oxide-production/omega-process.html.

9 Whiteley, K.S., Geoffrey Heggs, T., Koch, H. et al. (2005). Polyolefins. In: *Ullmann's Encyclopedia of Industrial Chemistry* (ed. Claudia Ley), pp. 36–69. Wiley-VCH: Weinheim.

10 Devesh, Tripathi (2002). Practical guide to polypropylene. UK: Shawbury. https://www.worldcat.org/oclc/568032693

11 Gahleitner, M. and Paulik, C. (2014). Polypropylene. In: *Ullmann's Encyclopedia of Industrial Chemistry*, 1–44. Wiley-VCH Verlag GmbH & Co. KGaA. https://doi.org/10.1002/14356007.021_o04.pub2. IBBN 9783527306732.

12 Nijhuis, T.A., Makkee, M.M., Moulijn, J.A., and Weckhuysen, B.M. (2006). The production of propene oxide: catalytic processes and recent developments. *Industrial & Engineering Chemistry Research* 45 (10): 3447.

13 Brazdil, J.F. Acylonitrile. In: *Ullmann's Encyclopedia of Industrial Chemistry* (ed. Claudia Ley), pp. 1–10. Weinheim: Wiley-VCH.

14 Ngoh, S.K. and Njomo, D. (2012). An overview of hydrogen gas production from solar energy. *Renewable and Sustainable Energy Reviews* 16: 6782–6792.

15 Cho, W.C., Lee, D.Y., Seo, M.W. et al. (2014). Continuous operation characteristics of chemical looping hydrogen production system. *Applied Energy* 113: 1667–1674.

16 Appl, M. (2006). Ammonia. In: *Ullmann's Encyclopedia of Industrial Chemistry* (ed. Claudia Ley), pp. 1–33. Weinheim: Wiley-VCH. https://doi.org/10.1002/14356007.a02_143.pub2.

Section III

Aromatics

5 Aromatic Production Technologies

5.1 Introduction

Aromatics are cyclic hydrocarbons with delocalized π electrons forming a planar ring system with alternating double and single bonds. With the resonance of the conjugated bonds in a ring system, aromatics are stable molecules and yet highly reactive with excellent ability for electrophilic and nucleophilic aromatic substitutions. Historically, aromatics were originally found as a lighting gas in the eighteenth century. Benzene was later discovered in 1825 by Michael Faraday by recovering the condensed part of a lighting gas found in whale oil. Benzene in coal tar was later found in 1845 by August Wilhelm von Hofmann. The early findings show that coal tar possesses a variety of aromatic components that can be used as fuels. In the 1920s, the primary sources of aromatics came from coal tar and coke-oven benzole, which derive from thermal processing of coal. With excellent octane as a fuel property, aromatics were in high demand for fuel supply during World War II. In 1940, Standard Oil Company and M.W. Kellogg Limited developed the hydroforming process to upgrade virgin naphthas. Toluene was extracted to produce explosives (trinitrotoluene). The process used molybdenum oxide on alumina as a catalyst. In 1949, UOP innovated and commercialized the first platinum-based reforming process, the Platforming™ process. It provided better yields and higher-octane products for motor fuels. The catalytic reforming process also has contributed to the petrochemical industry as the primary source of industrial aromatics, overtaking conventional coal tar and coke-oven production.

5.2 Aromatic Production

Aromatics can be produced via thermal or catalytic processes. While thermal conversion processes involve higher temperature and nonselective cracking/dealkylation making various byproducts, catalytic processes make use of selective catalysts for on-purpose production of aromatics and tailor rearrangements of the aromatic molecules to improve *para*-xylene yield. Both thermal and catalytic processes complement each other with mature technologies and reliability to meet market demands.

Modern Petrochemical Technology: Methods, Manufacturing and Applications, First Edition.
Santi Kulprathipanja, James E. Rekoske, Daniel Wei, Robert V. Slone, Trung Pham, and Chunqing Liu.

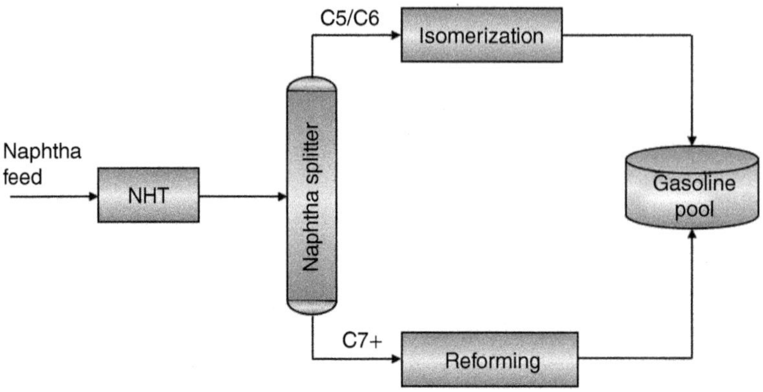

Figure 5.1 Typical naphtha complex configuration.

5.2.1 Reforming Technology

One of the major technologies that contribute to the production and consumption of BTX in the world is catalytic reforming. Historically, catalytic reforming was first developed in 1930s by UOP to produce high-octane aviation and commercial gasoline replacing lead additives for octane booster. The UOP Platforming process can produce high-octane gasoline, aromatics that are raw materials for further processing downstream in petrochemicals, and valuable byproducts hydrogen and liquefied petroleum gas (LPG). For gasoline production, naphtha-reforming product is a great octane contributor with high concentration of aromatics directly correlated with high-octane value. A typical reformate (C5+ liquid) includes 50–85 vol% aromatics, 10–40 vol% paraffins, less than 1 vol% olefins, and less than 1 vol% naphthenes.

For a traditional naphtha complex, a naphtha feed stream is hydrogenated in the naphtha hydrotreating (NHT) unit and the hydrotreated naphtha is sent to a naphtha splitter (Figure 5.1). The light naphtha components C5/C6 paraffins are routed to the isomerization unit, while the heavier C_7^+ are sent to the Platforming unit. For fuel production and blending requirements, the benzene level in the gasoline pool is an important parameter that impacts the naphtha complex configuration. Benzene and benzene precursors can be routed to the isomerization unit if the minimum benzene level is required in fuel specifications. On the other hand, if maximum benzene or aromatics is desired, then benzene precursors would be sent to the Platforming unit. Optimization of the C6 components can be considered when sending these components to the isomerization to offload the Platforming unit. There are several advantages: more capacity can be pushed through the Platforming unit with much less C6 components resulting in increased aromatic production with reduced benzene concentration in the reformate. Another important advantage is that the Platforming unit can be operated at lower severity since C6 components are among the hardest to convert.

The isomerization unit couples hand in hand with a Platforming unit to help convert C5 and C6 components to higher-octane components or isomerate while reducing benzene level at the same time. This allows the refineries to recover lost octane if benzene is removed elsewhere by other processes. If high level of Benzene is present (~10–20% volume) in the isomerization feed, the UOP Penex Plus® process has been a workhorse to convert high-level benzene to isomerate product. This process has a Bensat® reactor, where benzene is hydrogenated, followed by Penex® isomerization reactors. Note that Euro-V gasoline requirements allow maximum content of aromatic at 35 vol%, with 1.0 vol% max benzene.

The role of the NHT is key to protect and maintain the life of catalysts in the isomerization and Platforming units. Many types of naphtha can be processed in the NHT unit, including crude straight-run naphtha (SRN), fluid catalytic cracking (FCC) naphtha, hydrocracker, and/or coker naphtha. While shale oil or tight oil crudes have low contaminants, Canadian bitumen or heavier crudes possess high levels of sulfur, nitrogen, and metals. In the NHT unit, the sulfur and nitrogen impurities in the naphtha feed are converted to hydrogen sulfide (H_2S) and ammonia (NH_3) in the presence of the hydrotreating catalyst. Olefin and diolefin components are also hydrogenated to paraffins. The hydrogenation reaction exothermicity is generally small for straight-run naphtha while higher with coker naphtha or naphtha processed from heavy vacuum residues. The operating severities can be quite different and need to be evaluated for increased amount of coker naphtha in the feed. The hydrotreating catalyst is typical bimetallic catalyst, including group VIII metals such as nickel, cobalt, and molybdenum. For straight-run, low N naphtha, UOP's HYT-1118 catalyst is optimized to increase hydrodesulfurization (HDS) and hydrodenitrification (HDN) activity while being more than 30% less dense than the proven performance S-120 catalyst. For cracked naphthas with high nitrogen level in the feed, UOP's H-1119 is a more robust catalyst for selective sulfur and nitrogen removal. The catalyst can also be regenerated with standard regeneration methods to minimize catalyst changeout costs.

Some features and benefits [2] of HYT-1119:

- Improved metal dispersion
- Increased surface area (20% higher for higher silicon pickup) and stability
- Lower aromatic saturation
- Lower fill cost with reduced loaded density
- Higher HDS and HDN activity

The heart of the naphtha complex is the reforming unit where a heavy naphtha cut is converted to a reformate product that is rich in aromatics. The feed naphtha generally contains C6 to C8 to target the production of benzene, toluene, and xylenes. The feed is typically characterized by hydrocarbon types (paraffins, naphthenes, and aromatics – P, N, A). The naphthene-rich feed can easily convert to aromatics due to the ease of aromatic ring formation from already-cyclic hydrocarbons. Thus, naphthene dehydrogenation and dehydrocyclization are important pathways to get to the desired aromatics requiring both metal and acid functions (Figures 5.2 and 5.3). The metal is mainly for dehydrogenation, while the acid function is for isomerization or ring cyclization. The reforming catalyst needs to be optimized to have the balance of metal and acid functions to preserve yields and avoid byproducts such as light ends from cracking or dealkylation. Paraffin hydrocarbon is the most difficult type in the feed to convert to a ring due to competitive pathways between hydrocracking and isomerization and dehydrocyclization. While certain steps or reactions can occur reversibly due to equilibrium, hydrocracking is not reversible. Once the light ends or LPGs are made, they cannot recombine back to form higher chain length hydrocarbons. The dealkylation and demethylation reactions are also not reversible (Figure 5.4). Over the years, Platforming catalysts have been developed and optimized to make maximum liquid yield (or C5+ yield) and to maintain stability even at high operating severities.

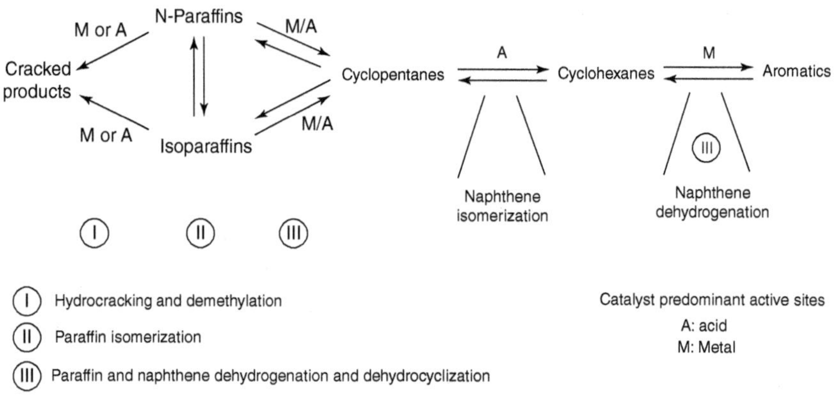

Figure 5.2 General platforming reaction mechanism.

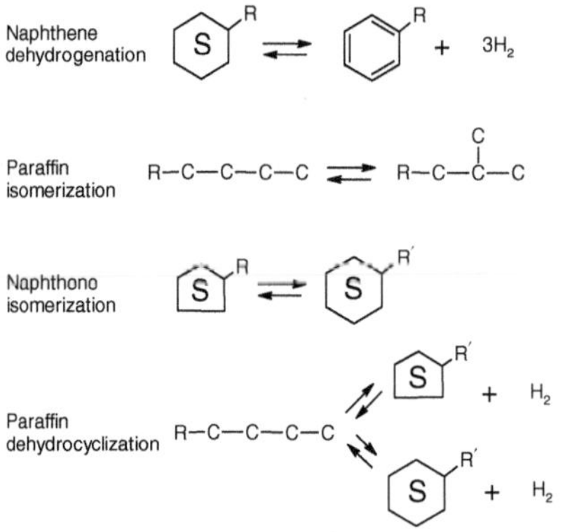

Figure 5.3 Major reactions happening in the reforming reactor.

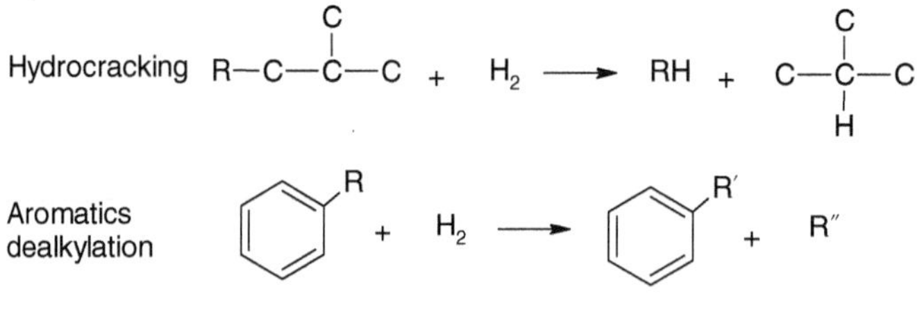

Figure 5.4 Undesirable reactions in the reforming reactor.

5.2.1.1 Semi-regenerative Reforming

Catalytic reforming process has been a go-to process for refineries around the world to upgrade low-octane naphtha streams to higher-octane motor fuel–blending components. The naphtha materials can be from various sources, including straight-run naphtha from crude units and cracked naphtha from thermal processes, coking, or hydrocracking units. Significant milestones in catalyst development and commercialization were achieved in the late 1960s with the advent of bimetallic Platforming catalysts. UOP developed R-16, R-18, R-50, R-60, and R-80 series catalyst with optimization of activity and stability over the years. With further development by adding a proprietary promoter metal to enhance catalyst activity and selectivity, UOP introduced R-98 and R-500. The most recent addition is R-560 (best in class for stability), which is a modified base catalyst that is more robust to harsh operation conditions and has a longer cycle length. These catalysts have been in operation and proven to give better activity, stability, and selectivity than the all-platinum catalysts. They enabled refiners to extend their Platforming operations to higher charge rates and higher product octane levels with existing constraints and equipment.

As of today, more than 600 UOP fixed-bed Platforming units are onstream. For motor fuel production, catalytic reforming can be coupled with isomerization units to produce targeted octane grades to meet gasoline-blending specification.

5.2.1.2 Cyclic Reforming

Cyclic reforming also includes fixed beds of catalytic reactors. In addition, there is a spare same maximum size reactor, which allows swing operation when one of the reactors is taken out for service or regeneration. The cyclic operation is more complex with valve switches and other safety precautions; however, the reforming process can be maintained continuous and may allow operators to run at lower or moderate pressures to achieve higher reformate yield and hydrogen production.

5.2.1.3 CCR Reforming

In 1971, the first continuous catalyst regeneration (CCR) Platforming unit was started up, which revolutionized the reforming process. Extremely high severities can be achieved without frequent shutdowns with catalyst deactivation. As of 1999, there were 156 CCR Platforming units onstream. Since the introduction of the UOP CCR Platforming™ process in 1971, more than 250 units have been commissioned and in operation. With lower-pressure operation and higher severity, more hydrogen yield can be achieved. The hydrogen co-product in the reforming process plays a critical role for other processes in the refinery to produce other key products such as high-quality, low-sulfur diesel that meets stringent regulations.

Over the years, UOP has developed a wide range of CCR Platforming catalysts to meet customers' demands on better stability and higher performance, especially on yields, octane, or maximization of aromatics production. The evolution of various catalysts is summarized in Figure 5.5 [2]. The R-130 series catalyst was introduced in 1993 with high activity (lower reactor inlet temperatures), good surface area stability, and chloride retention. Low-coke R-234 catalyst came about in 2000 that gives 20–25% lower coke make and higher C5+ yield vs. R-134 catalyst while maintaining

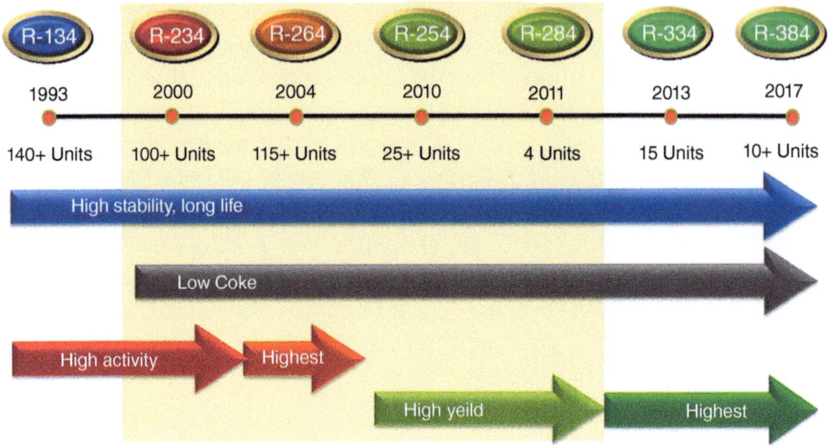

Figure 5.5 UOP CCR platforming catalysts in commercial operations.

Table 5.1 UOP CCR Platforming catalyst properties.

Catalyst	**R-234**	**R-254**	**R-334**
Shape	Sphere		
Pill diameter	1/16 in. (1.6 mm)		
Surface area	180 m^2/g		
Density (ABD)	35 lb/ft^3 (560 kg/m^3)		
Pt level	0.29 wt%		
Crush strength	50+ N		
Promoter	No	Yes	No

Source: Wier et al. [2]. © 2014, Curran Associates.

excellent surface area stability and chloride retention as the R-130 series. R-264 catalyst was introduced to markets that offer high yield, so customers can increase their throughput. With higher density, this catalyst has increased reactor pinning margin. With high-throughput benefit, the refiners can dial up their feed rate, and with higher activity, maximum BTX aromatics production can be achieved. R-254 and R-284 are promoted catalysts that can offer higher yields. UOP introduced 300 series products with R-334 catalyst in 2013 and the latest R-364 catalyst in 2017 providing highest yield, long life, and low coke. These catalysts, however, do not include a promoter and are primarily optimized for high performance by using proprietary base and manufacturing technique (Table 5.1). The latest-generation R-364 builds on the success of UOP's proven R-264 catalyst with over 100 loads sold. R-364 provides great flexibility in operation for the refiner with two different modes, high activity or high yield.

The first catalytic reforming unit was installed in 1949. The unit was the first commercial UOP designed semi-regenerative process, which includes three fixed-bed reactors. The use of a noble metal catalyst has enabled catalytic reforming a workhorse to produce high-octane gasoline. Typically, the semi-regenerative reforming unit can operate from 12 to 24 months until the catalyst loses its activity and selectivity. Catalyst is then regenerated in situ. Thus, higher-pressure operation generally favors cycle length, keeping the unit onstream longer.

The catalytic process has evolved over times from semi-regenerative, cyclic reforming to continuous catalytic regeneration (https://www.uop.com/processing-solutions/refining/gasoline/ https://www.uop.com/processing-solutions/refining/gasoline/#naphtha-reforming).

Process Variables:

- Reactor temperature
- Reactor pressure
- Space velocity
- Hydrogen/hydrocarbon ratio (H_2/HC ratio)
- Feedstock

Reactor Temperature

This is an important variable that directly controls the severity of the process to affect activity, overall yields, or product quality (RON). High severity is referred to as high-temperature operation while keeping other conditions constant. At high severity, high-octane reformate can be achieved at the expense of catalyst activity. Typical range for reactor inlet temperature is 490–550 °C. Higher-range temperature increases the rate of catalyst deactivated with rapid coke laydown leading to more frequent regeneration.

Reactor Pressure

A reforming reactor can typically run at 3.5–30 bar pressure. Lower pressures favor higher reformate and H2 yield, while higher pressures increase longer life of catalyst with lesser coke buildup with lower reformate yield compared to lower-pressure operation. The lowest operating pressures are only run in reforming units with CCR. Note that high-pressure operations tend to give lower reformate yield and more light end byproducts. This is a direct result of undesirable hydrocracking of paraffin components favoring light end formation rather than aromatization or ring formation.

Space Velocity

Space velocity is defined as the ratio of the volumetric feed rate to the amount of catalyst. A common acronym for space velocity is liquid hourly space velocity (LHSV) in units of h^{-1}. It measures how fast the feed passes the catalyst. Lower space velocity means larger catalyst volumes (more active sites available), and severity or temperature can be lower to achieve the same activity or product quality. However, higher space velocity (by pushing more feed flow) requires higher temperature to increase reaction rates for the same product target. With fixed design of the reactor for a given capacity, LHSV can only be varied by changing the naphtha charge rate to the unit.

Hydrogen/Hydrocarbon Ratio (H_2/HC Ratio)

Hydrogen is required to maintain catalyst stability, prolong catalyst life by keeping coke formation under control. Thus, hydrogen/hydrocarbon ratio, or H_2/HC, is defined as the ratio of the moles of hydrogen co-fed with the moles of feed naphtha. The H_2/HC can be adjusted by varying the rate of the recycled gas going back to the reactor. The hydrogen purity from the separator overhead gas also has a direct impact on the H_2/HC. To maintain a low catalyst deactivation rate (or less severity), higher partial pressure of hydrogen is required and thus higher H_2/HC. However, utility consideration must be considered when recycling more recycled gas to maintain H_2/HC.

Feedstock

Feedstocks to a reforming unit are characterized by distillate cut points and components, including paraffins, naphthenes, and aromatics (PNA). Feeds with high paraffin components are harder to convert to aromatics due to slow reaction rates. Naphthene components convert at a much faster rate and get almost consumed in the very first reactor. Thus, rich naphthene feedstocks would require less severity to reach a product octane target and aromatic concentration. On the other hand, lean naphthene feedstocks (more paraffin concentration) require more severe operating conditions to achieve the product quality target. Regarding boiling range, feedstocks with a higher end point or a wide boiling range require more severe operating conditions. If the target product is fed to an aromatic complex, the boiling range of the feedstock to the reformer is limited to a more narrow cut range so that the desired aromatics, such as xylenes (and less C9 aromatics), can be obtained.

5.2.1.4 High-pressure vs. Low-pressure Consideration

Without continuous catalytic regeneration, reforming units were operated at higher pressure with high H2 partial pressure to suppress coke formation and increase run length. However, the reaction equilibrium is limited at higher pressure, and thus lower overall yield of C5+ product is typically observed. With UOP introduction of Continuous Catalytic Regeneration in 1971, CCR Platforming process allows units to operate at lower pressure with higher yields of aromatics while maintaining catalytic activity by continuous catalyst circulation. CCR helps Platforming units achieve better yields approaching toward equilibrium limit, producing high quantities of aromatics with maximum octane barrels (Figure 5.6).

It is noticeable to see the reverse effect of pressure on C_{5+} reformate yield ranging from 75–91 LV% (Figure 5.8). While CCR Platforming allows low-pressure operation without having to shut down the unit for regeneration, semi-regenerative reforming or fixed-bed reforming cannot afford to operate at high severity and such low pressure. An ideal reformer is the one that can operate at highest severity and lowest pressure as possible to maximize yields. The catalyst should have high activity for max octane product or aromatic concentration. Besides product quality, the stability or onstream factor needs to be high, so the units continuously operate without the need for shutdown losing productivity. In this direction, CCR reforming > Cyclic reforming > Semi-regenerative reforming (Figure 5.7).

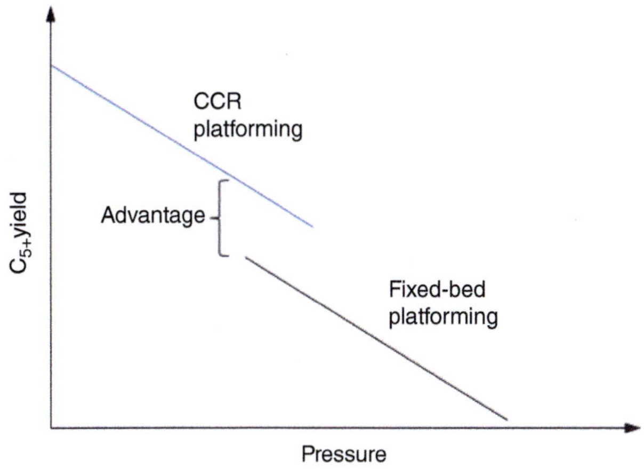

Figure 5.6 Reformate yield-pressure relationship for CCR and fixed-bed platforming.

UOP CCR Platforming is the leading catalytic technology with more than 250 units licensed in the world, producing rich aromatic reformates for either refining or petrochemical uses. UOP RZ Platforming is another technology that converts C6 and C7 materials in the raffinate stream rejected from extractive distillation to produce additional aromatics. With RZ-100 catalyst used in a fixed-bed reactor system, RZ Platforming offers a way to convert difficult components to aromatics, thus reducing the naphtha feed requirements for a target aromatics capacity.

5.2.2 Aromatic Transalkylation and Disproportionation Technology

A catalytic reforming process can produce relatively high yields of BTX aromatics. However, the selectivity or distribution of B, T, X aromatics cannot be controlled or manipulated easily in a reformer alone but rather depending on the feed composition or cut points allowed to enter the unit.

Among benzene, toluene, and xylene aromatics, benzene and xylene are of stronger demand with higher-value aromatic derivatives than toluene. Aromatic transalkylation offers the ability to convert toluene or C9 aromatics to more xylene product. Thus, this technology allows specific target conversion of C7 and C9 aromatics components to more desired xylene to meet the market demand while unloading the C7 and higher aromatics traffic that can build up in an aromatic complex. In older aromatic complexes, thermal dealkylation (THDA) process is very common for direct dealkylation of toluene to produce more benzene. This process can be inefficient in that the methyl group is rejected as methane, and thus the methyl group loss cannot be recovered. However, with transalkylation technology, the methyl group is retained and not wasted by alkylating with another aromatic molecule to further produce xylene, which is a highly desired product. Furthermore, the methyl group can also be introduced from an outside source such

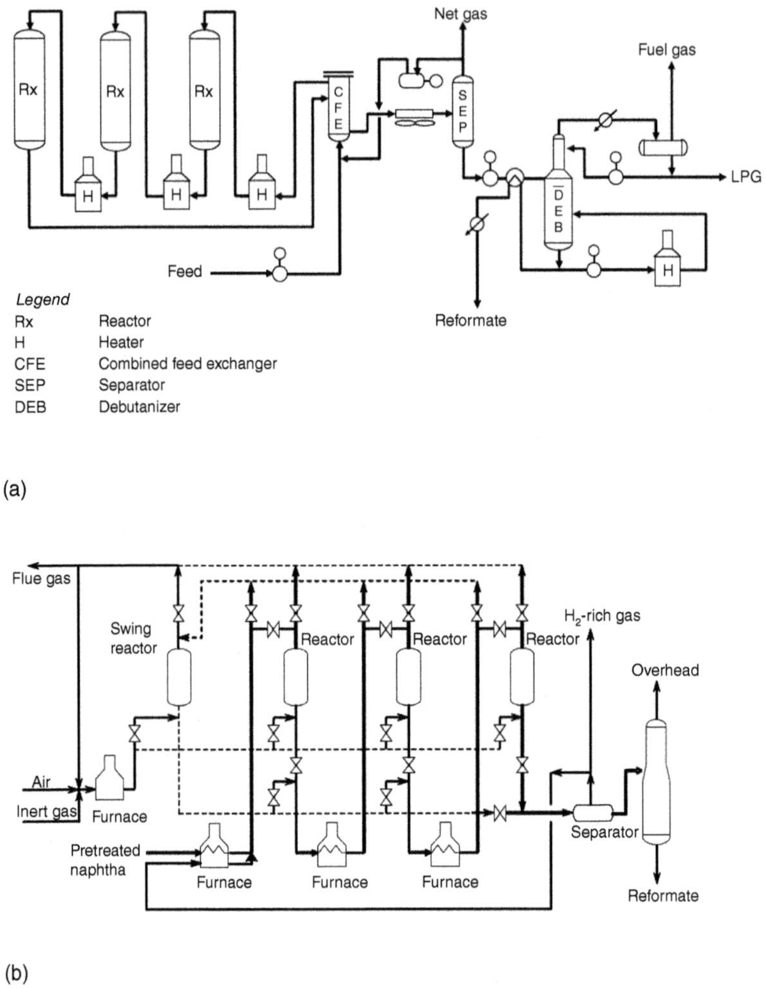

Figure 5.7 Typical process diagrams of commercial reforming processes, including (a) semi-regenerative, (b) cyclic regenerative, and (c) continuous catalyst regeneration.

as methanol injection or in a dedicated process unit such as toluene methylation process with the same purpose of increasing xylene yield.

In 1969, the UOP Tatoray™ transalkylation process was commercialized from the joint development of UOP with the inventor, Toray company [4]. UOP then became the licensor of the process. Figure 5.9 shows the process-flow diagram for the UOP Tatoray unit. ExxonMobil also developed a similar transalkylation process in 1976. The reactor is a fixed-bed type reactor converting toluene or a mixture of toluene and C9 aromatics in the presence of a catalyst and H_2 at a ratio of 3–12 H_2/toluene. Hydrogen flow is required or co-fed in the reactor to maintain stable operation preventing deactivation or coke formation.

The catalyst for this reaction was developed in the form of extrudates with a supported noble metal on alumina or zeolite, or a rare-earth catalyst [4]. Operating

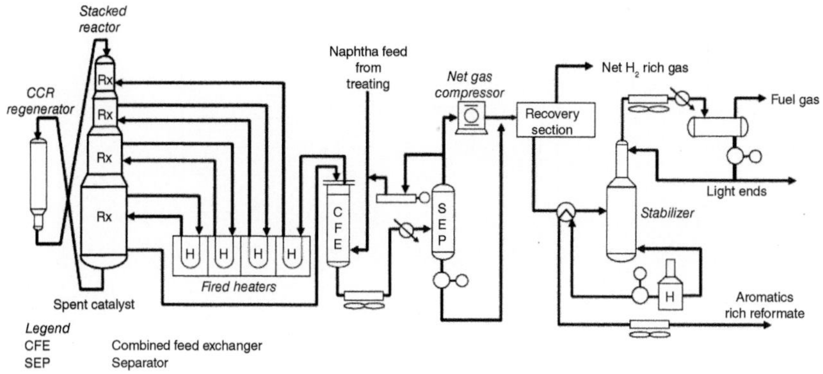

(c)

Figure 5.7 (*Continued*)

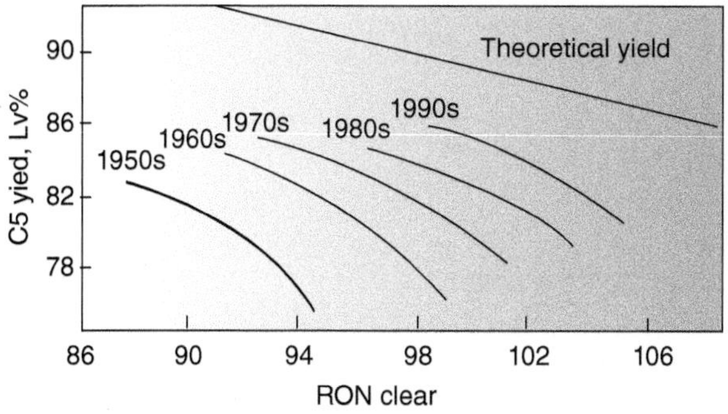

Figure 5.8 Reformate – octane relationship

conditions are reported at 350–530 °C, and 20–50 bar [4]. The xylene products obtained are in equilibrium with 23–25% *p*-xylene, 50–55% *meta*-xylene, and 23–25% *ortho*-xylene. Also, note that the equilibrium xylenes produced in the Tatoray process contain low ethyl benzene (EB). This low EB concentration product makes Tatoray process attractive and ready for isomer separation using Parex™ unit or *para*-xylene crystallization unit.

For reaction mechanism, the two main pathways are toluene disproportionation and transalkylation of toluene and C9 aromatics. With methyl group ability to jump over from one ring molecule to another in the presence of the catalyst, the methylated aromatic products are limited by equilibrium. Figure 5.10 illustrates the equilibrium compositions of aromatic products for any given feed methyl/phenyl ratio. For 100% feed, the methyl to phenyl ratio is 1, while all C9 aromatic feed would have a 2 : 1 of methyl to phenyl ratio. The more methyl groups available in the feed, the more chances or favorable equilibrium to proceed toward alkylated aromatic products. However, two high methyl/to methyl ratios also lead to production of heavier alkyl aromatics such as C10 and C11 aromatics. Thus, a typical 50 : 50

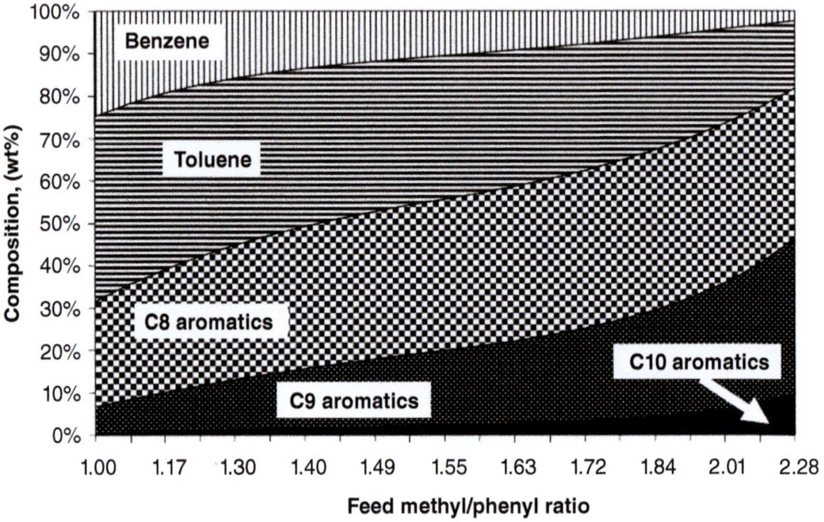

Figure 5.9 Equilibrium compositions of aromatics with various feed methyl/phenyl ratios at 700 K.

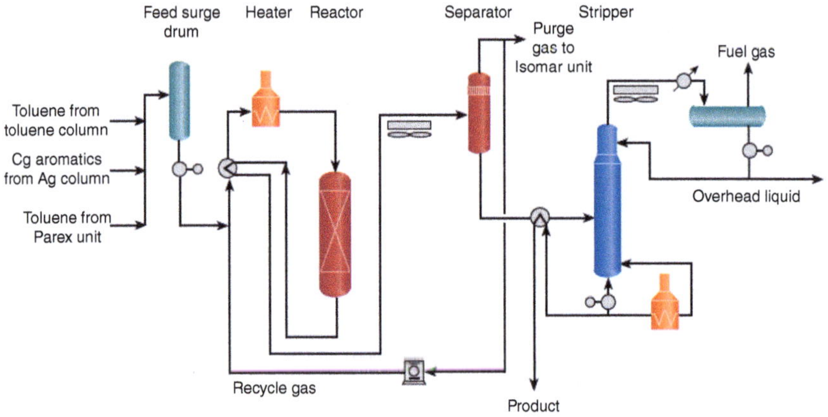

Figure 5.10 UOP Tatoray flow diagram.

mixture feed of toluene and C9 aromatics is a good feed to produce xylenes. Conversion per pass is from 46% to 59% [4]. Higher conversion reduces the amount of recycle materials through the benzene-toluene fractionation section of the complex for equipment cost and utility consumption considerations. However, pushing high conversion may result in benzene and xylene selectivity drop, faster catalyst deactivation, and increase rates of ring loss or formation of heavy aromatic by products.

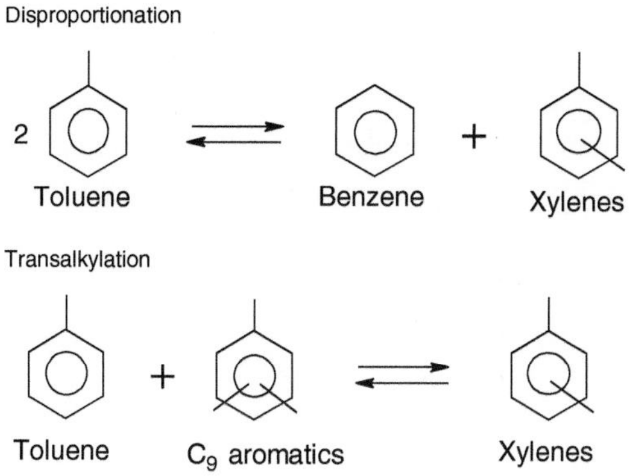

Transalkylation mechanism shows a conversion pathway for C9 aromatics to additional xylenes. This allows more C9 aromatic precursors to be fed to a reforming unit, and thus the end cut point for the naphtha feed is not restricted to exclude C9 components. Also, having C9 components together with toluene increases the methyl to phenyl ratio that favors equilibrium formation of xylenes. For older petrochemical complex, thermal hydrodealkylation provides a simple solution to convert toluene to additional benzene. However, this process does not allow additional production of xylenes, which may be more valuable than benzene. Thus, the amount of C9 components or precursor materials is strictly limited in the original naphtha feed entering the reforming unit and subsequent aromatic complex since circling these components around in the complex consumes utilities and is thus generally not favored.

As of 2013, UOP has licensed a total of 60 Tatoray units with a wide range of catalyst portfolio, including TA-4, TA-5, TA-9, TA-20, TA-30, and TA-32. The latest generation of catalysts provides high-stability catalyst with improved performance and lower overall ring loss to light ends or benzene co-boiler products. The feedstocks can vary widely from light feed 100 wt% toluene to heavy feed mixture of 30 wt% toluene and 70% C9 aromatics depending on feedstock availability upstream of the reforming unit and overall complex capacity.

5.2.2.1 Selective Toluene Disproportionation Process

While transalkylation process can produce xylenes at equilibrium composition, with about 25% *p*-xylene, selective toluene disproportionation can produce highly selective product, *p*-xylene, at about 90%, much higher than equilibrium composition. UOP PX-Plus selectively disproportionates toluene to benzene and xylenes with a selective catalyst that includes a porous structure allowing benzene and *p*-xylene to escape while inhibiting the diffusion of *o*-xylene and *m*-xylene. The reaction mechanism of disproportionation of toluene occurs via a biomolecular intermediate. Upon cleaving of the intermediate to benzene and xylene, the shifting of methyl groups can occur on the xylene molecule in the presence of the catalyst.

The process equipment for selective toluene disproportionation are similar to those for the transalkylation process; however, the feed is more exclusively toluene. The stoichiometry of the disproportionation reaction shows equal molar of benzene and xylene made. Thus, the selective toluene disproportionation process is highly suitable for producers who desire to make significant quantities of benzene together with *p*-xylene.

5.2.3 Isomerization of Aromatics

Xylene mixture produced from a reforming unit typically includes equilibrium isomers such as *para*-xylene, *ortho*-xylene, *meta*-xylene, and ethylbenzene. With equilibrium composition of xylenes, it is desirable to isomerize ethylbenzene and other xylenes to *para*-xylene, which is highly desired to produce polyethylene terephthalate (PET). The isomerization process is usually accompanied with a selective separation process, either by crystallization of selective adsorption to remove *para*-xylene from the equilibrium mixture. For adsorptive separation, UOP Parex process can separate out highly pure *para*-xylene product up to 99.9 wt% purity and 97 wt% recovery. The *para*-xylene product is referred to as the extract, while the rest of the xylene isomers are retained in the raffinate stream. The raffinate stream can then be contacted with an isomerization unit, such as a UOP Isomar™ unit, to convert to more *para*-xylene and the equilibrium product can be again separated in the downstream fractionation and Parex unit. The side-by-side Parex and Isomar units are typically referred to as Parex–Isomar loop and the primary intention is to break the equilibrium composition of xylenes by reaction coupling with separation. Having the separation unit (Parex) is convenient for product removal (*para*-xylene) to get it out of the equilibrium system and unload the material flows or traffic in the reaction-separation loop. Theoretically, all the xylene isomers can be cycled to extinction and converted to all *para*-xylene with such as a system like this. However, for such highly integrated aromatic complex shown in Figure 5.11, *ortho*- and *meta*-xylene can also be produced or extracted by boiling-point fractionation or separate adsorptive process (MX Sorbex), respectively.

5.2.3.1 Reaction Mechanism

The reactions for C8 aromatic isomerization process proceed with two main categories: ethylbenzene dealkylation and ethylbenzene isomerization, the first pathway provides valuable byproduct benzene, while the second pathway preferably isomerizes ethylbenzene to xylenes. Other xylene isomers can isomerize by re-rearranging the methyl group to an equilibrium composition the reaction condition. The process chemistry requires both acid site and metal site for the reactions to proceed. Xylene isomerization takes place on acid sites, while metal sites are required for aromatic ring hydrogenation and naphtha ring dehydrogenation. To a certain extent or increased operating severities, ring loss may occur by ring opening followed with cracking to light ends. Dealkylation also occurs, but this does not lead to ring loss. The byproduct again can be valuable such as benzene or toluene that can be recovered and converted in the Tatoray unit.

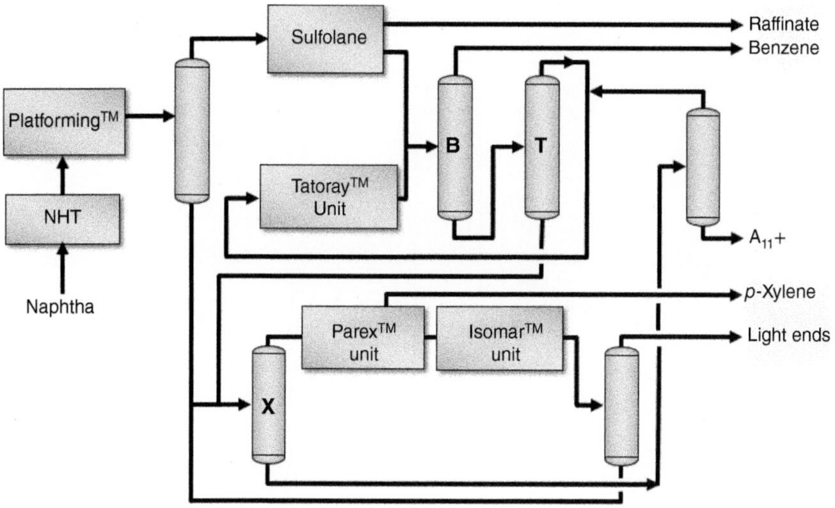

Figure 5.11 Typical UOP aromatic complex.

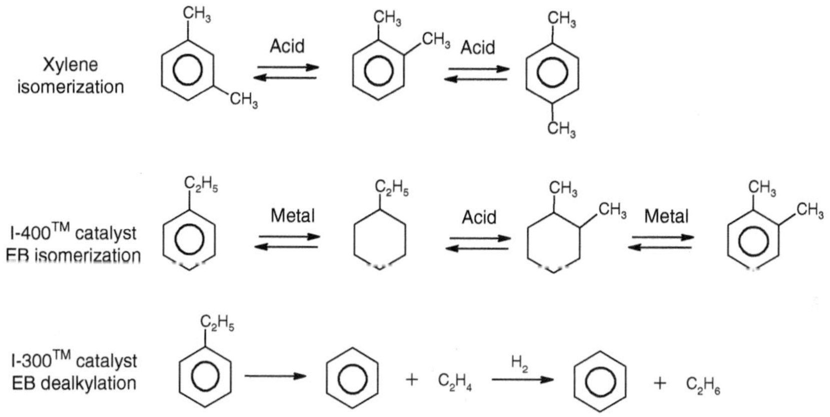

The Isomar flow diagram is shown in Figure 5.12. The feed usually comes from the Parex unit or referred to as Parex raffinate with depleted *para*-xylene. The feed is then combined with hydrogen-rich recycle gas and make-up gas. Small amount of hydrogen is consumed across Isomar reactor. The combined feed is heated and vaporized to reactor operating temperature. The reactor is a fixed-bed radial-flow reactor where hot feed is passed radially through the catalyst bed. The reactor effluent stream is heat exchanged with the feed and sent to the product separator where liquid and gas are recovered. A small gas stream is purged to avoid light ends buildup in the recycle gas. The liquid effluent is stabilized in the deheptanizer where overhead gas is taken as fuel gas to supply fuel in the complex. The overhead liquid is usually sent back to sulfolane unit for benzene recovery. The deheptanizer bottoms product is sent to xylene splitters for separation and eventually makes its way back to the Parex and Isomar units.

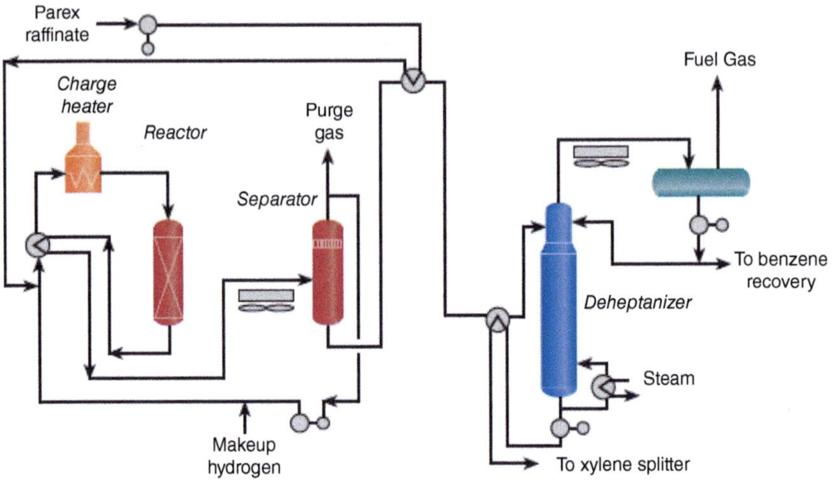

Figure 5.12 UOP Isomar flow diagram.

Based on the reaction mechanism highlighted for C8 aromatics, the catalysts can be tailored to maximize xylene product vs. xylene and benzene co-product. UOP Isomar catalysts I-350 and I-400 can be different in that I-350 can selectively convert EB to benzene with a selectivity more than 90 mol%, while ring retention is high with more than 99 mol% of aromatic rings conserved. Also for I-350 catalyst, byproduct such as toluene can be re-converted back to xylenes with the presence of a Tatoray unit in the complex. The EB dealkylation pathway is not equilibrium limited and thus conversion can be as high as 65–90 wt% per pass that can substantially unload the recycle materials in the Parex–Isomar loop. The I-400 type catalyst is specifically designed for isomerization of xylenes, which is equilibrium limited reaction. The conversion of EB is limited to 30 wt% per pass due to overall xylene equilibrium. The overall *para*-xylene yield is slightly higher for the I-400 catalyst as shown in Figure 5.13 due to more selective catalyst and lower operation severity. Both catalysts have commercially shown very high aromatic ring retention corresponding with high yields of xylene and benzene and low hydrogen consumption.

With classical chemical engineering reaction and separation to overcome equilibrium, UOP Isomar and Parex continue to be workhorse for the petrochemical industry. With highly energy-efficient design for aromatic complexes, UOP introduced the Isomar and Parex, which is a selective adsorption process that can deliver *para*-xylene with high recovery, 97 wt% and high purity, up to 99.9 wt%. The high recovery of the adsorption process tremendously helps debottleneck the flow traffic and recycle from the distillation and Isomar–Parex loop. The energy consumption or utilities are thus lowered compared to other separation process, including crystallization, which is based on the difference in melting points of xylene isomers. The eutectic composition of mixed xylene has 13 wt% of *p*-xylene at −52 °C, which limits the recovery of the separation, and thus can be as high as 60–70% per pass. For a

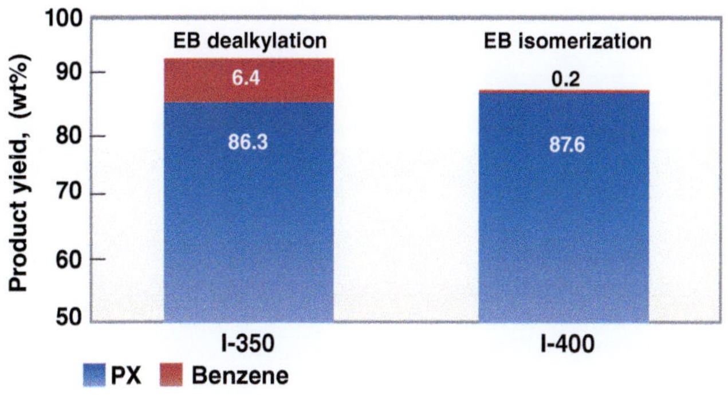

Figure 5.13 Overall Parex–Isomar yields for various isomerization catalysts.

given xylene column and isomerization unit, Parex process can deliver more than 50% more *p*-xylene with higher recovery of separation per pass.

5.2.4 Toluene Methylation

Toluene methylation has recently gained traction as an emerging technology for on-purpose *para*-xylene production. While other technologies revolve around re-arranging methyl group from an internal feed or mixture to increase overall yields of xylenes, toluene methylation directly introduces a methyl group addition from an external source to a C7 aromatic. The methyl group is introduced via a methanol molecule, which is a flexible intermediate for petrochemical production either for olefins or aromatics.

Many licensors in the world have developed and commercialized this technology, including UOP, ExxonMobil, SABIC, GTC, and Sinopec. The catalyst choice is typical MFI zeolite or modified MFI base for selective *para*-xylene production. The reaction pathway is direct alkylation of toluene with methanol in the presence of a catalyst. While the main desired reaction can selectively produce *para*-xylene with 85–95% concentration in mixed xylene (>23% equilibrium composition), methanol and toluene also participate in side reactions. This leads to a certain limit on methanol utilization since methanol is highly reactive and can self-react to form dimethylether (DME) or olefins such as ethylene and propylene. The methanol to olefins can occur at the operating condition with an acid catalyst. Once the olefins are formed, they can further alkylate with aromatic rings present in the system to form alkylated aromatics such as ethylbenzene, propylbenzene, and heavy aromatics. To limit the side reactions from taking place, toluene and methanol feed ratio may be controlled at higher than 1 : 1 stoichiometric molar ratio, thereby increasing the rate of methanol utilization toward methylation reaction. Typical conversion of toluene per pass is less than 36%, while methanol overall conversion per pass is less than 95%. Note that toluene can undergo subsequent methylation to form heavier aromatics such as trimethylbenzene and tetramethylbenzene. The operating temperatures for toluene methylation can be from 400 to 700 °C depending on the operating target and choice of reactor configurations.

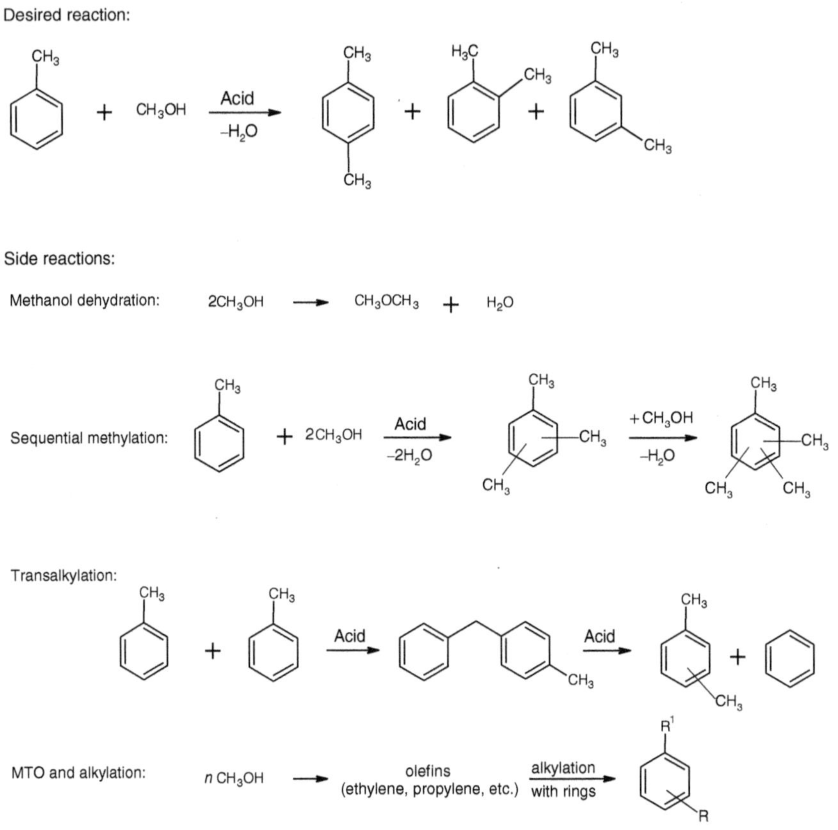

Toluene methylation unit can be integrated into existing aromatic complex or new complex to increase *para*-xylene production capacity. For a new aromatic complex, certain design considerations can be made to take advantage of toluene methylation for its selective *para*-xylene production with reactor effluent xylene composition, including more than 90% *para*-xylene, whereas typical equilibrium xylenes only include 22–24% of *para*-xylene. One of the design considerations is to maintain a high-quality toluene feed stream dedicated to the toluene methylation unit. Separate benzene and toluene columns can be combined into a single dividing wall column where toluene can be separated in a side draw product. Capital and operating costs savings can be realized without compromising the overall target and performance for the whole complex. A typical UOP design for new aromatic complex with toluene methylation is shown in Figure 5.14. It is also desirable to drive benzene conversion to more profitable products. Benzene can also methylate to toluene and subsequently *para*-xylene. A competitive pathway is to have benzene transalkylate with heavier aromatics to directly produce toluene and xylenes. UOP has considered and designed different scenarios to optimize the equipment investment, profit margins, and internal rate of return for respective options. The common benefit with toluene methylation unit is that the stream going to the Parax and Isomar loop is concentrated in *para*-xylene resulting in reduction in equipment size

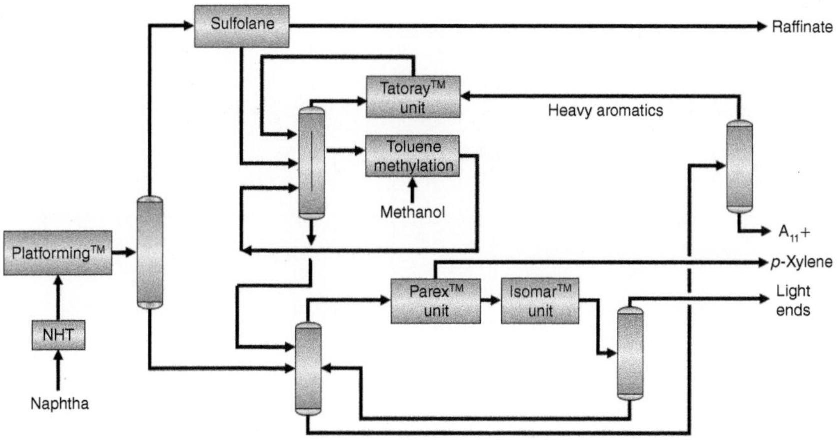

Figure 5.14 Aromatic complex with toluene methylation integration.

and utility consumption. The same can be said for existing units using crystallizer for *para*-xylene separation instead of adsorptive separation. For existing aromatic complexes, more than 20% capacity of *para*-xylene can be achieved using toluene methylation with little changes in the process units. Toluene methylation is thus a viable option for on-purpose *para*-xylene production particularly for regions with cheap and abundant feedstocks such as China and Middle East.

5.2.5 Aromatization of Light Naphtha/Hydrocarbons

BP-UOP Cyclar process was first developed by British Petroleum (BP) in 1975. The purpose is to valorize LPG or light hydrocarbons to valuable petrochemical products. As tight oil and lighter crudes become more abundant nowadays, cheap or stranded feedstocks are readily available for viable conversion to petrochemical base aromatics. Leveraging UOP's expertise in CCR system and Platforming technology, BP turned to UOP to jointly develop the technology and a high-strength formulation of catalyst that can be used in the radial-flow or stack-bed reactors as previously designed for UOP Platforming process. The catalyst activity can be maintained and regenerated using the UOP CCR process.

In contrast to production methods via thermal or catalytic cracking, aromatics can be re-constructed from smaller-chain molecules (C3/C4 hydrocarbons). Thus, LPG is expanded from conventional use as fuels to modern use of turning to petrochemical base aromatic production. Process-flow diagram for Cyclar is shown in Figure 5.15. The overall process and equipment are very similar to a typical catalytic reforming unit. The process has three major sections. The reactor section, which consists of several adiabatic, radial-flow reactors in series (reactors can also be stacked), charge heater, and feed-effluent heat exchanger. Next to reactor section is the CCR section, including the regenerator equipment and catalyst transfer system. And the last section is product recovery section, which consists of product

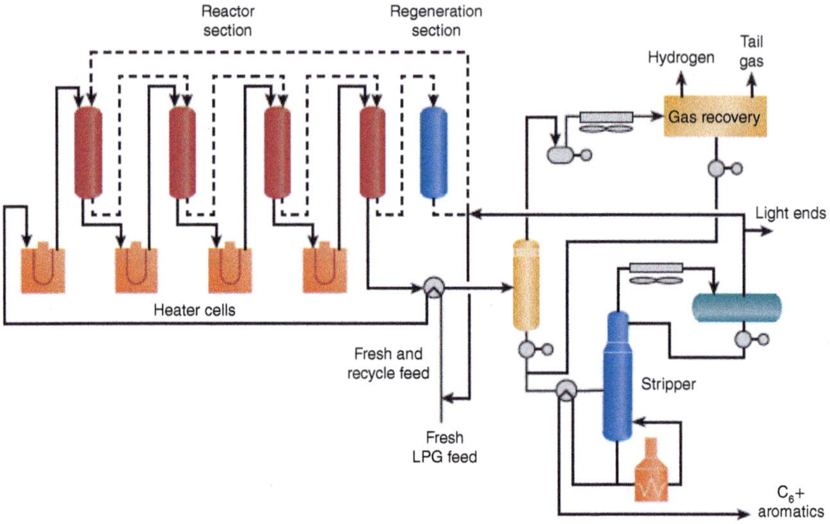

Figure 5.15 Cyclar flow diagram.

separators, the main compressor, hydrogen recovery equipment (pressure swing adsorption – PSA, or cold box, or absorber-stripper system), and product stripper.

The overall reaction is endothermic. The process stream leaving a reactor is thus re-heated to raise to the reaction temperature between reactors. At normal operating conditions, the rate of carbon deposition is slow and relatively small, accumulating less than 0.02 wt% of the feed processed. This low level of carbon on spent catalyst results in relatively mild regeneration conditions that prolong the life of the catalyst. The catalyst is formulated to have high thermal stability and good resistance to common feedstock contaminants. Proper regeneration can fully restore catalyst activity and selectivity to the original performance as observed with fresh catalyst. Good mechanical strength and low attrition property translates to less fines made and less makeup cost when circulating the catalyst in the system.

5.2.5.1 Mechanism

The reaction mechanism for LPG and condensate aromatization process is driven by dehydrocyclodimerization chemistry, which favors at 425–500 °C temperature. The reaction pathway includes the following steps:

- Dehydrogenation of paraffin to olefin – this is also the rate-limiting step to generate reactive olefin intermediates
- Dimerization of the olefin
- Cyclization of the dimer to a naphthenic ring
- Dehydrogenation of the naphthene to an aromatic

Beside aromatics, a significant amount of valuable byproduct present is hydrogen that is generated from the dehydrogenation step. A small amount of heavy hydrocarbons and coke is also present a result of side reactions from olefin dimerization and ring formation. This can be managed by coke burns and catalyst regeneration in the CCR system.

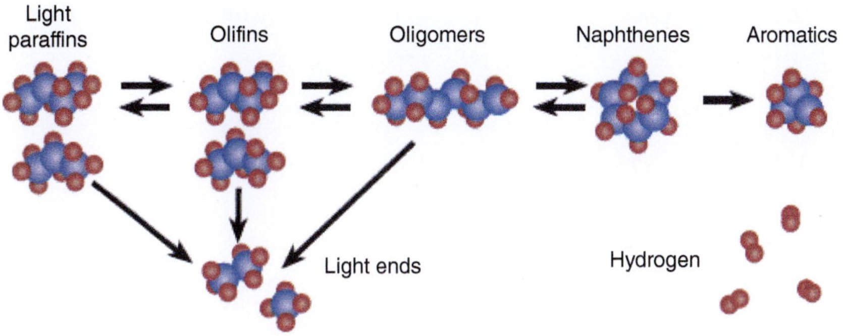

The aromatics yield increases with the carbon number in the original feedstock (butane vs. propane). Low-pressure operation with high temperature also favors high yield of aromatics. The liquid product contains greater than 90% BTX aromatics depending on the feed composition. The overall aromatics yield for propane or butane-rich feed can be 60–67%. Again, butane-rich feed would produce higher overall aromatics yield. Hydrogen is a significant byproduct and up to 5 wt% of the feed can be recovered as high-purity H_2 with standard PSA technology

Also, with design flexibility, Cyclar unit can be configured or revamped to turn to an olefin production unit such as propane dehydrogenation unit (PDH or UOP Oleflex) or repurposed to a naphtha-reforming unit. Starting from a small-chain-length paraffin (C3–C4 in LPG), the molecule dehydrogenates to an olefin, which has more active functional group that can then oligomerize to a dimer olefin. The ring closure can result in an aromatic molecule and at the same time produce hydrogen as a byproduct. The aromatization of the dimer molecule is an acid-catalyzed reaction. The commonly used catalyst is ZSM-5 as a base with a promoted metal.

5.2.6 Catalytic Cracking

Fluid catalytic cracking also produces a major amount of mono-aromatics that are present in the gasoline cut. With the use of zeolite targeting high production of olefins (propylene and butylenes), the gasoline cut from the main fractionation contains high concentration of aromatics, which makes it a great gasoline blend due to its high-octane value. However, with strict regulation on aromatic limits in the final gasoline product, refiners need a way to extract the high-value aromatics out from the FCC unit. Thus, more and more refiners are integrated with their petrochemical counterparts to expand their aromatics production to maintain high profits, reduce the price swing vulnerability in fuel market, and still meet all environment constraints at the same time.

Why is LCO a good stream or intermediate to upgrade? For component analysis in Figure 5.16, LCO has a very high aromatics content from 75 to 85 wt% depending on the crude source or the feed and the operating conditions in the FCC unit. However,

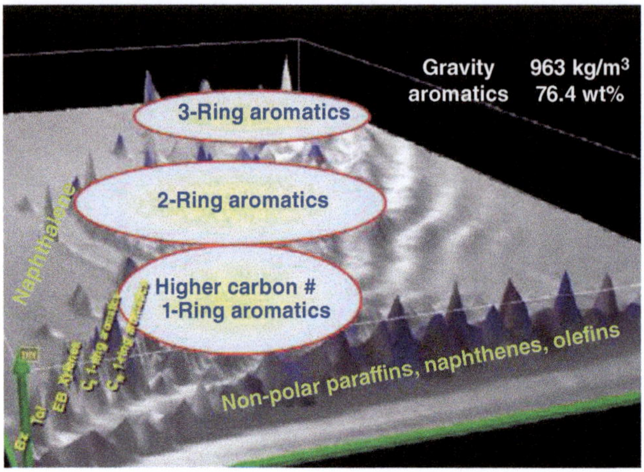

Figure 5.16 LCO characterization showing rich concentration of multiring aromatics.

high aromatic content (usually means good octane) makes LCO a low cetane value stream, with typical cetane from 15 to 25. In the US, significant amounts of LCO can be blended to the diesel pool in up to 20% with similar boiling range with diesel – 95% cut point at 360 °C. LCO also has high sulfur and nitrogen content, up to 1.5 wt% and 750 wt ppm, respectively. Note that major aromatic components are multiring aromatics, 2-ring, and 3-ring components, shown by quantitative two-dimensional GC × GC analysis. For aromatics production, these multiring components need to be selectively converted or ring opened to single-ring aromatics. This is a suitable strategy for the current trend of highly integrated refinery and petrochemical complexes that can operate their FCC to maximize not only propylene but also aromatics to maximize profits.

LCO upgrading mainly includes hydroprocessing for contaminant removals, cetane improvement, or cut-point adjustments to meet fuel specifications.

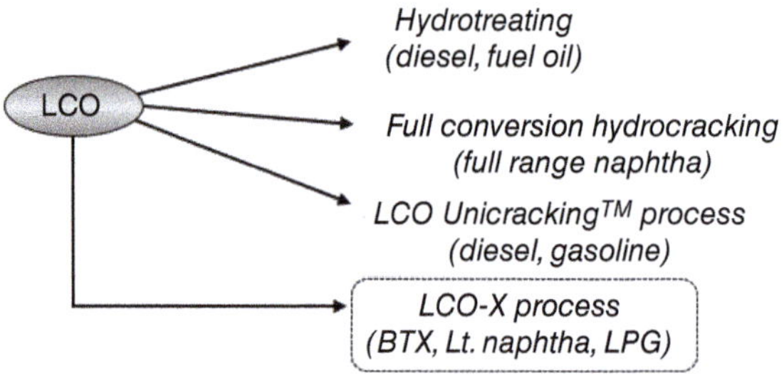

For a typical 60 000-barrel-per-day (bpd) FCC feed capacity, LCO yield can be obtained at 14 000 bpd and sent to LCO-X process, which can produce 215 000 tons/year of mixed xylenes and 80 000 tons/year of benzene as petrochemical

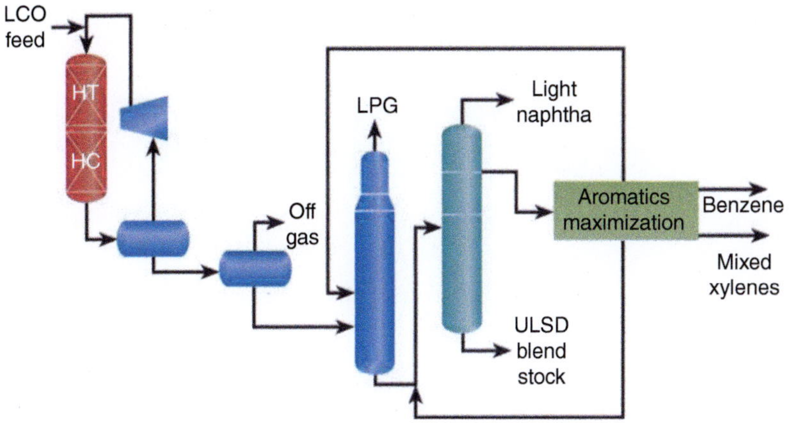

Figure 5.17 UOP LCO-X process for maximization of aromatics.

products. Other byproducts include fuel gas, LPG, light naphtha, and ultra-low sulfur diesel (ULSD). The amount of xylenes produced can easily be picked up by the demand growth of *para*-xylene. For economy-of-scale considerations, LCO intermediates from other FCCs can be routed together to a larger LCO-X unit to produce more xylenes to gain healthy margin profits and quick return on investment.

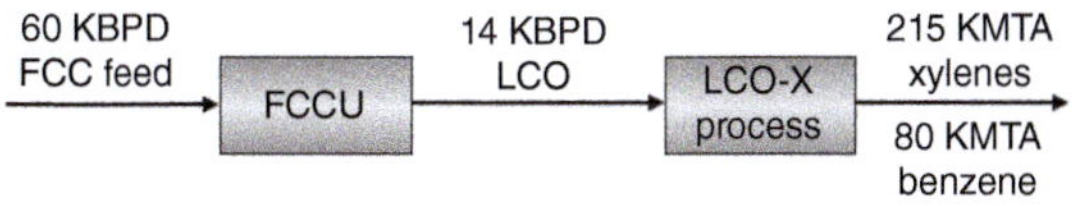

A general process diagram for UOP LCO-X process is shown in Figure 5.17. The reactor includes a multizone hydrotreating combined with hydrocracking catalysts. The reactor effluent is sent to a hot high-pressure separator where recycle gas is taken off in the overhead and returned to the reactor and the hot liquid is sent to a low-pressure separator. Offgas is separated out and the separator liquid is routed to fractionation section for product recovery. Several products can be obtained from LCO-X process, including fuel gas, LPG, light naphtha, aromatics, and ultra-low sulfur diesel.

From process chemistry point of view, the catalyst is designed to convert multiring aromatics to single-ring components. The simplified reaction pathway is depicted in Figure 5.18. Di-alkyl naphthalene is desired to cleave to 2 monoaromatic molecules. Some saturation of the ring, whether complete or partial saturation, can occur at a certain extent to accommodate the subsequent ring opening or dealkylation. Thus, some naphtha range materials may be formed as intermediate products and aromatic maximization reactor can further convert the alkyl-cyclohexanes to corresponding aromatics (Figure 5.18).

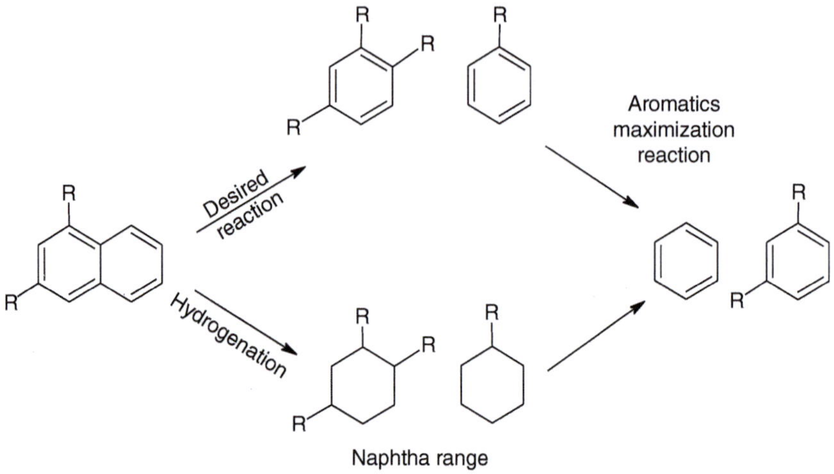

Figure 5.18 Simplified pathway for multiring conversion in LCO-X process.

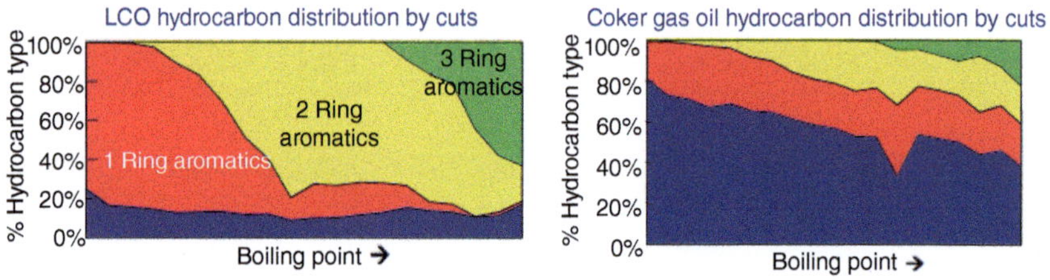

Product quality	
Benzene	• 90–99.9%
Mixed xylenes	• Suitable for Parex unit feed for *p*-xylene • Exceeds ASTM D-5211 specification • ~1% ethylbenzene
Light naphtha	• 80–90% C_5–C_7 paraffins, 10–20% naphthenes • 76–82 RON, high Iso content
LPG	• <0.5% olefin • 60% C_4, 40% C_3
Diesel	• <10 ppm sulfur, Ultra-Low-Sulfur Diesel (ULSD) • 10–15 cetane increase over LCO

Refiners can re-configure or integrate their FCC to export aromatics for petrochemical production. BTX can be extracted using extractive distillation processes with commercial proprietary solvents such as Carom, Udex, or Sulfolane (developed by UOP and Shell).

UOP licensed LCO-X technology levering Unicracking™ technology to convert the byproduct LCO from FCC to convert multiring aromatics to single aromatics.

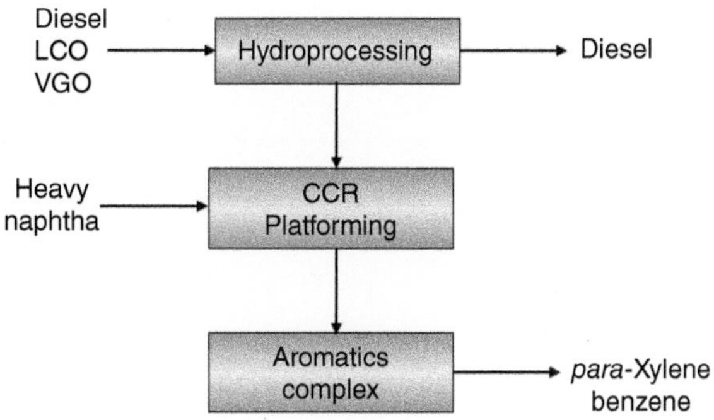

Figure 5.19 Hydrocracking integration with reforming for petrochemicals production.

As future trend of gasoline decreases with the emergence of higher-efficiency engines, hybrid and electric vehicles, refineries will adjust and re-purpose their FCC units to convert or extract more aromatics out of their intermediate product streams. The LCO stream is rich in aromatics, including both single and double rings. UOP LCO-X process offers a pathway to convert double-ring aromatic to single-ring aromatics to maximize the aromatic yields, thus improving the value of LCO stream, which can typically be hydrotreated and blended with diesel fuels (Figure 5.19).

For hydrocracking-based refineries, as the fuel demand decreases and petrochemical demand increases in the near future, the key factor to remain profitable while keeping process flexibility is to integrate with reforming process to produce aromatics for aromatic complex. The hydrocracking unit will adjust to run at higher severity and produce higher yields of naphtha at the expense of diesel. The naphtha in turn is fed to the reformer to produce aromatics. The hydrocracking unit also needs to be able to convert to naphtha with a variety of feedstocks ranging from straight-run diesel to LCO and VGO. The hydrogen management will also be critical and balanced prudently among various units with the main source of hydrogen coming from the reformer beside any hydrogen from steam reforming of natural gas.

5.2.7 Production from Pyrolysis and Coking

5.2.7.1 Steam Cracking/Pyrolysis Gasoline

Steam cracking has matured over the years and is the main technology to produce olefins in the world. With the simple design, high reliability, and flexibility to process a wide range of feed, steam cracking continues to be chosen for high-demand, high-capacity petrochemical complexes. Historically, with naphtha range and heavier feeds, beside olefin products, steam cracking also produces pyrolysis gasoline, a valuable aromatic-rich product.

In the United States, lighter feedstocks such as LPG or ethane are preferred feed for steam cracking. As a result, high yield of ethylene and olefins is produced from light feedstocks. Very low yields of higher carbon chain products are produced. Pyrolysis gasoline yield, including benzene, toluene, and xylene, is thus very low.

With the evolution of shale gas production in the United States, more and more steam crackers will come onstream to produce olefins from light feedstocks such as cheap ethane recovered from shale gas.

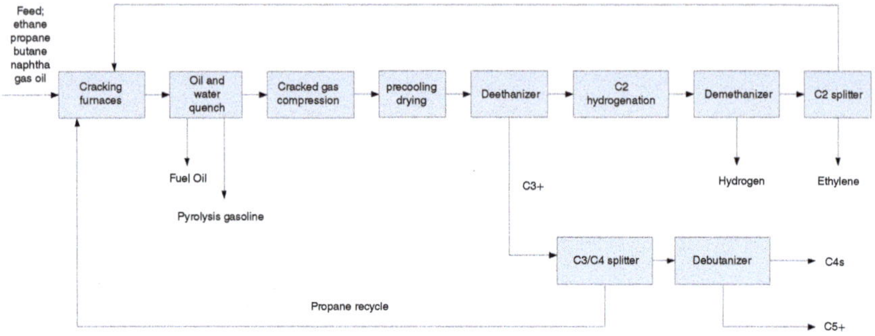

Figure 5.20 Steam-cracking flow diagram for petrochemical production.

Steam cracking is a highly energy-intensive process. Naphtha feed enters the coiled tubes and vaporized at the convection section of the furnace and is preheated with steam to 600 °C. About half a ton of steam is added per ton of naphtha feed. The mixed feed is then passed to the cracking tubes, which are positioned vertically in a cracking furnace. The cracking or pyrolysis reaction temperature is high at 800–900 °C. The reactions take place in 0.1–0.5 seconds. The effluent gas leaving the cracking furnace at 850 °C is rapidly quenched to stop further reactions or undesired secondary products. The high heat associated with the product is recovered to produce high-pressure steam (before the quench unit), which is used to drive the cracked gas compressors. At the quench unit, pyrolysis tar or gasoline can be recovered in the main column and the effluent gas is further washed in a water wash column. The cracked gas is compressed to 30–40 bar pressure, treated and dried to remove contaminants such as CO_2, H_2S, and H_2O before entering olefin recovery section. High purity of ethylene and propylene, 99.9% polymer grade can be achieved using C2 and C3 splitters with 90–150 trays with high overhead reflux rate. Ethane and propane at the bottom of the columns are recycled back to the cracking furnaces. Methane and hydrogen can also be recovered by using cryogenic separation or low-temperature distillation (<-100 °C) with refrigeration and power recovery system using Joule–Thompson effect. Note that the deethanizer overhead, including mainly ethylene and ethane, is contacted with a selective hydrogenation reactor to remove acetylene from the system. Ethane is further separated and recycled. The pyrolysis gasoline (pygas) can be recovered from the quench oil in a decanter after the cracking reactor and from the debutanizer column bottom (Figure 5.20).

Steam-cracking Yields

Steam crackers are a major workhorse to produce ethylene, a critical monomer for the production polyethylene and other polymer products. Other olefins such as propylene and butenes are byproducts. Table 5.2 summarizes typical yields from steam cracking from different feedstocks with the basis on the same amount of ethylene production. Pyrolysis gasoline includes high-octane aromatic components such as benzene, toluene, and xylene. Pyrolysis gasoline yields increase with heavier hydrocarbon feeds. However, due to the nature of thermal nonselective cracking,

Table 5.2 Typical steam cracker yields for different feedstocks.

Product	Feedstock				
	Ethane	Propane	Butane	Naphtha	Gas oil
Ethylene	0.79	0.43	0.40	0.34	0.31
Propylene	0.03	0.17	0.22	0.16	0.19
Crude C4s	0.03	0.04	0.10	0.10	0.11
Pygas	0.02	0.07	0.07	0.20	0.19
Gas oil (6 oil)	0.00	0.01	0.01	0.03	0.11
Hydrogen	0.02	0.01	0.01	0.01	0.01
Fuel	0.27	0.11	0.19	0.15	0.08

benzene concentration tends to be higher than toluene and xylene concentration in pyrolysis gasoline. Thus, if xylene is the preferred product or component extract, a more selective production method such as catalytic reforming of naphtha should be used. With shale gas evolution in the US, lighter feedstocks become much cheaper and petrochemical producers tend to reserve more expansive naphtha feedstocks for production of aromatics, while lighter feedstocks are for production of olefins.

Product yields from a steam cracker can vary widely depending on the type of feeds. Typical yields for steam crackers are summarized from Platts Analytics [5] in Table 5.2. Ethane is particularly good and selective for ethylene production. With cheap and abundant feedstock, ethane steam cracking is the most efficient and economical way to produce ethylene for polymer and petrochemical intermediates. However, the yields of other byproducts are reasonably low when ethane is the feed. The higher yields of byproducts such as gas pyrolysis gasoline (pygas) increased with increased carbon chain length in the feedstock. However, the pygas yield tops out at 20% when naphtha is fed to the steam cracker. Thus, steam crackers are not considered as the base aromatic production engine, whereas naphtha reformers continue to be the main driver for BTX aromatics.

5.2.7.2 Coke-oven Benzole

The industrial aromatics started in early 1920s with tar coal and coke-oven benzole being the main sources for aromatics production. Coal tar has a wide range of aromatic components, including benzene, toluene, naphthalene, anthracene, pyrene, styrene, and indene. Other valuable compounds that can be recovered from coal consist of phenols, anilines, pyridines, and quinoline.

Heating or pyrolysis of hard coal can produce a mixture of rich aromatics consisting of high concentration of benzene and toluene. Coal pyrolysis process occurs without air and product effluents may vary with different degrees of heating. At low heating temperatures, less than 150°C, light components such as carbon dioxide, water, and C_2 to C_4 hydrocarbons are emitted. Above 180 °C, higher molecular weights such as aromatics start to appear. At 350–550 °C, degasification

Table 5.3 Composition of a typical coke-oven benzole.

Composition	%/ppm
Benzene	67%
Toluene	16%
pX, mX, oX, EB	6%
Styrene	1.3%
C_{9+} aromatics	7%
Nonaromatics	2%

rate increases to produce semicoke. Higher temperatures, 600–800 °C, lead to further coke formation accompanied by methane and hydrogen evolution. The volatile aromatic product from coal pyrolysis is referred to as coke-oven benzole. However, the yield of xylene compared to benzene and toluene is very low. Because of high-temperature carbonization of coal, benzene produced is much higher than the catalytic reforming and pyrolysis gasoline. The typical composition obtained from coke-oven benzole is shown in Table 5.3.

Industrial scale of hard coal pyrolysis or carbonization is operated at very high temperatures of 1000–1200 °C [1]. The main product is coke. Typical yield from one ton of coal is 750 kg of coke, 35 kg crude tar, 11 kg benzole, 2.4 kg ammonia, 150 kg water, and 370 m^3 coke-oven gas. In fact, coal tar and benzole are just byproducts from coal pyrolysis with very low overall yields, total less than 5 wt%. As the demand for industrial aromatics ramped up, pyrolysis of petroleum and other production methods such as naphtha reforming became viable to supply the market with cheap abundant feedstocks.

Figure 5.21 shows the coal pyrolysis process where coal is heated up in the coke ovens at high temperature (1000–1200°C). The main product includes coke, but many valuable co-products can be obtained, including tar, ammonium sulfate, and coke-oven benzol. The crude benzole is extracted from the coke-oven gas by counter-current absorption in a column using a sponge oil that selectively absorbs benzol. The sponge oil is a heavy tar-oil fraction with boiling range 220–330°C and can be recovered in a stripper column and returned to the absorption tower. The coke-oven benzol stream can be further treated in a hydroprocessing unit and extractive distillation in the refinery for BTX aromatics production.

Figure 5.22 shows the relative distribution of benzene, toluene, and xylene using different technologies. While catalytic reforming produces lowest yield of benzene, coke-oven benzole has highest yield for benzene. Catalytic reforming selectively produces highest xylene compared to steam cracking or coal pyrolysis processes. Thus, catalytic reforming is the most efficient production method of xylenes. In fact, catalytic reforming is highly integrated with aromatic complex for extraction or further conversion to maximize *para*-xylene, which is a building block for PET.

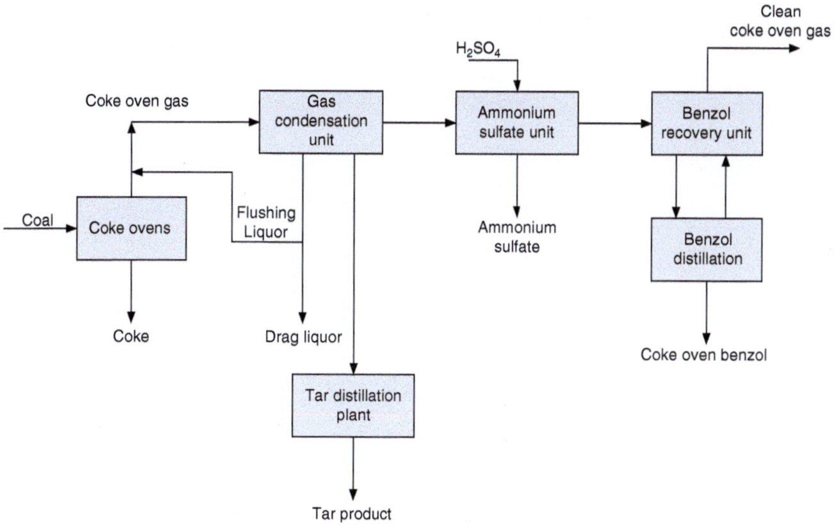

Figure 5.21 Coke-oven flow diagram with coke-oven benzol recovery unit.

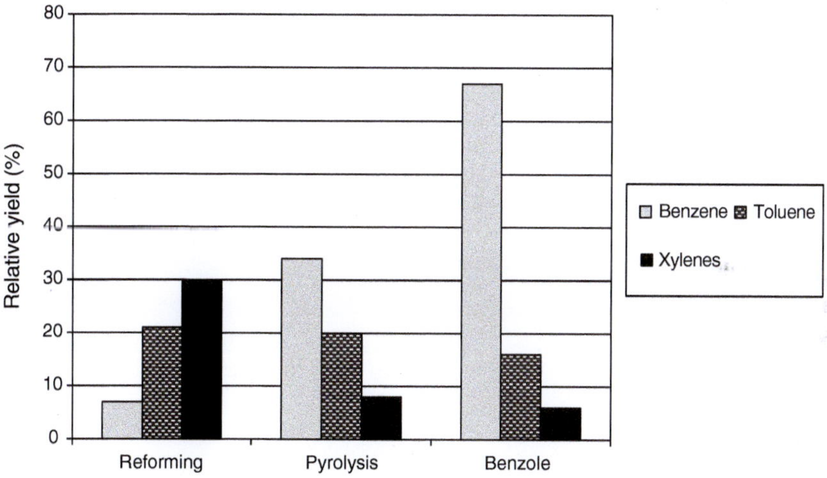

Figure 5.22 Relative yields of aromatics for various technologies.

5.2.7.3 Biomass Pyrolysis

One of the most important uses of biomass is the production of wood derivatives or chemicals such as natural rubber, cellulose and ethanol and organic chemicals such as acetic acid or phenolic compounds. It is also well known that biodiesel and ethanol products can be produced from renewable resources and these can be used in fuel blend applications. From woody biomass, thermal pyrolysis or torrefaction can yield a wide range of organic compounds that can be similarly obtained from coal pyrolysis. The liquid product yield obtained from a fast pyrolysis can be as high

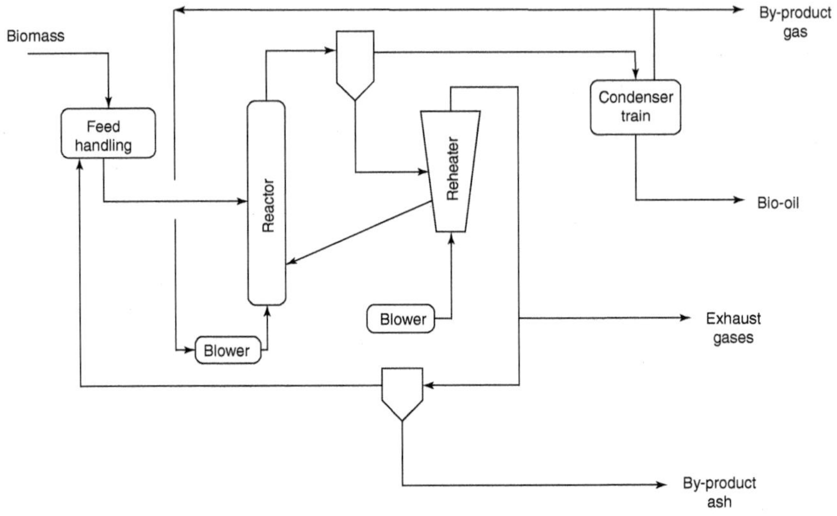

Figure 5.23 RTP™ flow diagram. Source: Based on RTP™ [8].

Table 5.4 RTP green fuel typical properties.

Property	Typical values
Gross calorific value, MJ/kg	16–19
Net calorific value, MJ/kg	15–17
Water content, wt%	15–30
Solids content, wt%	0.5–2.5
Ph	2–3
Specific gravity	1.1–1.3
Ash content, wt%	0.1–0.25
Flash point, °C	45–99
Pour point, °C	−23 to −30

Table 5.5 Comparative fuel properties for various fuels.

Fuel	MJ (l)	Btu (US Gallon)
Heavy fuel oil (No. 6)	40.4	144 800
Light fuel oil (No. 2)	39.3	141 000
Ethanol	23.5	84 000
RTP green fuel	19.9	71 500

as 70–75 wt%. Other products include char and light gas. The liquid product contains high concentration of oxygenated hydrocarbons. Nonoxygenated hydrocarbon products or BTX aromatics are also present but in much small quantities and can be very difficult to recover. The liquid product obtained from biomass pyrolysis can be referred to as pyrolysis oil, bio-oil, or bio-crude.

A UOP joint venture commercial pyrolysis process is shown in Figure 5.23. This is a rapid thermal process referred as UOP RTP™ leveraging experiences from UOP fluid catalytic cracking (FCC) technology. Biomass, in the absence of oxygen, is pyrolyzed in a fluidized reactor at 500 °C and quickly quenched to yield 65–75 wt% RTP green fuel (Table 5.4). Utility is minimized by circulating hot sand inside the fluidized reactor. Char and sand are disengaged in a cyclone and can be reheated and cycled back to the reactor. Lighter char is lifted and disengaged in another cyclone and discharged as byproduct ash.

The energy content per volume basis for RTP green fuel is about half of typical fuel oil (No. 6) (Table 5.5). Thus, for a boiler or furnace requirement, about twice the volume of RTP green oil is needed to substitute for conventional fuel oil. This can be economical if the biomass feedstocks are readily abundant and cheap. However, for transportation fuel market, the bio-oil needs to be further processed or refined to lower the oxygen content while meeting the other fuel specifications required for motor fuels.

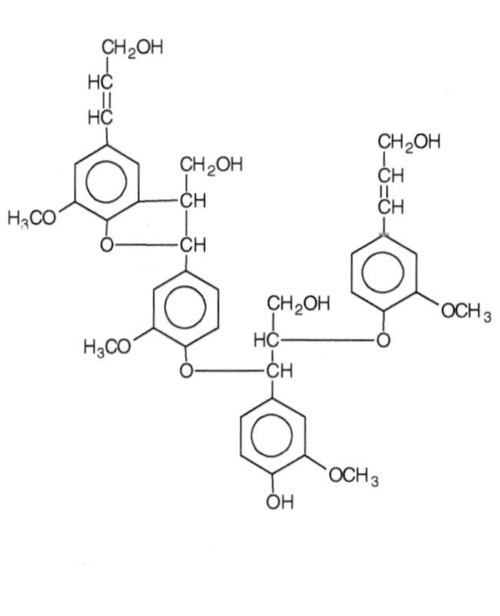

Tetralignol

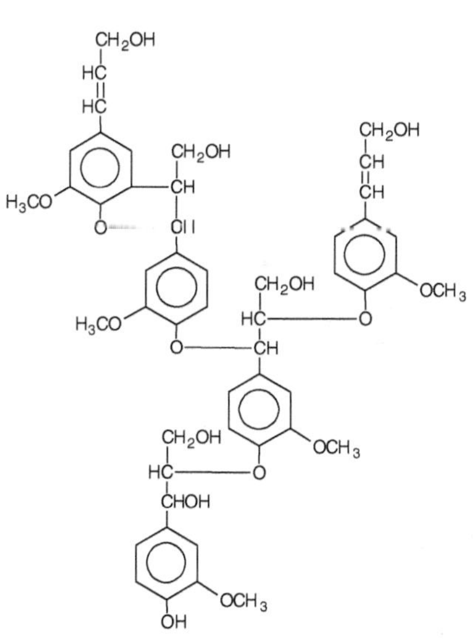

Pentalignol

Typical woody biomass contains about 50% cellulose, hemicellulose, and lignin. Lignin component (as high as 25%) has high aromaticity and complex structures, which include multiple linkages of three basic monolignol building-block molecules: *p*-coumaryl alcohol, coniferyl alcohol, and sinapyl alcohol. These monomers have a common phenolic backbone with alkyl- or oxygen groups

Table 5.6 Product yield summary for catalytic hydrogenation of lignin.

Product	Weight percentage of organic lignin (%)
H_2O	17.9
C1–C5	25.2
C6–150 °C	8.3
150–240 °C	5.7
150–240 °C phenols	37.5
240–260 °C	8.7
Residue	2.4
Hydrogen consumption	5.7

attached to it. Theoretically, these monomers can be deoxygenated or dealkylated to yield corresponding aromatics or phenols.

Catalytic upgrading of bio-oils has been recently focused on to produce higher-value products such as aromatics or specialty chemicals. The Hydrocarbon Research Institute had shown that catalytic hydrogenation of lignin can yield very high yield of phenolic molecules (37.5 wt%) as shown in Table 5.6 [7]. The aromatics and naphthene components in the neutral oils can be as high as 14 wt%. However, the overall yield on the biomass feed is low since the typical lignin fraction in woody biomass is about 25 wt%.

The total aromatic hydrocarbon yield is quite low with less than 1 wt%, detected and quantified by GC × GC-FID/TOF-MS for detailed composition of the bio-oil [6]. The aromatic oxygenates concentration is higher, but these molecules are highly oxygenated. Economic considerations need to be evaluated whether to recover the aromatic oxygenates (phenolic branch compounds) by groups for specialty chemicals or to process the molecules by deoxygenating them in hydrogen atmosphere with mild or severe operating conditions. The catalysts can be quickly deactivated by strong adsorption of oxygenates that further repolymerize and form heavier species that depress the catalyst activity.

Thermal pyrolysis or fast pyrolysis of biomass produces a mixture of hydrocarbons and oxygenates. The product from the fast pyrolysis process is also referred to as bio-oil or bio-crude due to the nature of highly oxygenated compounds. The low boiling components may contain high concentration of light oxygenates such as aldehydes, alcohols, and ketones. The higher boiling point fraction includes furanic compounds and phenolics. The phenolic fraction includes low concentration of aromatic hydrocarbons, including benzene, toluene, and xylene.

5.3 Summary

The aromatic production technologies continue to evolve. Catalytic reforming provides the major production of aromatic building blocks. Industrial licensors continue to develop catalysts with better performance and selectivity for aromatics while minimizing byproducts such as light ends or LPG. At the same time, design improvements, data analytics, and process intensification will help refineries utilize their resources at the highest efficiency, increase profits/margins, and minimize wastes and downtime.

On-purpose and secondary conversion processes continue to provide pathways to fine-tune the distribution of products to high yields of desired products. To achieve this, reaction and separation engineering will continue to be integrated to overcome equilibrium limitation. Novel production methods taking advantage of in situ reactive separation will emerge as a breakthrough challenging the current/traditional technologies that have been used over the past 70 years. This will require new technologies or hybrid systems making use of new catalysts, adsorbents, and membrane materials to improve the performance and yields while lowering costs and investment barriers.

References

1 Franck, H.-G., Stadelhofer, J.W., (1988) *Industrial Aromatic Chemistry*. Springer.

2 Wier, J.J., Sioui, D., Metro, S. et al. (2014). Optimizing naphtha complexes in the tight oil boom, AFPM Annual Meeting AM-14-35 (March 2014).

3 https://www.uop.com/processing-solutions/refining/gasoline/#naphtha-reforming

4 Rase, H.F. (2000). *Handbook of Commercial Catalysts: Heterogeneous Catalysts*. CRC Press.

5 Johnson, J.A., Frey, S., Thakkar, V. *"Unlocking high value xylenes from light cycle oil", AM-07-40, Annual NPRA Meeting* 2007.

6 Negahdar, L., Gonzalez-Quiroga, A., Otyuskaya, D. et al. (2016). Characterization and comparison of fast pyrolysis bio-oils from pinewood, rapeseed cake, and wheat straw using ^{13}C NMR and comprehensive GC × GC. *ACS Sustainable Chemistry & Engineering* 4 (9): 4974–4985.

7 Derk, T.A.H. and Huge J.P. (1983). Lignin hydrocracking process to produce phenol and benzene, US4420644.

8 RTP™ (18 May 2015). Technology Overview, RTP Process. http://www.ensyn.com/technology.html (accessed 23 October 2020).

6
Uses of Aromatics

6.1 Introduction

Aromatics make excellent monomers that can be crosslinked to synthesize valuable polymers. With excellent substitution ability, high reactivity, good solvency, and increased reactivity with substituent groups, aromatics continue to be the key materials for polymer syntheses and development in material science. Essential materials such as nylons, polyesters, plastics, pesticides, coatings, and resins are primarily derived from basic aromatic building blocks, including benzene, toluene, and xylene.

6.2 Polymer Production from Basic Aromatic Building Blocks

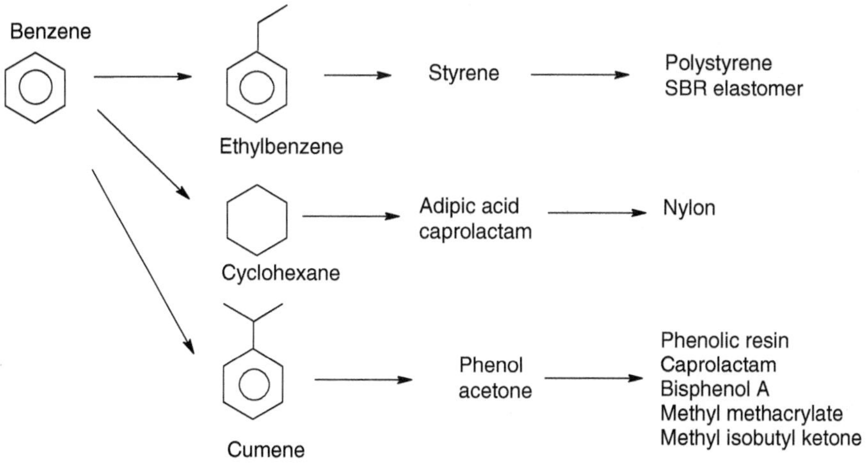

Caprolactam - Nylon-6 production; BPA - polycarbonate plastics

Modern Petrochemical Technology: Methods, Manufacturing and Applications, First Edition.
Santi Kulprathipanja, James E. Rekoske, Daniel Wei, Robert V. Slone, Trung Pham, and Chunqing Liu.

Table 6.1 World consumption of benzene by major end use – 2016.

Derivatives	Percentage of use
Ethylbenzene	49
Cumene	21
Cyclohexane	12
Nitrobenzene	9
Detergent alkylation	3
Chlorobenzene	2
Maleic anhydride	2
Other	2

Source: Data from IHS Markit [2].

The world consumption of benzene in 2016 was 44 206 thousand metric tons and summarized in Table 6.1 (IHS Markit [2]). It is projected to grow at 2.6% average annual growth rate.

6.2.1 Ethylbenzene and Styrene Production

Currently, direct or straight recovery of ethylbenzene from an aromatic mixture, including mixed xylenes and ethylbenzene, is not economical due to high required energy and equipment/process cost. The production of ethylbenzene is typically carried out via Friedel-Crafts process with direct alkylation of benzene and ethylene using $AlCl_3$ with co-catalysts such as ethyl chloride, BF_3, $FeCl_3$, $ZrCl_4$, $SnCl_4$, and H_3PO_4. The typical process performance includes 40% conversion and 98 mol% selectivity [1]. The reactant ratio of benzene and ethylene is typically maintained more than required stoichiometric ratio so that ethylene does not self-react and polymerize to other products.

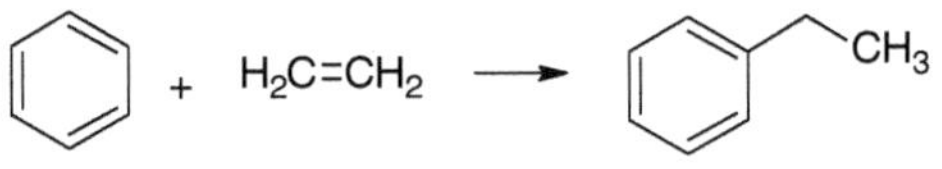

Styrene and Polystyrene:
Next to ethylene and vinyl chloride, styrene is the most important monomer building block in the production of plastic materials and synthetic rubber such as styrene-butadiene (SB) copolymer and polystyrene-divinylbenzene. Styrene production is almost exclusively via catalytic dehydrogenation of ethylbenzene. Similarly, divinylbenzene can be produced from diethylbenzene by dehydrogenation at a high temperature of 600 °C with superheated steam. The main application of divinylbenzene is the manufacture of ion-exchange resins by co-polymerization with styrene.

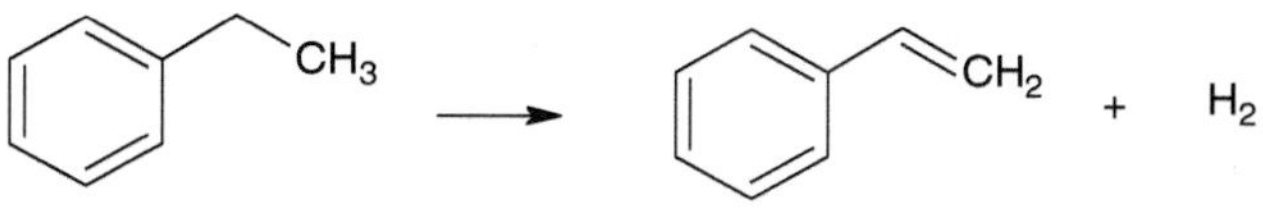

Polymer from Styrene:
Polystyrene:

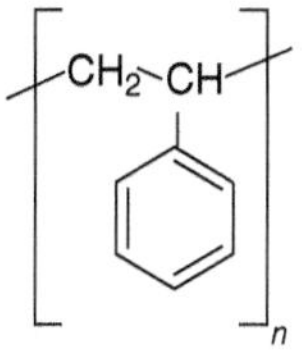

Cross-linked polystyrene-divinylbenzene

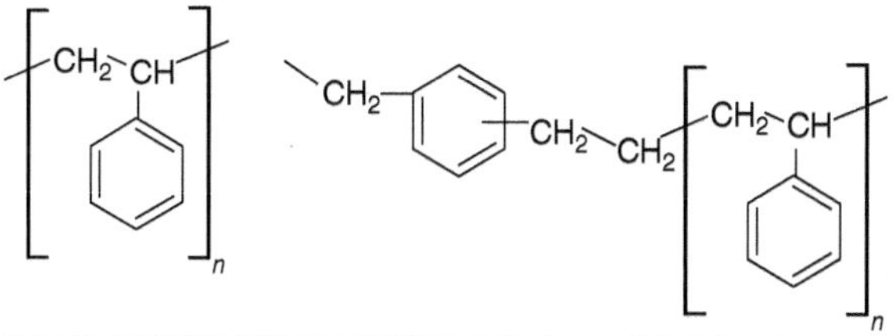

Nylon 6, Nylon 6,6, PET, Polyurethane, Polystyrene, Polycarbonate...

6.2.2 Cyclohexane to Nylon 6, Nylon 6,6

Cyclohexane is an important intermediate for nylon production. Cyclohexane can be produced from catalytic hydrogenation of benzene. Note that the benzene purity is very critical since trace amount of sulfur can poison the catalyst.

From cyclohexane, different derivatives can be produced to monomer molecule for different nylons, namely nylon 6 and nylon 6,6 [1].

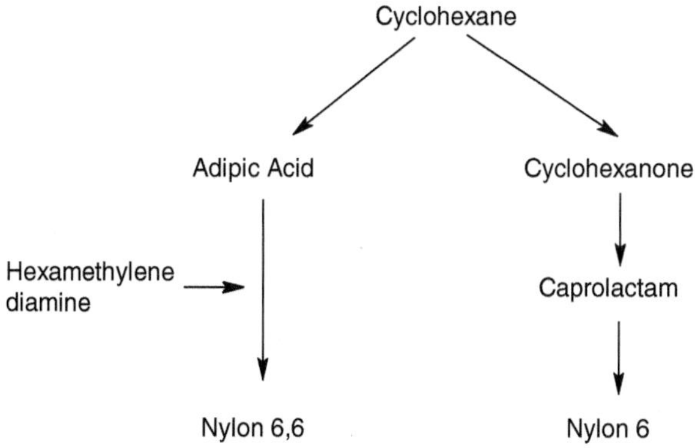

Table 6.2 World consumption of cyclohexane in 2017.

Derivatives	Percentage in use
Caprolactam	55
Adipic acid	22
Cyclohexanol/cyclohexanone	20
Other	3

Source: Data from IHS Markit [2].

The total world consumption of cyclohexane in 2017 was 5223 thousand metric ton. The major use is for nylon synthesis. The relative demand is summarized in Table 6.2.

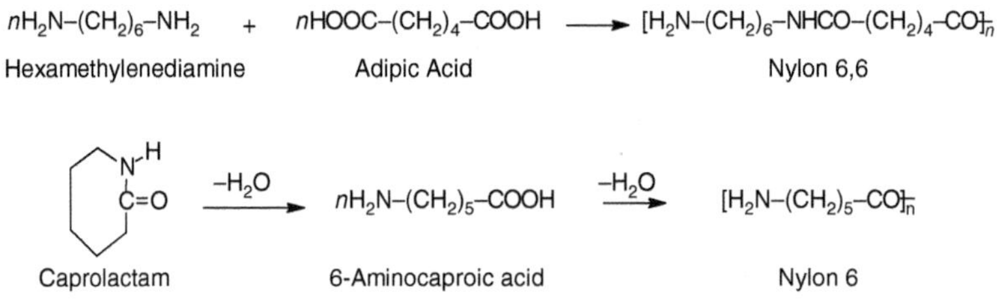

6.2.2.1 Adipic Acid Production

Adipic acid can be synthesized from cyclohexane oxidation or from benzene via phenol intermediate.

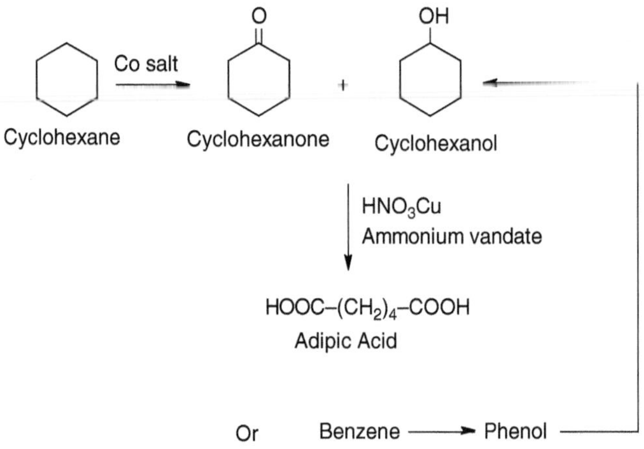

Production of Hexamethylenediamine

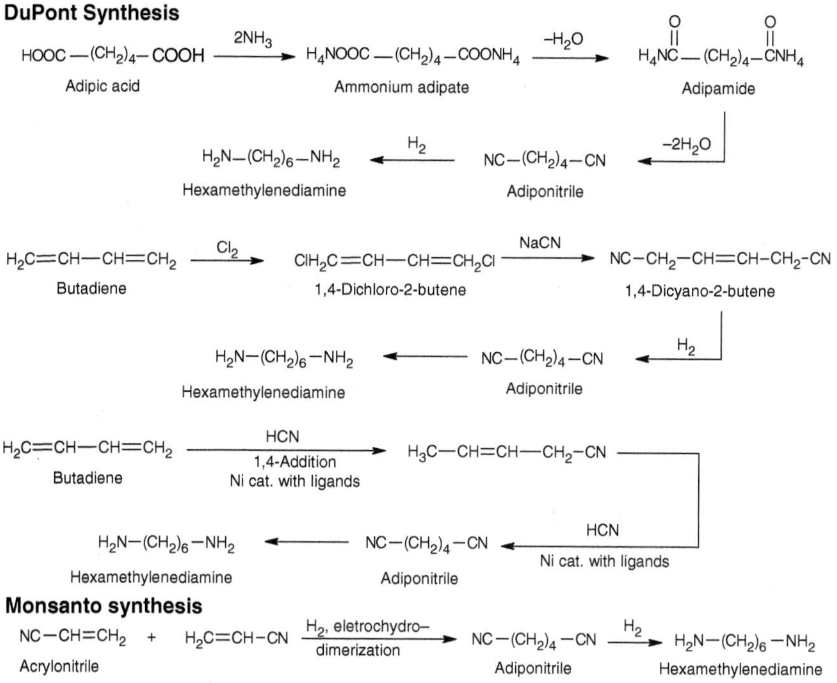

6.2.2.2 Caprolactam Synthesis

There are seven major methods to synthesize caprolactam:

- Conventional
- Toray
- DuPont
- Union Carbide
- SNIA Viscosa
- DSM
- ENI

Conventional

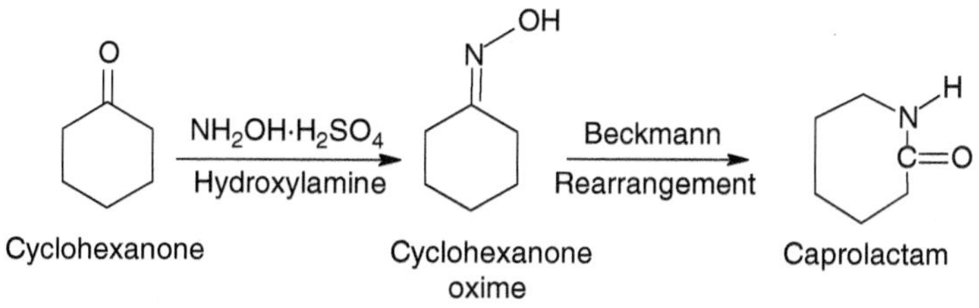

$$NH_4NO_2 + NH_3 + H_2O \longrightarrow HON(SO_3NH_4)_2$$

$$HON(SO_3NH_4)_2 + 2H_2O \longrightarrow NH_2OH{\cdot}H_2SO_4$$

Note that for the conventional method, the byproduct is ammonium sulfate. For every pound of caprolactam produced, 4.4 pounds of ammonium sulfate is obtained. The high amount comes from sequential steps: manufacture of hydroxylamine sulfate, manufacture of the oxime, and Beckman rearrangement.

Toray

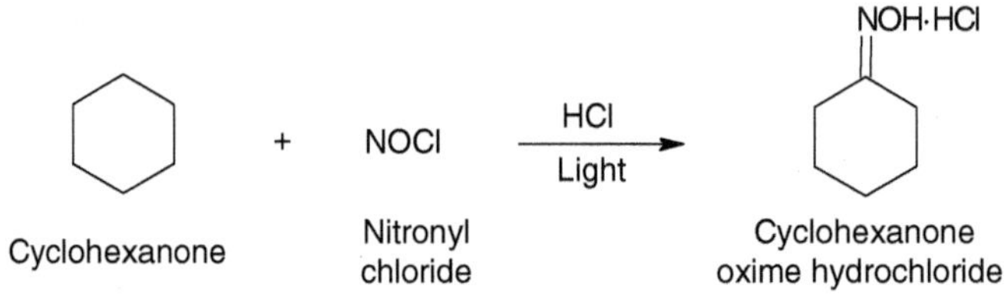

$$2H_2SO_4 + N_2O_3 \longrightarrow 2HNOSO_4 + H_2O$$

$$HNOSO_4 + HCl \longrightarrow NOCl + H_2SO_4$$

DuPont

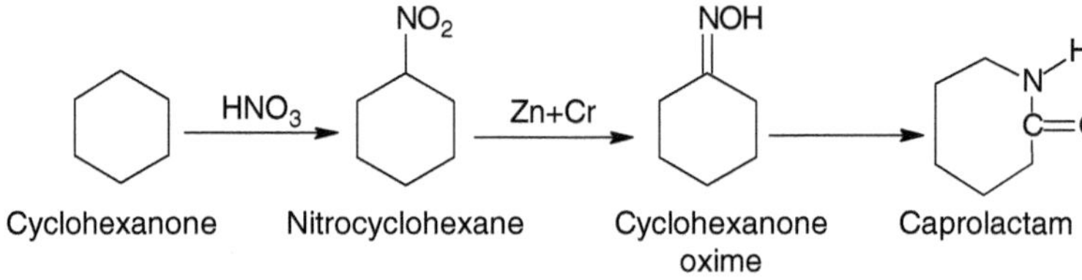

Union Carbide

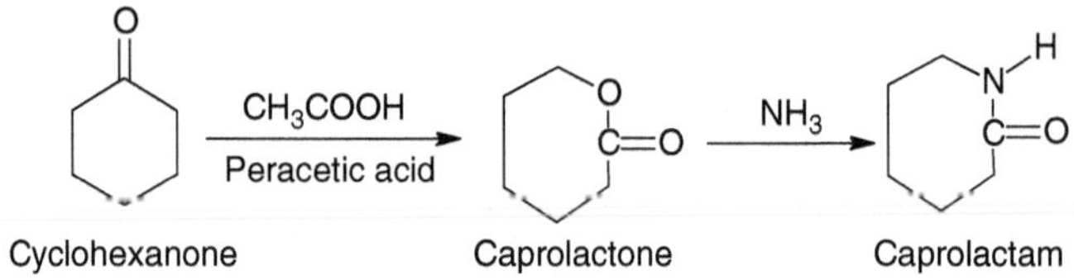

SNIA Viscosa

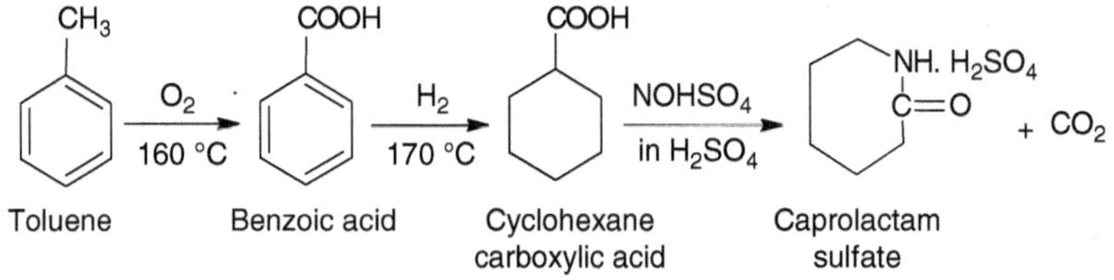

DSM

A buffered solution of NH_2OH, $NO_3^- + 2H^+ \rightarrow NH_3OH^+ + 2H_2O$

The buffered NH_3OH^+ in H_2O is reacted with cyclohexanone in toluene to produce the oxime, which undergoes catalytic Beckmann rearrangement to yield final caprolactam product. The spent aqueous solution is recycled back to previous reaction steps.

ENI

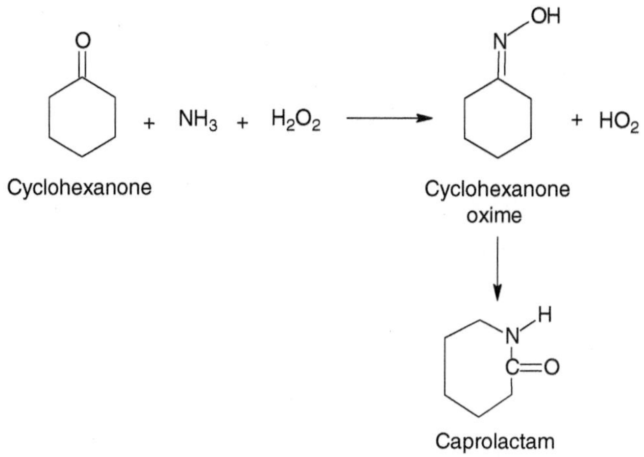

DuPont – DSM process eliminates completely ammonium sulfate formation. The caprolactam is produced from a different starting material, butadiene, rather than cyclohexane or benzene derivative.

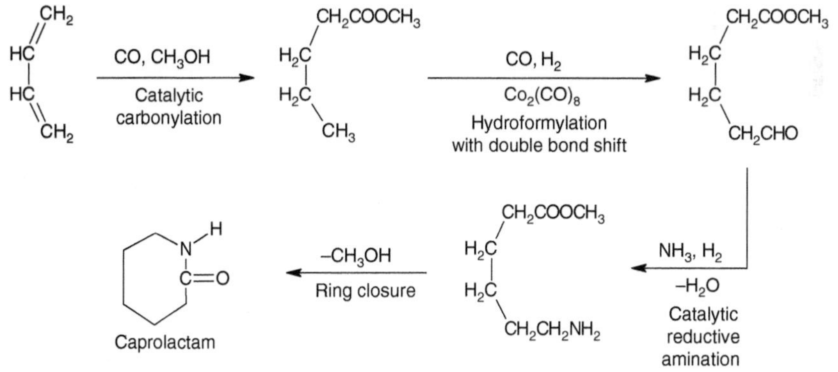

A different approach for nylon synthesis from butadiene: DuPont-BASF HDMA-Caprolactam 2-for-1 synthesis process

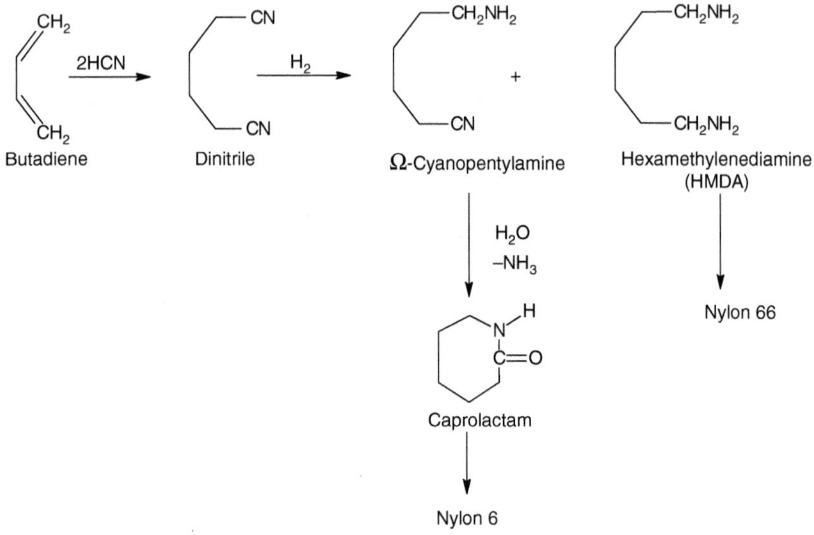

6.2.2.3 Properties of Nylon 6 vs. Nylon 6,6

Both nylon 6 and nylon 6,6 are compatible materials. The major differences are:

- Nylon 6,6 has higher melting point. It is preferred for some engineering thermoplastics applications and also temperature-resistant packing films.
- Nylon 6 is used where flexibility and oxygen barrier properties are required or more critical.

6.2.3 Production of Cumene

Isopropylbenzene or cumene can be produced by benzene alkylation with propylene in either gas phase or liquid phase. The alkylation reaction is the classic Friedel–Craft reaction, which takes place in the presence of sulfuric acid or $AlCl_3$ as a catalyst at 30–40 °C. The benzene/propylene reaction molar ratio is kept at about 5-to-1 to control the selectivity and keep propylene from polymerization.

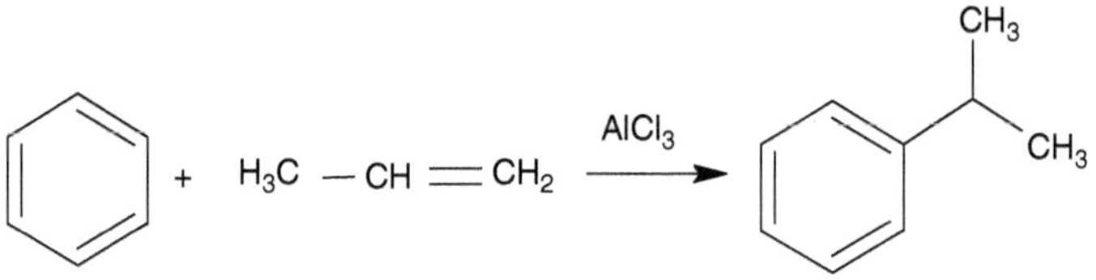

Cumene to Phenol

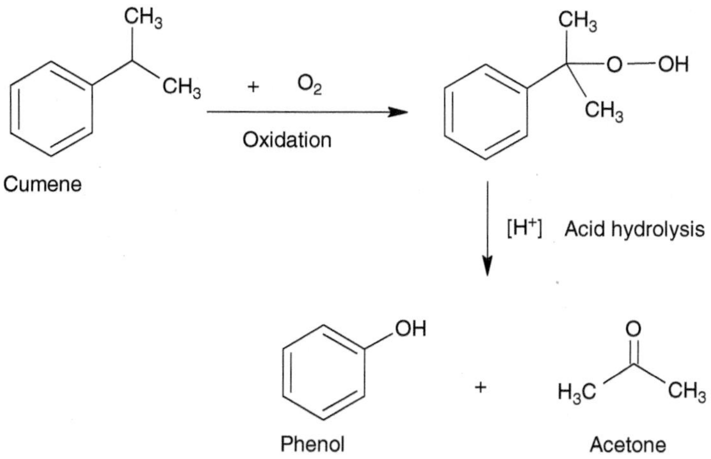

The phenol market is globally driven by the demand for bisphenol A (BPA), which consumed 47% of global phenol production in 2017 (Table 6.4). BPA is driven by strong demand for polycarbonate products, which outpaced the GDP growth rate and is expected to continue in the future. Polycarbonate is used in automotive, OEM, construction, optical media, and appliance industries. It directly competes with glass and acrylic resins in glazing/sheet and acrylonitrile-butadiene-styrene (ABS) applications. BPA is also driven by strong demand of epoxy resins, which are used to make surface coatings, circuit boards, and adhesives. The global use for phenol is summarized in Table 6.3, which is adapted from 2018 IHS Markit report [3].

Table 6.3 Uses of nylon.

Application	Examples
Carpet fibers	Commercial , residential carpets
Textiles	Hosiery, lingerie, wearing apparel, sport wear
Tire cords	Truck, automobile, airplane tires
Industrial fibers	Soft luggage, backpacks, marine ropes, conveyor belts, military webbings, airbags
Engineering thermoplastics	Various automobile components, wire % cable coating, molded components
Packaging films	Flexible food packaging

Table 6.4 World consumption of phenol by end use – 2017.

Derivatives	Percentage of use
Bisphenol	47
Phenol formaldehyde (PF) resin	27
Nylon	15
Alkylphenol	3
Xylenol	2
Other	6

6.2.3.1 Bisphenol A production

Bisphenol A is produced by acid-catalyzed reaction of acetone with phenol using hydrochloric acid (HCl) or sulfonated crosslinked polystyrene resin (H^+) at 50–90 °C. The reaction requires excess phenol with a high molar ratio of 15 : 1 of phenol to acetone to control product selectivity and prevent acetone from dimerization or trimerization reaction in the presence of acid catalyst. Bisphenol A product is separated from the reaction mixture by crystallization.

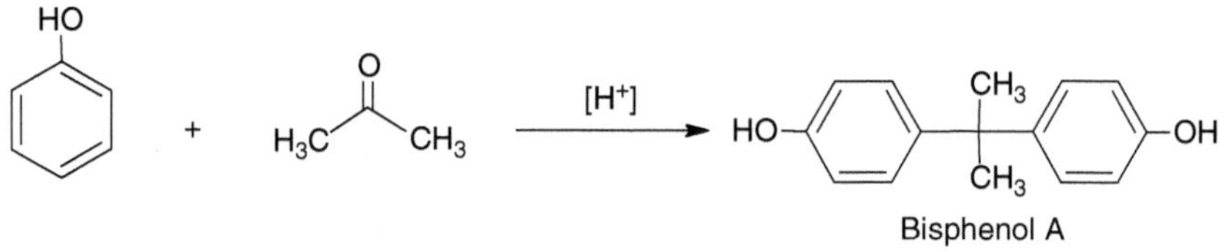

6.2.3.2 Production of Nitrobenzene

The industrial production of nitrobenzene is achieved by isothermal nitration of benzene. The nitration medium consists of 40% HNO_3, 40% H_2SO_4, and 20% water. Due to safety standpoint, stainless steel equipment is used due to high corrosive nature of the catalyst or nitration medium. The reaction proceeds heterogeneously between aqueous and organic phases with nitration occurring in acid phase.

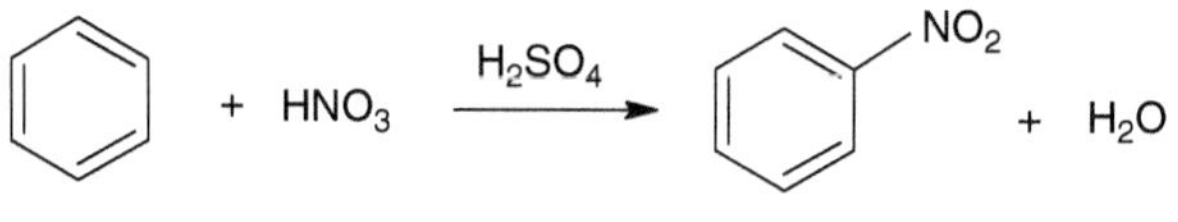

Nitrobenzene is mainly used for the production of aniline to further produce 4,4-diphenylmethane diisocyanate (MDI) and poly(methylene diphenylene isocyanate) (PMDI). Nitrobenzene is also used for the production of *m*-nitrochlorobenzene or *m*-chloroaniline for pharmaceutical applications. Other uses of nitrobenzene include the production of *m*-nitrobenzenesulfonic acid and *m*-aminophenol also for pharmaceutical applications.

6.2.3.3 Production of Aniline

The most important method to produce aniline is from gas-phase reaction of nitrobenzene using iron (Fe) and HCl catalyst that gives very high yield of 99% aniline.

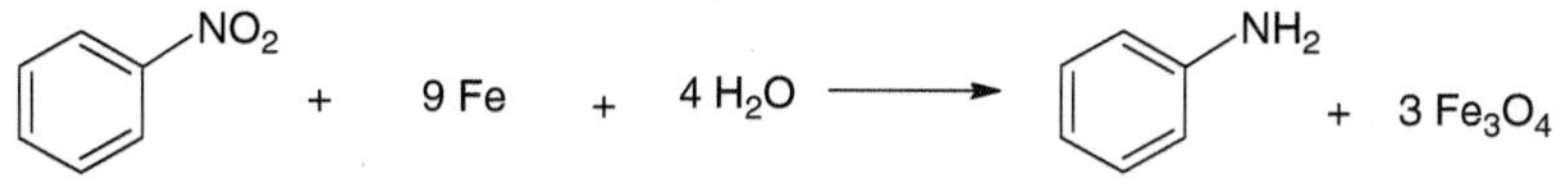

Another method to produce aniline is from gas-phase ammonolysis of phenol at 400 °C, 200 bar using SiO_2/Al_2O_3 with tungsten and vanadium.

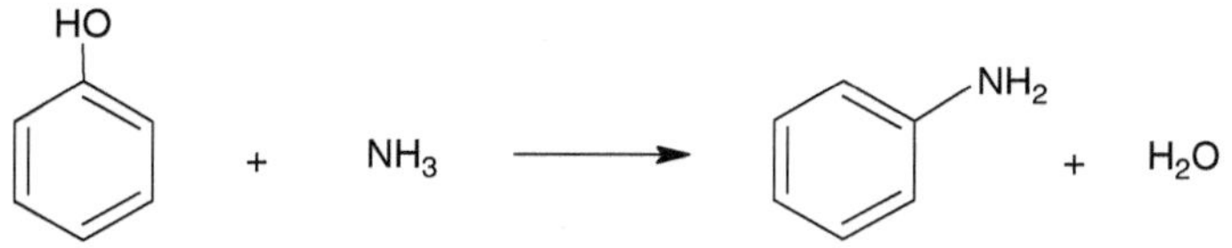

6.2.3.4 Production of MDI and Poly-MDI (PMDI)

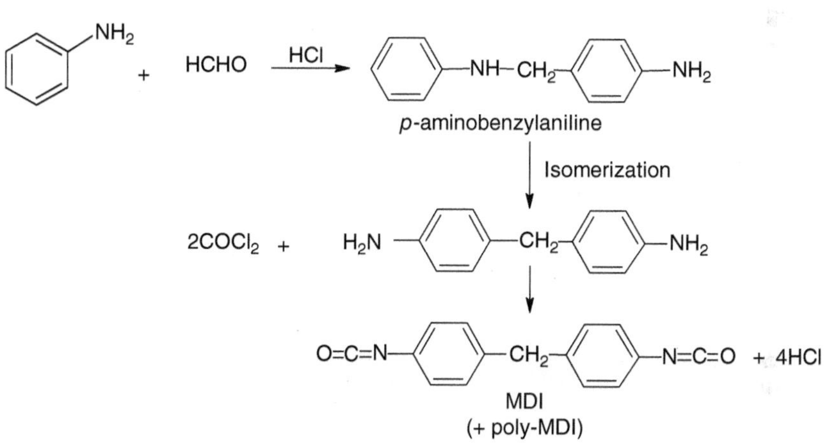

6.2.3.5 Production of Toluene Diisocyanate (TDI)

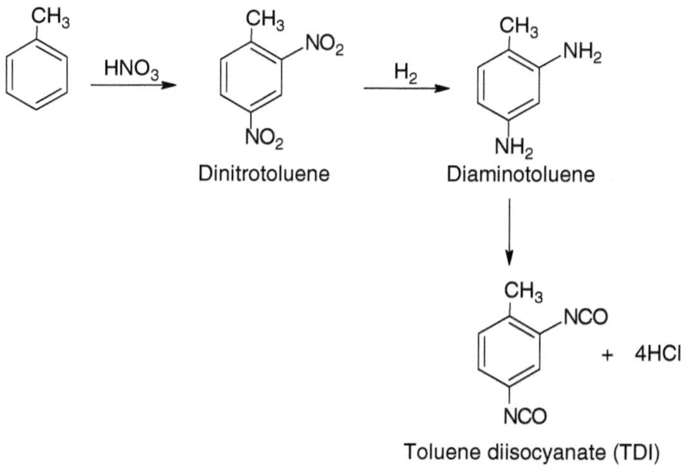

Table 6.5 Total US 2014 demand for TDI, pure MDI, polymeric MDI (2603.5 million pounds).

Application	Percentage-%
Rigid foam	40.7
Binders	18.5
Flexible foam slabstock	16.8
Flexible foam molded	7.5
Elastomers	4.8
Coatings	3.8
Adhesives	3.5
Others	4.4

6.2.3.6 Production of Polyurethane

Toluene diisocyanate is used to produce polyurethane by reacting with MDI or PMDI and polyols

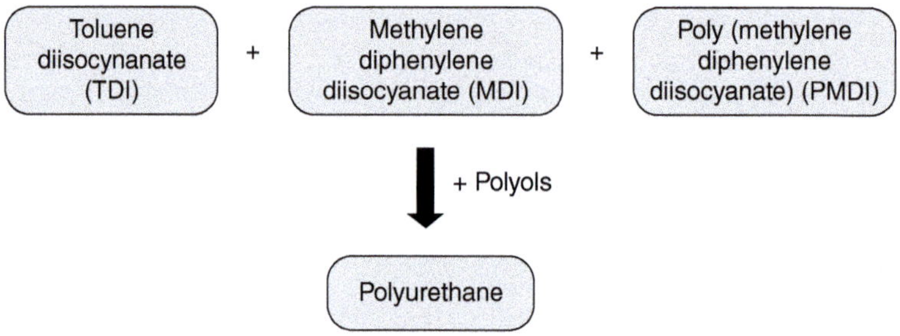

Polyurethane or TDI, MDI/PMDI are used in a variety of applications, from furniture, automobiles, building insulation to bedding, refrigeration/packaging, and nonfoam uses. The total US 2014 market demand for TDI, MDI, and PMDI is summarized in Table 6.5 [4].

6.2.3.7 Alkylbenzenes and Alkylbenzene Sulfonates

Monoalkylbenzenes with long alkyl chains (>C8 hydrocarbons) in the form of alkylbenzene sulfonates are the most important raw materials for the production of ionic detergents. There are two types of alkylbenzene sulfonates: linear-alkylbenzene and branched-alkylbenzene

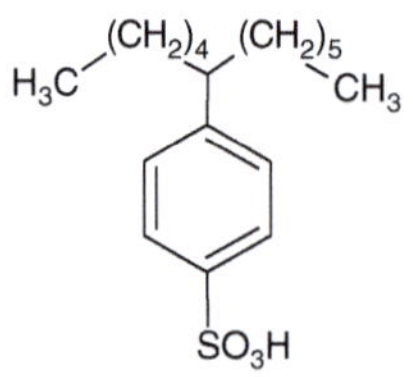

Linear alkylbenzene sulfonate (LAS)

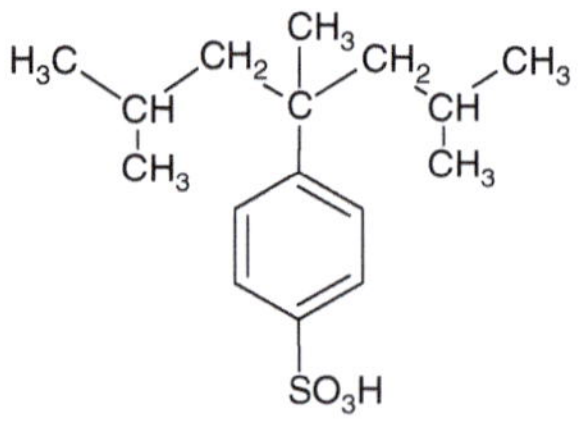

Branched alkylbenzene sulfonate (BAS)

6.2.3.8 Maleic Anhydride from Benzene

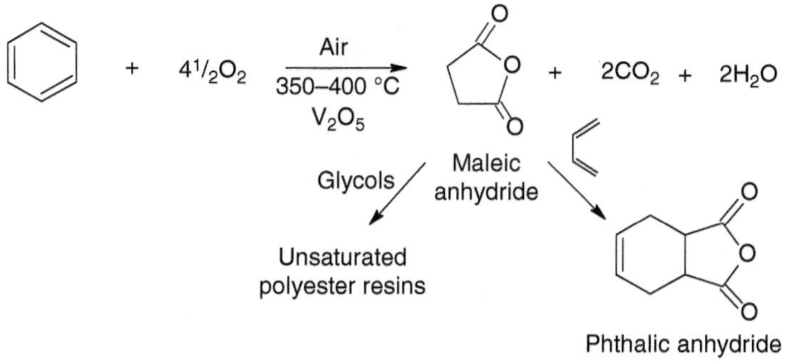

6.2.3.9 Dichlorodiphenyltrichloroethane (DDT) from Benzene

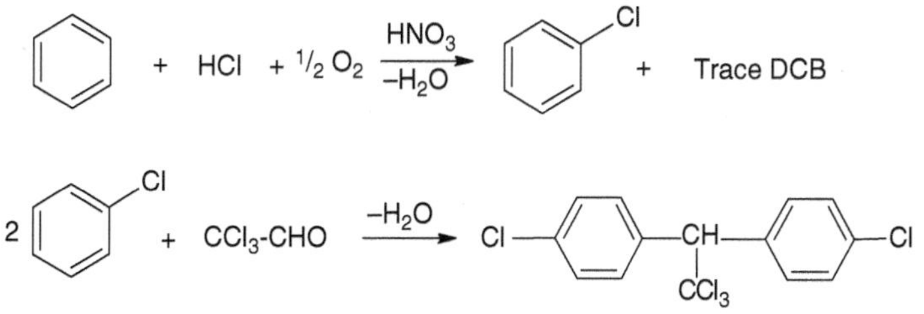

Dichlorodiphenyltrichloroethane (DDT) was one of the most important insecticides. It is no longer used due to its poor biodegradability and is further banned from usage due to links with cancer and birth defects (Figure 6.1). In summary, Benzene and Toluene are original feedstocks to produce intermediates and polymers which are shown in Figure 6.1.

6.3 Resins and Chemicals Production

Plastics and chemicals are produced from xylene derivatives. The important intermediates include phthalic anhydride, isophthalic acid, and terephthalic acid, which are derived from *o*-xylene, *m*-xylene, and *p*-xylene via oxidation reactions [1]. The xylene derivatives are used to manufacture important chemicals such as plasticizers, unsaturated polyester resins, alkyd resins, and synthetic fibers. Most of the xylenes used in the United States and Western Europe come from catalytic reformate. The world supply and demand for p-xylene are shown in Figure 6.2.

Significant investments in xylene production have been made over the past decades to increase overall capacity due to high demand of mixed xylenes. These investments have been focused in Northeast Asia region with highly developed and integrated polyester industry.

Among xylene isomers, *p*-xylene is the most widely used and produced product, accounting for about 83% of total mixed xylenes' consumption in 2017 (IHS Markit [3]). PX is the raw material to produce purified terephthalic acid (PTA) and dimethyl

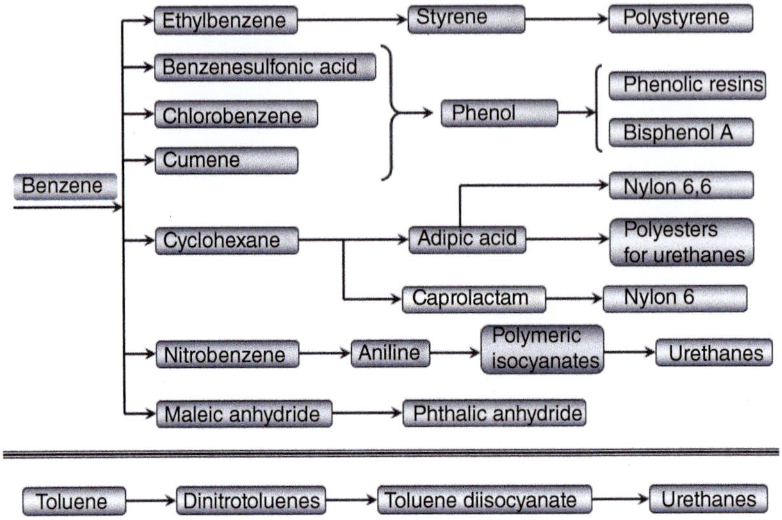

Figure 6.1 Polymers from benzene and toluene.

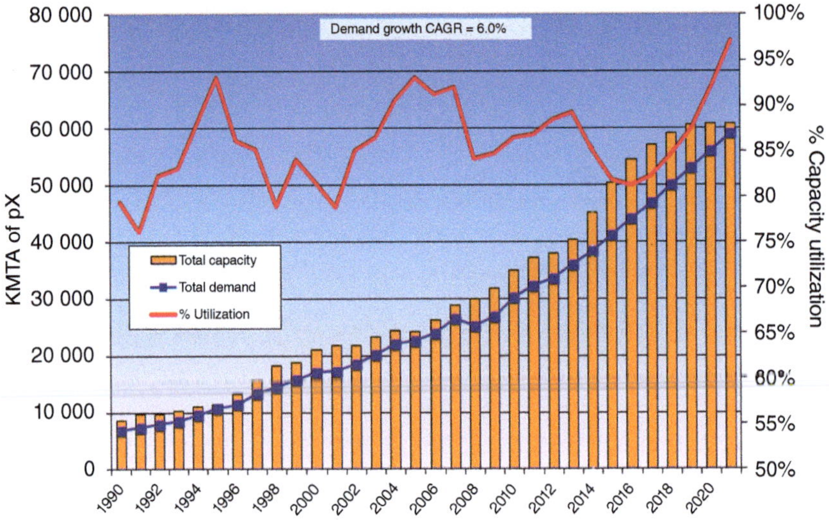

Figure 6.2 World supply and demand balance for *p*-xylene.

terephthalate (DMT), which are key ingredients for polyethylene terephthalate (PET) polymer, which is, in turn, used for the production of polyester fibers, PET resins, and PET film. *o*-Xylene is about 5.5% global consumption of mixed xylenes and used for production of phthalic anhydride. This intermediate is key ingredient for alkyd resin or plasticizer to be used in the production of PVC products, including pipe, flooring, cable, and packaging film. *m*-Xylene is only about 1% of mixed xylenes, which is used to make isophthalic acid, a key ingredient to produce plastic bottles (Figure 6.2).

Table 6.6 Properties of xylene isomers.

C_8-aromatics	Freezing point (°C)	Boiling point (°C)	Catalytic reformate content range (%)
p-Xylene	13.0	138	18–22
m-Xylene	−47.9	139	40–45
o-Xylene	−25.2	144	17–22
Ethylbenzene	−95.0	136	14–21

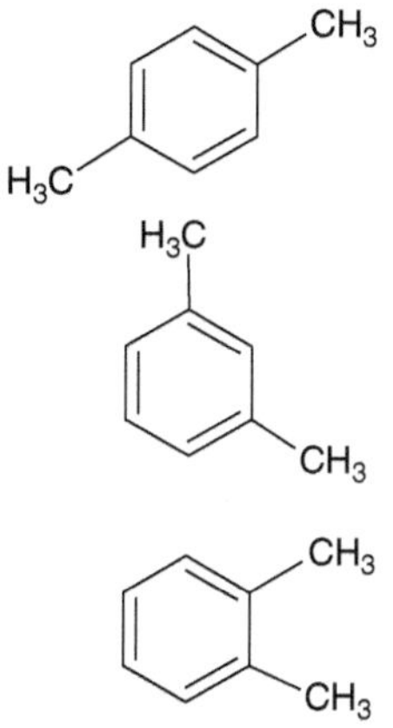

The freezing points among xylene isomers are different enough that certain processes can take advantage for separation, for example, crystallization process. Other separation processes such as adsorptive separation take advantage of different adsorptive properties of the isomers (Table 6.6). Once the isomers are separated, they can be oxidized to give acid or anhydride intermediates to produce polyester fiber, plastic, plasticizer, and resin materials (Figure 6.3). The worldwide consumption of mixed xylenes in 2017 is shown in Table 6.7 with predominantly high percentage use of *p*-xylene, 83% of global xylene demand (IHS Markit [3]).

Table 6.7 World consumption of mixed xylenes in 2017 (thousands of metric tons).

Isomer consumption	Capacity (thousands of metric tons)	Percentage of use
p-Xylene	46 523	83
m-Xylene	598	1
o-Xylene	3 077	5
Ethylbenzene	—	0
Solvents and others	6 118	11

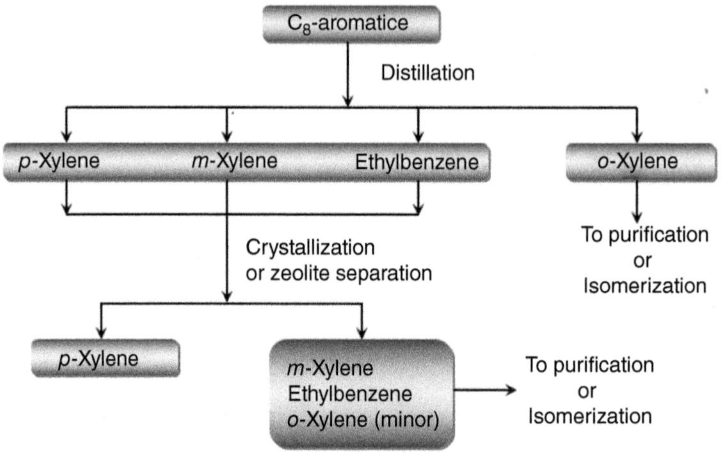

Figure 6.3 Overall C_8-aromatics separation.

The major use of *p*-xylene is to produce PTA, which is then used to produce PET polymer. The PET polymer is in turn used to make PET solid-state resins and PET films. The overall consumption for mixed xylenes is strongly tied with the global polyester industry by the high growth of demand, consumption, and development for polyester.

The total world producers for mixed xylenes consist of about 170 companies; among those, the top 10 largest companies account for 45% of the market with major region based in Asia (Figure 6.3). The top 10 producers are listed in Table 6.8.

Table 6.8 Top 10 world producers of mixed xylenes in 2017.

Ranking	Producer	Average annual capacity (thousands of metric tons)	Percentage of total
1	SINOPEC	7126	9.2
2	ExxonMobil Corp.	5575	7.2
3	CNPC	4386	5.7
4	Reliance Industries	4217	5.4
5	JXTG Nippon Oil & Energy	3075	4.0
6	Total	2597	3.3
7	SK Innovation	2531	3.3
8	Saudi Aramco	2047	2.6
9	CNOOC	1943	2.5
10	Formosa Group	1815	2.3
	Others	42320	54.5
	Total	77633	100

Oxidation reactions of Xylenes

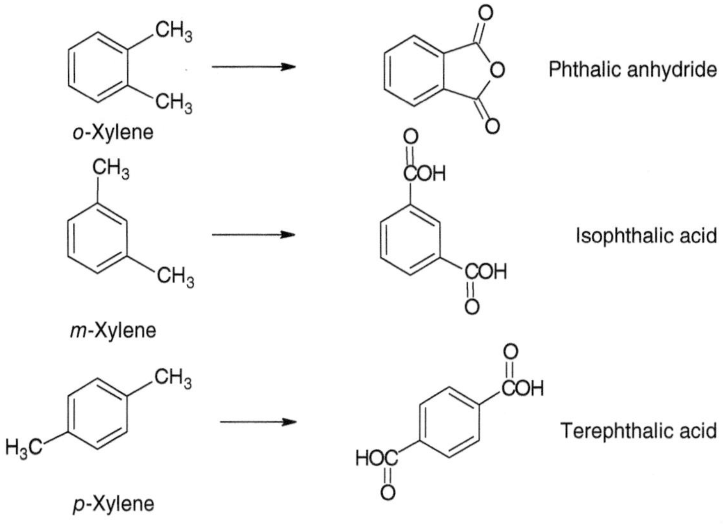

o-Xylene oxidation to phthalic anhydride

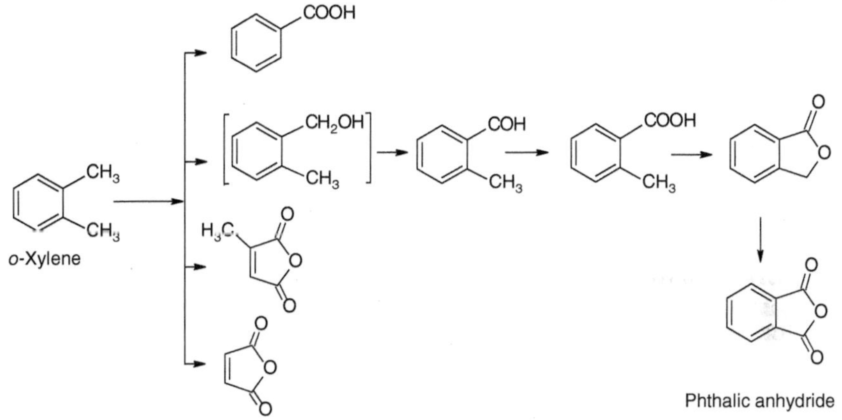

The oxidation reaction of *o*-xylene is in gas phase using V_2O_5 catalyst and $K_2SO_4 + TiO_2$ as modifier at 350–450 °C in air. The reaction mechanism is complicated with risk of explosion. The presence of byproducts requires distillation for final purification. Note that phthalic anhydride can also be produced by direct oxidation of naphthalene using V_2O_5, which is a strong oxidizing agent.

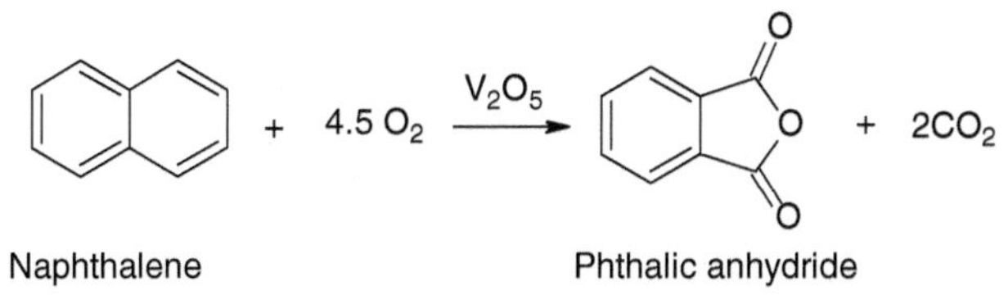

Table 6.9 World consumption of phthalic anhydride in 2016.

Applications	Capacity, thousands of metric tons	Percentage of use
Plasticizers	1,931	50.3
Unsaturated polyesters	507	13.2
Alkyd resins	842	21.9
Others	561	14.6

The largest consumption component for phthalic anhydride (PA) is for the production of phthalate plasticizers, which consumed about 50% of total use of PA in 2016 (IHS Markit [2]) (Table 6.9). China is the world's biggest consumer of PA for plasticizers followed by the Western Europe and the United States.

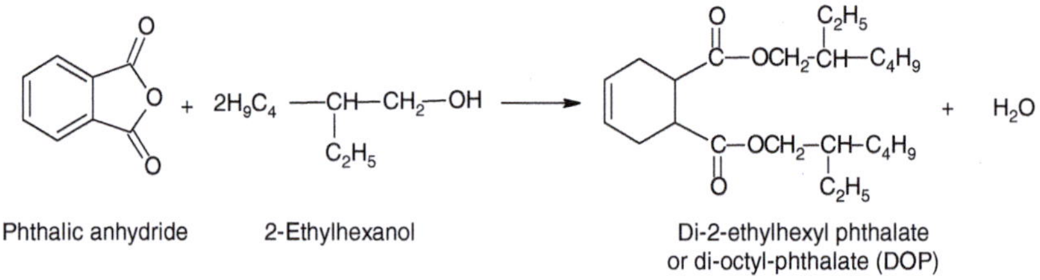

Di-2-ethylhexyl phthalate or di-octyl-phthalate (DOP) is widely used (80%) in PVC. DOP is an increasingly important plasticizer for auto manufacturing industry. DOP tested positive in rats for carcinogenicity and has been phased out from toy manufacturing industry.

6.3.1 Unsaturated Polyester Resins Synthesis

Unsaturated polyester resins can be produced by reacting phthalic anhydride with maleic anhydride and diethylene glycol. The polymer product is then cross-linked with styrene. Unsaturated polyester resins can be used in reinforced and nonreinforced applications, including glass, cloth, fiber, or metal replacement for boat, storage tanks, and automobile manufacturing.

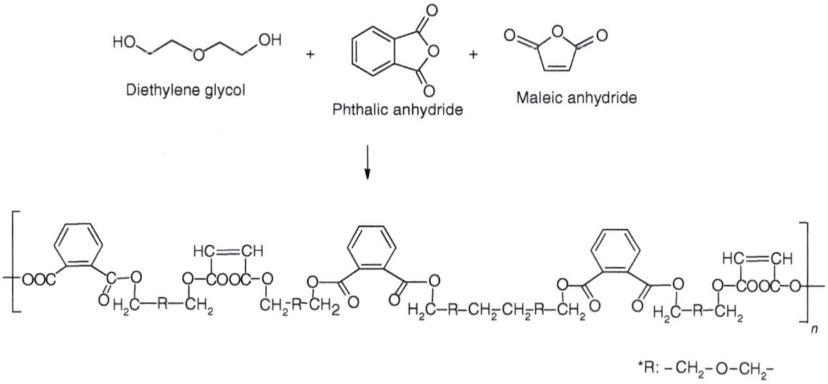

6.3.2 Alkyd Resin Synthesis:

Alkyd resins are the most important vehicles for oil-based paints. However, their growth has been static for years because more resistant coatings based on vinyl, epoxy, and urethane resins have come into use. Also, the growth of alkyd resins has been inhibited by the popularity of water-based or latex paints derived from poly(vinyl acetate) and acrylic resins.

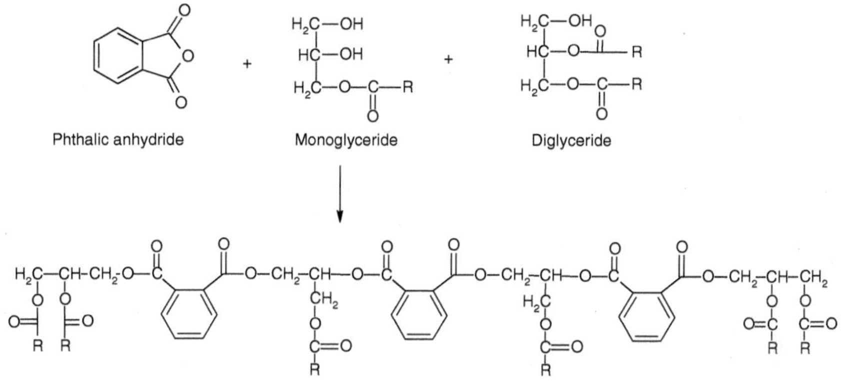

*R: alkyl, e.g. $CH_3-(CH_2)_4-CH=CH-CH_2-CH=CH-(CH_2)_7-$

6.3.3 *p*-Xylene Oxidation to Terephthalic Acid:

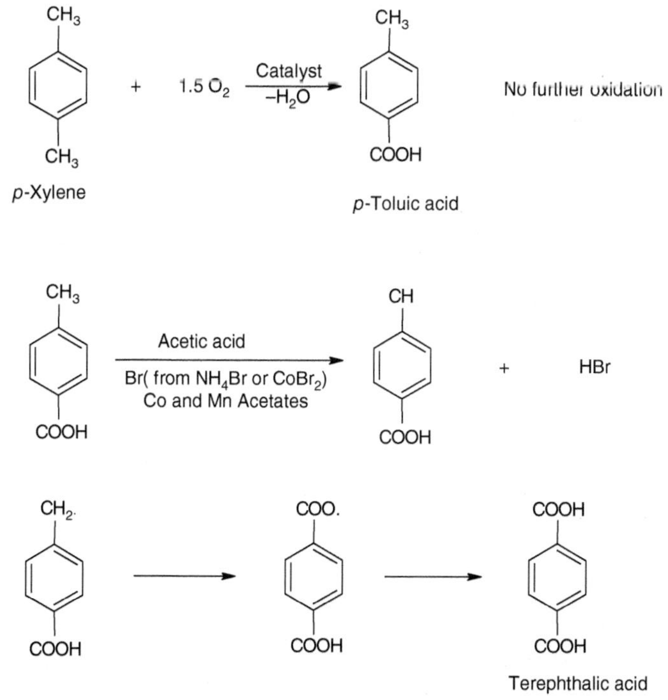

Table 6.10 Global supply of PET polymer in 2017.

Region	Percentage of distribution
Asia/Pacific	80
North America	7
Europe/Africa/Middle East	11
Latin America	2

Table 6.11 Worldwide PET consumption by end-use applications.

End-use applications	Percentage of distribution
Polyester fibers	63
Polyester solid-state resin	19
Polyester film	4
PET chip	12
Other	2

For byproduct removal, metal-catalyzed hydrogenolysis is carried out followed by *p*-toluic acid removal by crystallization.

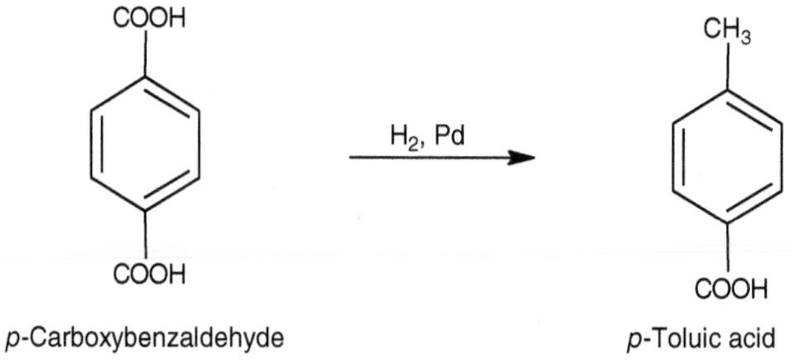

Poly(ethylene terephthalate) is produced from reacting terephthalic acid with polyol

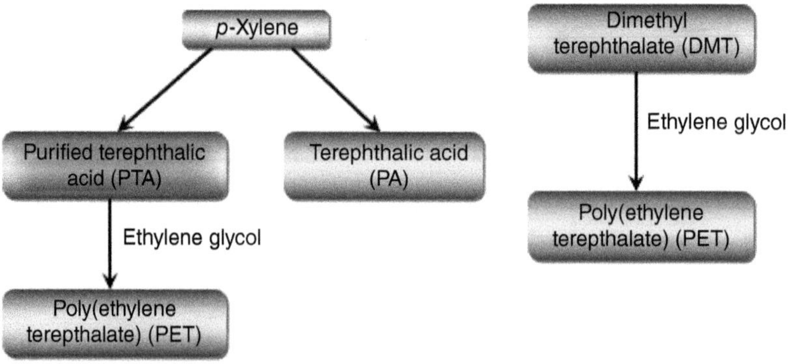

6.3.4 Poly(ethylene terephthalate), PET, and Poly(ethylene naphthalate), PEN

PET is one of the most important and widely used commercial polyesters. Tables 6.10 and 6.11 show global supply and worldwide consumption of PET in 2017 [2]. It is an amorphous thermoplastic polymer that is solidified upon rapid cooling and can be easily molded for various applications. PET has high strength, good abrasion and heat resistance, and excellent dimensional stability that is good when reinforced in glass-fiber applications. Another important polyester is poly(ethylene naphthalate) (PEN). PEN polyester has lower oxygen permeability than PET, making it more oxygen resistant for high-barrier performance applications (Table 6.12). The glass transition temperature (T_g), a critical property – temperature region where the polymer transitions from a hard, glassy material to a soft, rubbery material, is higher for PEN (120 °C) than for PET (75 °C) (https://polymerdatabase.com/polymer%20index/polyesters.html). Thus, PEN has higher tensile strength and service temperature with reduced elongation and shrinkage due to higher glass transition temperature (T_g). However, PEN resin is more expensive, about 3–4 times the cost of PET (Table 6.12). To recap, xylenes are the major feedstock to produce chemical intermediates and polymers that are shown in Figure 6.4.

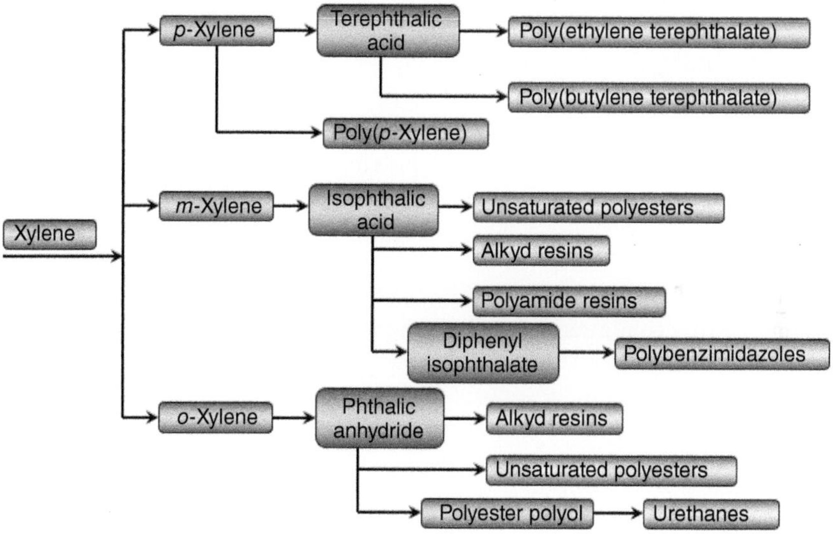

Figure 6.4 Polymers and chemical resins from xylenes.

Table 6.12 PET and PEN polymer general characteristics.

Material	Temperature resistance	O_2 permeation resistance	Uses
PET	Up to 75 °C	Low to medium	Water, liquid, dressing, peanut butter, soft drink
PET–PEN copolymer	Up to 100 °C	Low to medium	Jam, tomato sauces, sport drinks
PET–PEN blends	Up to 100 °C	Medium to high	Cutups, carbonated drinks
PEN	Up to 120 °C	High	Vegetables, meats, baby food, beer, wine

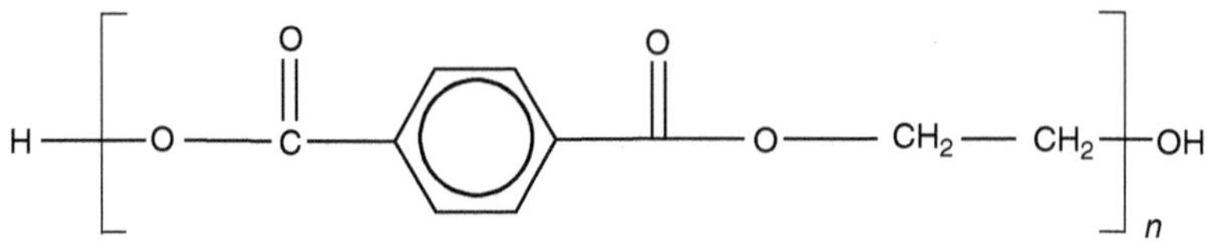

Poly(ethylene terephthalate) – PET

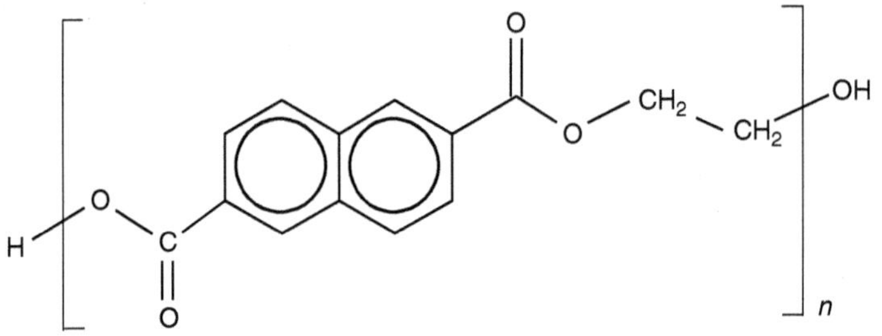

Poly(ethylene naphthalate) – PEN

6.4 Summary

Aromatic applications have been contributing to the society by improving the quality and every aspect of life. Industrial organic chemistry has evolved from the field of aromatic chemistry giving ways to the development of material science. From building-block BTX aromatics, many applications have been developed ranging from pharmaceuticals, dyes, detergents, pesticides, plastics, synthetic fibers, and chemical resins that can be used in construction and automotive industries.

Polymer science continues to develop with the fundamental chemistry using BTX aromatics. With liquid-crystallinity characteristic, aromatic-based polymers have been developed and continually invented to have better performance properties at lower costs and reduced environmental impacts. The more biodegradable chemicals or intermediates are the current challenges for development to combat wastes and environmental footprints. The same is true for pesticides and pharmaceutical products using aromatic-based molecules. They will be more environmentally benign, nontoxic and can be used effectively at minimal dosage to protect our environment and human wellbeing.

Acknowledgments

The author of Chapters 5 and 6 would like to dedicate the work to his wife, Linh Le, and his son, Bryan Pham, for their love and support. He deeply appreciates his wife's tolerance and sacrifice for the family with her everyday works, no matter how big or small. She is always there with him and his son to take care of them for the fun days and the bad days. The author, Trung Pham, is grateful and in debt to Dr. Santi Kulprathipanja for his leadership and professional advice. With Dr. Kulprathipanja's deep wealth of knowledge and expertise, he has taught a great deal over the years, and the author feels very fortunate to have him as a colleague and mentor. Trung thanks many colleagues at UOP, including Stanley Frey, Mary Joe Wier, Clayton Sadler, James Johnson, Patrick Whitchurch, Patrick Silady, Mark Lapinski, Dave Lafyatis, Steve Philoon, Patrick Bullen, Aronson Huebner, Gavin Towler, David Wegerer, and many others, who have discussed and contributed many ideas to him over the years. He would also like to thank Mark Goldberg and John Simley for reviewing these book chapters and providing valuable feedback.

References

1 Franck, H.-G., Stadelhofer, J.W. (1988). *Industrial Aromatic Chemistry*. Springer
2 (2017). *Chemical Economics Handbook*. IHS Markit.
3 (2018). *Chemical Economics Handbook*. IHS Markit.
4 (2014). *2014 End-Use Market Survey on the Polyurethanes Industry in the United States, Canada and Mexico*. American Chemistry Council.

Section IV

Industrial Separation

7
Adsorption

7.1 Principle of Adsorption

Adsorption process consists of two main pathways: adsorption and desorption. Adsorption is the process whereby a feed gas or feed liquid components adsorb at the surface of a microporous solid. The mixture of adsorbed feed components is called the "adsorbate," and the microporous solid is called the "adsorbent." Adsorption is dictated by the characteristics of the adsorbate and adsorbent interaction. If a certain species A has a greater affinity for the adsorbent than another species B in the mixture, the preferentially adsorbed species can in principle be separated from the other molecules in the gas or liquid mixture. Desorption is the process whereby the adsorbate is removed or recovered from adsorbent so that the adsorbent is available for reuse in the next cycle in the process. In gas-phase adsorption, the adsorbed material is most often removed by changing the temperature and/or the pressure of the system along with a carrier or sweeper gas. For liquid systems, a solvent must be found that preferentially displaces the desired product species from the adsorbent. The solvent that displaces the desired product from adsorbent is called "desorbent." Characteristics of desorption depend on specific interactions of desorbent with both the adsorbent and adsorbate. It is these interactions that dictate and complicate the chemistry of adsorption. However, because of the significant number of possible combinations of adsorbents and desorbents, there are few reasons why any two or more components cannot be separated by adsorption. The desorbent itself should be easily separated from the product in another separation step, usually distillation. This chapter will highlight the impact of adsorption technology for two industrial applications: impurity removal and liquid bulk separation.

There are two types of adsorption, chemical adsorption (chemisorption) and physical adsorption (physisorption). Chemisorption takes place when an adsorbed component reacts chemically with the adsorbent. Desorption or recovery of the adsorbed product is generally difficult or not possible. This type of adsorption is generally used for trace impurities' removal from liquid hydrocarbon or natural gas. Because of the nature of the chemical interaction between adsorbent and adsorbate, the adsorbent is not regenerable. On the other hand, physisorption does not involve chemical reaction between adsorbate and adsorbent. With this type of adsorption, the adsorbate is adsorbed through interaction with the adsorbent by electrostatic

Modern Petrochemical Technology: Methods, Manufacturing and Applications, First Edition.
Santi Kulprathipanja, James E. Rekoske, Daniel Wei, Robert V. Slone, Trung Pham, and Chunqing Liu.

or van der Waals forces. In liquid physisorption, the adsorbate can be desorbed by a desorbent as mentioned earlier. In gas-phase separation, the adsorbate can be recovered by raising the temperature, reducing the partial pressure of a component, or in some cases, by using a vapor-phase desorbent. This type of adsorption is generally used for bulk liquid or gas-phase separation. Liquid adsorption will be the main subject of this chapter's discussion. This is because bulk liquid adsorption has a great impact in refining and petrochemical separations. To demonstrate the impact of bulk liquid-adsorptive separation, two industrial hydrocarbon separation technologies, the UOP Parex™ and UOP MX Sorbex™ processes, will be discussed at the end of the chapter.

7.2 Adsorbents

An adsorbent is typically a solid porous material that allows molecules of a gas or liquid mixture to adhere to its surface, and this process is called adsorption. In separation processes via adsorption, the role of the adsorbent is to provide a large surface area for the adsorption of certain molecules. The strength with which these molecules are bonded to the surface is determined by the nature of the interaction between the molecule and the surface of the adsorbent. Adsorbents are made from many types of materials such as zeolites, silica gel, activated alumina, activated clay, polymeric resin, and activated carbon. General examples of adsorbent applications are listed in Table 7.1. The selection of adsorbents is dictated by the physical and chemical characteristics of adsorbents.

As shown in Table 7.1, zeolites are used in industrial bulk-scale separations such as xylene separation and separation of *n*-paraffins from *i*-paraffins. Zeolites are microporous crystalline aluminosilicate minerals with pores of uniform size from 3 to 10 Å, which are uniquely determined by the unit structure of the crystal. These pores will completely exclude molecules that are larger than their diameter. In contrast, activated carbon, activated alumina, silica gel, and activated clay do not possess an ordered crystal structure and consequently the pores are nonuniform. The distribution of the pore diameters within these adsorbent particles may be as narrow as 20–50 Å, or it may range as wide as 20 Å to several thousand Angstrom as is the case for some activated carbons. In this case, almost all molecular species may enter the pores. This is the reason why silica gel, activated alumina, activated clay, and activated carbons are not suitable for industrial bulk separation. Their uses are most beneficial in the areas of polar species removal from vapor or liquid. The pore size distribution for zeolite molecular sieve, a typical silica gel, and activated alumina are illustrated schematically in Figure 7.1.

7.2.1 Zeolite Adsorbents

Zeolites are microporous crystalline aluminosilicates with a three-dimensional microporous framework of $[SiO_4]^{4-}$ and $[AlO_4]^{5-}$ tetrahedra. Isomorphous substitution of Al^{3+} for Si^{4+} causes a negative charge of the zeolite framework, which

Table 7.1 General uses of adsorbents.

Adsorbent	Industrial applications
Zeolites	Separate individual PX, MX, or ethyl benzene (EB) from C8-aromatics
	Separate *n*-paraffins from *i*-paraffins and cyclic
	Separate olefins from paraffins
	Separate fructose from glucose in corn syrup
	Separate individual N_2 or O_2 from air
	Remove acid gas from natural gas
	Remove iodide from nuclear off-gas or acetic acid
	Remove water from azeotropes, or organic liquids
	Purify H_2
Silica gel	Desiccant in packings and double glazing
	Drying of gases, refrigerants, organic solvents
	Dew point control of natural gas
	Removal of polar impurities from hydrocarbons
Activated alumina	Removal of SO_x and NO_x
	Removal of vinyl chloride monomer (VCM) from air
	Removal of odors from gases
	Clean-up of nuclear off-gases
	Purification of helium
Activated clays	Removal of organic pigments
	Refining of mineral oils
	Removal of polychlorinated biphenyls (PCBs)
	Removal of dienes/trienes from hydrocarbons
Polymeric resins	Water purification
	Recover and purify steroids, amino acids
	Separate fatty acids from water and toluene
	Separate fructose from glucose in corn syrup
	Separate aromatics from aliphatic
	Recover proteins and enzymes
	Removal of colors from syrups
	Removal of organics from hydrogen peroxide
Activated carbon	Water treatment: portable, industrial, municipal
	Sweetener decolorization
	Chemical processing
	Pharmaceuticals
	Hydrocarbon recovery
	Industrial off-gas control

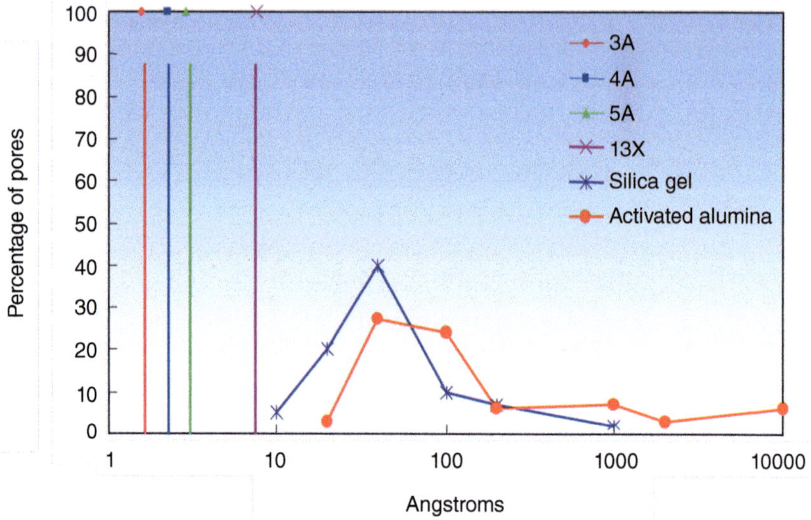

Figure 7.1 Distribution of pore sizes in adsorbents.

is compensated by charge balancing cations. Zeolites have uniform micropores, with size-selective adsorption and catalytic properties. Zeolite minerals were first described in 1756 by the Swedish mineralogist A.F. Cronstedt [1]. The zeolite structure may be represented by the formula $M_{x/n}[(SiO_2)_y(AlO_2)_x]\cdot wH_2O$, where M is a cation of valence n, y/x is the Si/Al ratio, and w is the number of water molecules. For a given framework, it is possible to tailor the properties of the zeolite for a desired application, by varying the Si/Al ratio and the counterion. Examples of zeolite frameworks that have an impact in industrial separations are LTA, MFI, and FAU. The zeolite A has an LTA framework. Both silicalite and ZSM-5 have an MFI framework but differ greatly in the Si/Al ratio. There is no formal convention about the Si/Al ratio of the two zeolites; however, silicalite has a very high Si/Al ratio (basically no Al in the structure), and ZSM-5 has a Si/Al ratio from several hundreds to 10 [2, 3]. The MFI framework has a three-dimensional pore structure with straight pores running in the b-direction with pore openings of 0.53 × 0.56 nm. and zigzag pores running in the a-direction with pore openings of 0.51 × 0.55 nm. The zeolites X and Y have the same FAU framework but differ only in the Si/Al ratio, which varies between 1 and 2.0 for zeolite X, and between 1.5 and 5.6 for zeolite Y [2, 3]. FAU has a three-dimensional pore structure with pores running perpendicular to each other in the a-, b-, and c-directions. The pore diameter is 0.7 nm and there is also a cavity with a diameter of 1.2 nm where the channels intersect. Structure of zeolites A, Y, and silicalite, along with the general characteristics of zeolites, are shown in Figure 7.2.

7.2.2 Zeolite Synthesis

Many high-quality books and monographs, as well as publications from the International Zeolite Association, have documented the important aspects of

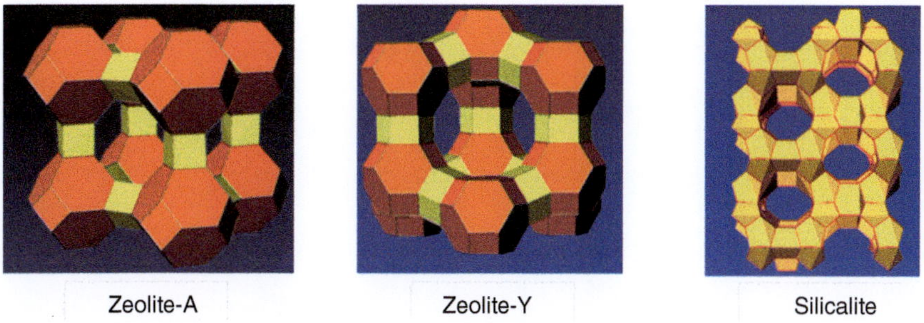

Figure 7.2 Distribution of pore sizes in adsorbents.

zeolite synthesis [4–7]. Additional information is available in the patent literature, although the experimental detail given in patents sometimes is lacking. Still, more formulation and synthesis details are present in the patent literature than the open literature, which makes these patent documents a necessity when a complete picture of this field is desired. Since there is enormous zeolite synthesis information documented in these sources, only a brief description of zeolite synthesis will be described here as a reference to this part of chapter.

The preparation of zeolite materials began in the late 1940s based on the pioneering work of R.M. Barrer. In 1948, Robert M. Milton at Union Carbide Corporation initiated research into zeolite production, building on the work of Barrer, and because of their efforts, Union Carbide entered the zeolite market in 1954. Since that time, Union Carbide, ExxonMobil, and numerous independent researchers have developed hundreds of zeolites. Despite the wide variety of zeolite microporous crystal structures that can be synthesized, virtually all zeolite synthesis procedures share the same basic steps as shown in Figure 7.3. That is, raw materials, which are composed of SiO_2, Al_2O_3, Na_2O, and water, are introduced to form a caustic aqua-gel solution of silica and alumina. The aqua-gel, containing the charge of nutrients necessary for zeolite synthesis, is sealed in a vessel and exposed to elevated temperatures (usually between 60 and 180 °C) in a batch process. The solubility of the gel phase is higher than the solubility of the newly formed zeolite phase and so the resulting equilibrium between the amorphous gel phase, supersaturated aqueous phase, and newly formed zeolite crystal phase at high temperature and autogenous pressure provides the driving force for zeolite homogeneous nucleation. Under these same conditions, subsequent crystal growth of formed nuclei competes with continued nucleation until all nutrients are depleted and the batch synthesis is stopped. Due to the variety of parameters involved in crystallization, batch times can vary from hours to days [8]. Several different silica sources can be used for hydrothermal synthesis such as colloidal silica, sodium silicate, or tetra ethyl ortho silicate. Aluminum can be from sources such as aluminum hydroxide, alumina, or sodium aluminate. Other parameters, which can be varied to produce different zeolite frameworks and crystal sizes, are the silica-to-aluminum ratio, the water content, the pH, and the presence of any cations in solution. Organic templates such as triethanolamine (TEA) and diethanolamine (DEA) can be used as structure-directing agents (SDAs). These

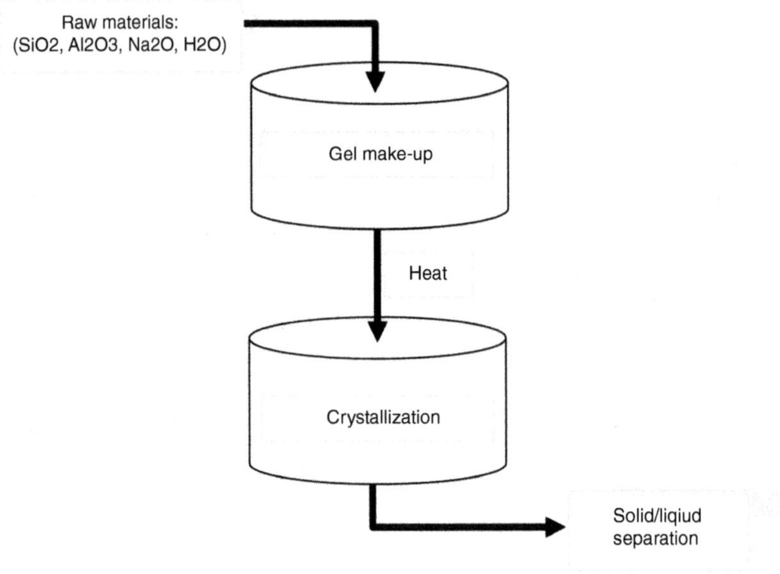

Figure 7.3 Zeolites synthesis process.

parameters have a strong influence on not only the identity of the framework which is synthesized, but also on the crystal size distribution, the yield of solids, and the crystallinity of the yielded solids.

7.2.3 Forming Zeolite Powders into Usable Shapes

Zeolites and other molecular sieves are important materials in catalysis, adsorption, and many other applications, both emerging and well established. Currently, no more than a few dozen crystalline microporous structures are widely manufactured for commercial use, in comparison to the hundreds of structures that have been made in the laboratory. The highest-volume zeolites manufactured are among the earliest-discovered materials: zeolite A (used extensively as ion exchangers in powdered laundry detergents and *n*-paraffin separation) and zeolites X and Y (used in xylene separation and catalytic cracking of gas oil).

Zeolites are generally in a powder form with a typical particle size ranging from 1 to 5 μm, and, in some cases, as small as a few nanometers. At this size, the zeolites are usually too small to be used as an adsorbent or catalyst. Therefore, forming zeolite powders into useful shapes is needed. There are several books and monographs that outline equipment and manufacturing procedures that are relevant to zeolitic adsorbents and catalysts [9, 10]. Many technologies are available to industry for manufacture of zeolitic adsorbents and catalysts. The unit operations and equipment that are used can therefore be found throughout the catalyst and adsorbent-manufacturing industries. There are two adsorbent shapes that industry normally produces – extrudates and beads. Pelletization and accretion

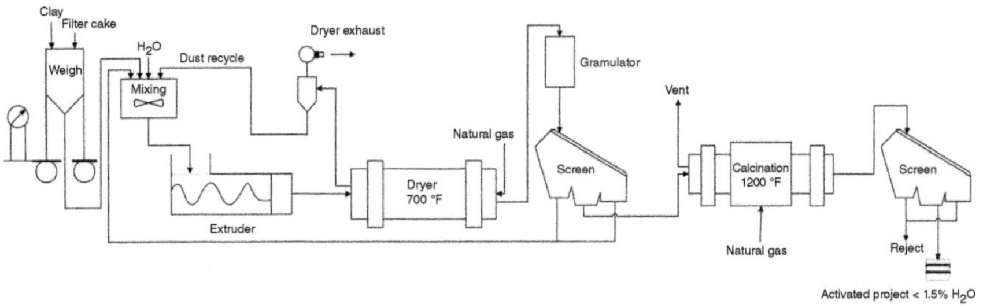

Figure 7.4 Adsorbent forming by extrusion process.

are the two techniques that are used to manufacture extrudate and bead adsorbents, respectively.

7.2.3.1 Pelletization of Zeolite Powder to Extrudate Adsorbent

The pelletization of zeolite powder usually requires the addition of inorganic particulate binders to make the paste moldable and to impart the adsorbent physical strength and attrition resistance after thermal treatment. Commonly used binders are clays (aluminosilicates), amorphous aluminophosphate, alumina, silica, titania, zirconia, or a combination of two or more of these components [11]. Clay binders are popular as they make the paste plastic and at the same time impart a relatively high mechanical strength to the extruded bodies enabling handling and post-treatment processes such as cutting. Kaolin, attapulgite, sepiolite, and bentonite clays are commonly used binders for zeolites [12, 13]. Various types of organic additives are added to the pastes to serve as thickening agents, lubricants, wetting agents, and temporary binders to improve the plasticity of the paste. Commonly used plasticizing agents for extrusion pastes are various types of modified water-soluble cellulose products, polyethylene glycol, polyvinyl alcohol, and glycerine [14–16].

A method of forming the zeolite powder into an adsorbent is shown in Figure 7.4. This includes the addition of zeolite (filter cake), clay, and water into the mixer. The blended clay zeolite mixture is extruded through the die. The extrudates are subsequently dried, granulated, screened, calcined, and activated to become the final adsorbent product. The calcination temperature of a zeolite-binder adsorbent is always lower than the thermal stability of the zeolite. The temperature and binder are selected to achieve a mechanically strong adsorbent while keeping the surface area and crystallinity of the zeolites intact. The thermally induced increase in strength is probably related to the creation of new and stronger bonds both between the binder phase and the porous zeolite powders and between the porous powders themselves.

7.2.3.2 Accretion of Zeolite Powder to Bead Adsorbent

As in the case of pellitization, the process of accretion of zeolite powder to bead adsorbent also requires the addition of inorganic particulate binders to make the paste moldable and to impart the adsorbent physical strength and attrition resistance after thermal treatment. The preparation of the zeolite/clay paste for accretion is like that which is used in palletization. Only minor differences are expected, such as the moisture content to suit to each system in making a strong adsorbent. The method of forming the zeolite powder into a bead adsorbent is shown in Figure 7.5. This includes adding zeolite (filter cake) and clay into a ribbon blender and transferring the mixture of zeolite and clay to an accretion drum with water addition. The blended clay zeolite mixture is accreted to form various sizes of bead particles. The beads are subsequently screened to the desired size particles as product, while the over- and undersize beads are recycled back to the accretion drum. The bead adsorbents are then calcined and finally activated to be the adsorbent product.

7.3 Bulk Liquid-Adsorptive Separation

Liquid adsorption includes two main events: adsorption and desorption. Adsorption of an adsorbate onto zeolite adsorbent is dictated by the characteristics of the adsorbate–adsorbent interaction. A zeolitic adsorbent is a crystalline porous solid. When adsorbent is immersed in a liquid mixture, the porous solid has the capability to adsorb specific species from the liquid. At equilibrium, the composition of material inside the pores differs from that of the liquid surrounding the porous solid. The amount of adsorbate that can be taken up by the adsorbent (i.e. the adsorbent capacity) is dependent on the nature of the adsorbate–adsorbent interaction, surface area, and pore volume of the adsorbent.[1]

In general, a higher adsorbent capacity is seen with higher adsorbent selectivity, higher adsorbent surface area, and higher pore volume. This phenomenon creates a higher concentration profile of the more selectively adsorbed component in the pores of the adsorbent than in the surrounding liquid. The process of adsorbing the adsorbate into the zeolitic adsorbent is known as the adsorption step.

To desorb the adsorbate from the zeolitic adsorbent, a desorbent is added. A desorbent is a suitable liquid that is capable of displacing – or desorbing – the adsorbate from the selective pores of the adsorbent. The process of recovering or desorbing the adsorbate from the adsorbent is known as the desorption step. In a chromatographic liquid separation process, the adsorption and desorption steps must ideally occur simultaneously. After the desorption step, both the rejected product (product with lower selectivity, resulting in less adsorption by adsorbent) and the extracted product (product with higher selectivity, resulting in strong adsorption by adsorbent) contain desorbent. In general, the desorbent is recovered by fractionation or evaporation and recycled back into the system.

7.3.1 Modes of Liquid-Adsorptive Separation

Zeolitic liquid separation mechanisms are complex, due to the interaction between the zeolitic adsorbent, adsorbate, and desorbent. To develop an adsorptive separation

1 Parts of 7.3 are reproduced from Kulprathipanja [14], Copyright Wiley-VCH Verlag GmbH & Co. KGaA. Reproduced with permission.

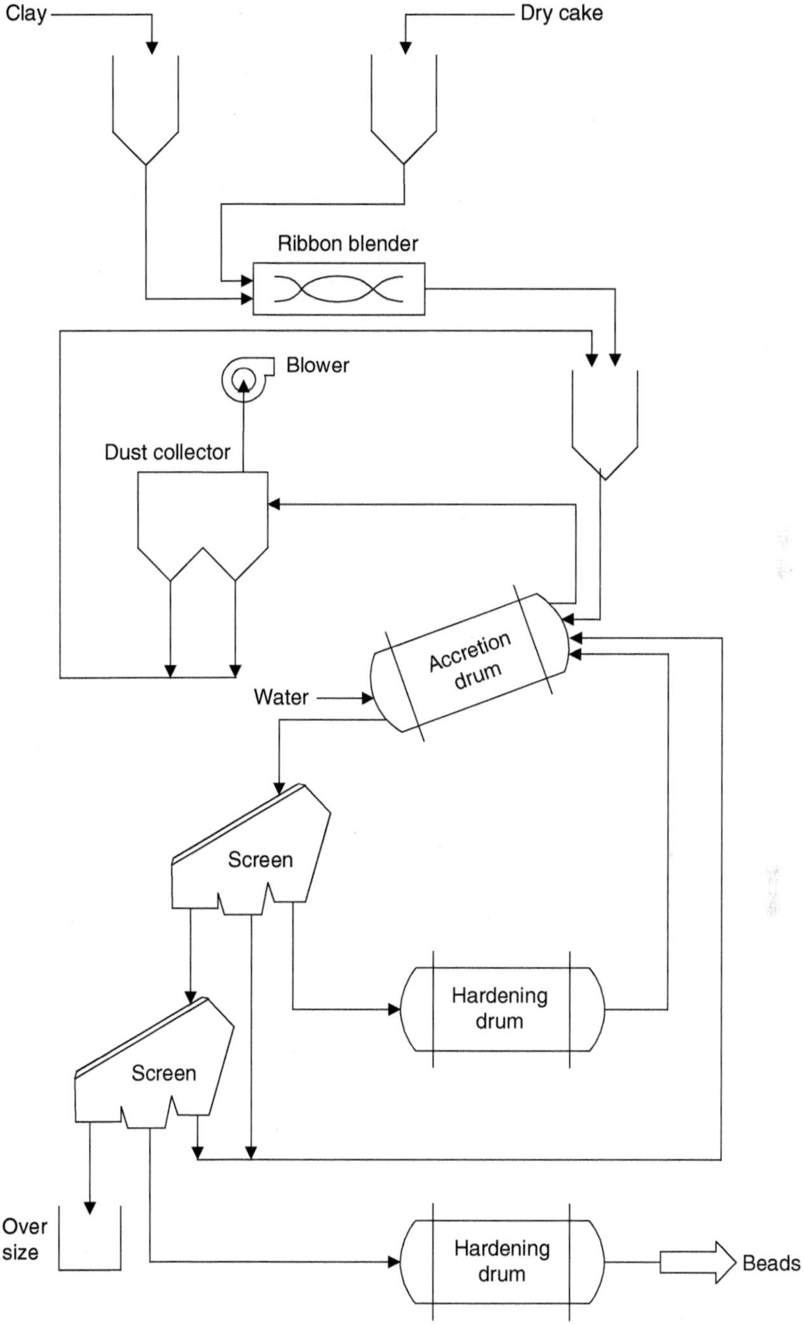

Figure 7.5 Accretion system for bead adsorbent manufacture.

process, an understanding of the zeolite adsorbent, adsorbate, and desorbent interaction is required. This has led to the development of appropriate techniques such as fundamental adsorption isotherms, pulse tests, and dynamic breakthrough tests. These techniques allow scientists and engineers to select the most suitable adsorbent and desorbent combinations, and optimize operating parameters such as temperature, liquid-flow circulation and adsorbent moisture. However, isotherms, pulse tests, and dynamic breakthrough tests do not offer enough information for scaling up the commercial unit. The scale-up information must be obtained from the simulated counter current – the Sorbex technology. Other very important considerations must be addressed: flow distribution across the chamber diameter, plug flow within the adsorbent bed, hydraulics, optimal configuration of adsorption, purification and desorption zones, flow and pressure measurement and control, desorbent recovery and energy integration, to name a few.

7.3.1.1 Adsorption Isotherms

In general, adsorption isotherms are generated using a batch experiment at a fixed temperature and a fixed feed composition. These experiments include exposing a known amount of adsorbent to a known concentration of adsorbate at a constant temperature. Once equilibrium is established, the net adsorbate concentration in the liquid is measured. This process is repeated at multiple adsorbate concentrations and temperatures. A plot of adsorbate loading (g adsorbate/g adsorbent) vs. adsorbate concentration reveals the adsorption isotherm with the shape of the isotherm determining the suitability of an adsorbent for a system [17].

7.3.1.2 Pulse-Test Procedure

The pulse test is a procedure used to search for a suitable adsorbent and desorbent combination for a liquid-adsorptive separation. The properties of the suitable adsorbent, such as a type of zeolite, exchange cation, and adsorbent moisture content, are a critical part of the study. The operating temperature and liquid-flow circulation are also critical parameters that can be set in the pulse test. A pulse-test procedure [18] begins with an injection of a small pulse of the feed mixture to be separated into a desorbent stream that is flowing through a packed adsorbent column at a fixed flow rate and temperature. The column effluent composition is then determined as a function of time or volume of desorbent passed through the adsorbent bed by using gas or liquid chromatography. Particularly important is the sequence and time when each of the feed components exits the packed adsorbent column because these characteristics describe the specific adsorbate and adsorbent interactions. By determining the interactions using the pulse test, the separation process can be optimized.

An example of results from a pulse test separating arbitrary components A and B along with a tracer is shown in Figure 7.6. An inert tracer is selected that will not be adsorbed by the system being studied. Each component elution time or volume, shown in Figure 7.6, is taken at the maximum peak height or concentration of a specific component. Furthermore, the net retention volume of each component is measured, based on the maximum peak height or concentration of the tracer as the zero origin. Because the net retention volume of any component is ideally

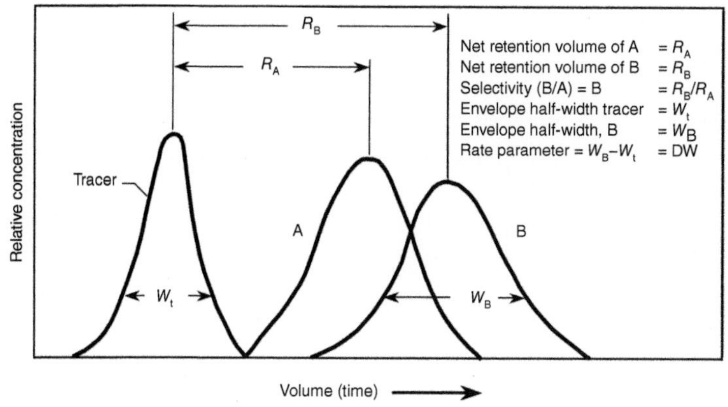

Figure 7.6 Schematic pulse test: two components and paraffin tracer.

proportional to its distribution coefficient (its concentration in the adsorbed phase divided by its concentration in the unadsorbed phase), the adsorbent selectivity for the more strongly adsorbed component B over component A can be calculated from the ratio of the net retention volumes of component B to component A. In Figure 7.6, component B is more strongly adsorbed in the adsorbent packed bed since it elutes after component A.

7.3.1.3 Breakthrough Procedure

The adsorbent and adsorbate interaction obtained using the pulse test takes place at a high level of dilution. Thus, the interaction information might not fully represent the actual interaction since the commercial feed concentration is normally much higher. For a more representative concentration, the dynamic breakthrough test technique is introduced. The breakthrough procedure is like the pulse-test procedure except a large amount of high concentration feed is used. The breakthrough procedure can be described as follows:

(1) Introduce solvent (such as desorbent) into a packed adsorbent column at a fixed flow rate, pressure, and temperature.
(2) Introduce a feed mixture to be separated until the effluent reaches the feed composition.
    The feed breakthrough comprises the components to be separated, the tracer, and the desorbent.
(3) Introduce solvent (such as adsorbent) until no component in the feed mixture is detected in the effluents.

A plot of column effluent composition as a function of time or volume of solvent and feed reveals the adsorption and desorption behaviors of the components in the feed mixture and illustrates whether the adsorption system is suitable for separating the components. Particularly important is the sequence and time when each of the feed components exits the packed adsorbent column because these characteristics describe the specific adsorbate and adsorbent interactions. From the breakthrough

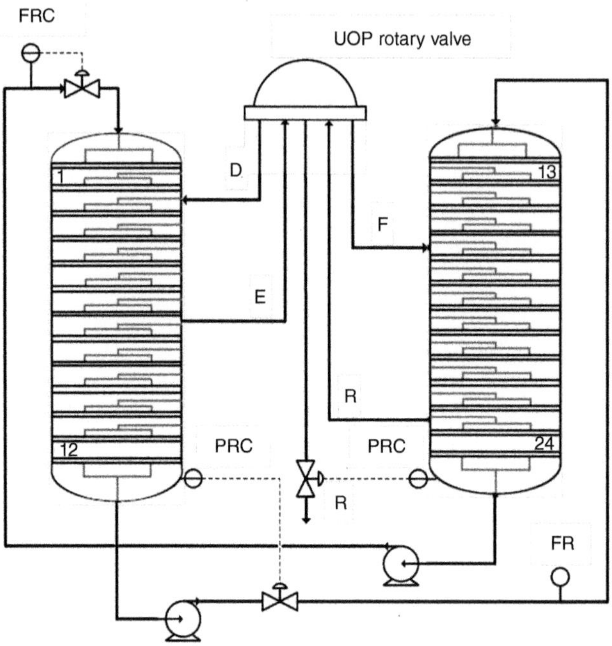

Figure 7.7 Sorbex: adsorption section.

plot, zeolitic adsorbent selectivity, capacity, mass transfer rate, and desorbent strength can be calculated.

7.3.1.4 Simulated Moving-Bed Operation: Sorbex Technology

Batch-type liquid-adsorptive separations can be carried out using the principle of pulse test, as mentioned earlier. The technique is still practiced in the field of small-scale separation such as in the pharmaceutical area. However, a batch-type liquid-adsorptive separation is not applicable to refining and petrochemical since it would require a very large volume of adsorbent and adsorbent. For this reason, a continuous liquid-adsorptive separation technology, the UOP Sorbex process, was developed. The Sorbex process operates on a liquid–solid countercurrent contacting principle. The basic layout for the Sorbex process is shown in Figure 7.7. The primary process consists of two adsorbent chambers, liquid-flow distributors, piping, and pumps directing the circulating flow through the chambers and from the bottom of one chamber to the top of the other. There is a bed line connecting each distributor grid to a port on the rotary valve. The rotary valve steps the net streams (feed and desorbent in, extract and raffinate out) around the 24 distributor grids. A control system regulates the flows, pressures, and rotary valve movement in the adsorption section. The adsorbent is contained in each of the 24 beds, with a flow distributor located above and below each bed. For practical reasons, in most cases, the adsorbent is divided between two chambers. There are some applications whereon 8 or 12 adsorbent beds are needed, and these will utilize a single chamber.

The UOP rotary valve has been used in more than two hundred Sorbex units across a variety of applications such as *p*-xylene (PX) separation, *m*-xylene (MX) separation, olefin separation, and *n*-paraffin separation. The purpose of the rotary valve is to move the inlet and outlet ports of the net streams (feed, desorbent, raffinate, extract) around the 24 beds in stepwise fashion, creating a semicontinuous simulated countercurrent flow of adsorbent relative to the liquid phase. This allows simultaneous adsorption, purification, and desorption to take place with more than an order of magnitude efficiency over batch adsorption [19].

7.3.2 Liquid-Adsorptive Separation Processes

The degrees of freedom in liquid-phase adsorption are very large due to the number of possible ways to modify the zeolite adsorbent characteristics. Key adsorbent variables include the zeolite framework structure, silica-to-alumina ratio, particle sizes, chemical composition, counter exchange ion, water content, and binder. These variables can be carefully modified to selectively adsorb one component over others. The adsorbed component can then be removed or desorbed from the adsorbent using a suitable solvent functioning as a desorbent. To optimize separation by liquid-phase adsorption, two opposing forces must be balanced: the adsorptive force of the adsorbent to the adsorbate and the desorptive force of the desorbent to the adsorbate. Ideally, the adsorbent should have a lower selectivity for the desorbent than the adsorbed component. However, the adsorbent should have a higher selectivity for the desorbent than the rejected components in the mixture. This is to ensure that the desorbent itself does not utilize most of the adsorbent capacity and, at the same time, can effectively desorb the adsorbate from the adsorbent. There are many adsorption mechanisms that zeolite separation processes can utilize. This depends on many factors such as zeolite structures, zeolite pore sizes, zeolite Si/Al ratios, exchanged cations, and the physical and chemical properties of adsorbates and desorbents. In general, the mechanisms present in zeolite adsorption can be classified as following:

(1) Equilibrium-selective adsorption;
(2) Rate-selective adsorption;
(3) Shape-selective adsorption;
(4) Ion exchange;
(5) Reactive adsorption.

7.3.2.1 Equilibrium-Selective Adsorption

The foundation of equilibrium-selective adsorption is based on differences in the equilibrium selectivity of the various adsorbates with the adsorbent. While all the adsorbates have access to the adsorbent sites, the specific adsorbate is selectively adsorbed, based on differences in the adsorbate–adsorbent interaction. This, in turn, results in higher adsorbent selectivity for one component than the others. One important parameter that affects the equilibrium-selective adsorption mechanism is the interaction between the acidic sites of the zeolite and basic sites of the adsorbate. Specific physical properties of zeolites, such as framework structure, choice of exchanged metal cations, SiO_2/Al_2O_3 ratio, and water content, can be manipulated to influence the acidity of zeolites, which affects separation performance. The degree of separation is characterized by the zeolite selectivity and capacity, as well as the choice of desorbent and operating conditions such as temperature, feed, and desorbent flow rates. Therefore, to achieve a meaningful separation using the equilibrium-selective adsorption mechanism, the following zeolite characteristics and operating conditions are to be considered:

(1) Zeolite framework structure;
(2) Metal cation exchanged in zeolite;
(3) Zeolite SiO_2/Al_2O_3 molar ratio;
(4) Moisture content in zeolite;
(5) Characteristic of desorbent;
(6) Operating temperature.

These variables are described in the following sections.

Zeolite Framework Structure

One of the most significant variables affecting zeolite adsorption properties is the framework structure. Each framework type (e.g. FAU, LTA, MFI, MOR) has its own unique topology, cage type (α, β), channel system (one-, two-, three-dimensional), free apertures, preferred cation locations, preferred water adsorption sites, and kinetic pore diameter. Some zeolite characteristics are shown in Table 7.2. More detailed information on zeolite framework structures can be found in Breck's book entitled *Zeolite Molecular Sieves* [11]. The variety of the different framework structures result in different adsorbent characteristics: acid strength, size of molecule adsorbed, adsorption/desorption rate of different molecules, capacity, and stability. Thus, these differences characterize the adsorbent's selectivity to a specific molecule and adsorbent–adsorbate interactions. Take, for example, the difference in selectivity of BaY and Ba-Mordenite [20] to PX, MX and *o*-xylene (OX):

BaY selectivity: PX > OX > MX
Ba-Mordenite selectivity: PX > MX > OX

Ba-Mordenite's selectivity to MX is higher than OX, but the opposite is true for BaY. This reversal in selectivity is a result of differences in adsorbent framework characteristics: mordenite has higher acid strength compared to Y zeolite. Adsorption and desorption rates of xylenes are expected to be faster in BaY compared to Ba-Mordenite because Mordenite is a one-dimensional channel system, while Y zeolite is a three-dimensional channel. With the reason stated, a three-dimensional channel zeolite is the preferred mass separating agent of choice compared to one- or two-dimensional channels for the liquid adsorption separation.

Metal Cation Exchanged in Zeolite

One of the parameters in the broad class of equilibrium-selective adsorption mechanisms is the interaction between the acidic and basic sites of the adsorbent and the adsorbate. Zeolites can be ion-exchanged with a variety of metal cations

Table 7.2 Typical properties of common zeolites.

Zeolite type	Channel system	Pore openings, Å (hydrated form)	Typical SiO_2/Al_2O_3 mole ratio	Theoretical ion exchange capacity, meq/g (Na form, anhydrous)
Analcime	1D	2.6	4	4.9
Chabazite	3D	3.7 × 4.2, and 2.6	4	4.9
Clinoptilolite	NK	4.0 × 5.5, 4.4 × 7.2, and 4.1 × 4.7	10	2.6
Erionite	3D	3.6 × 5.2	6	3.8
Ferrierite	2D	4.3 × 5.5, and 3.4 × 4.8	11	2.4
Phillipsite	3D	4.2 × 4.4, 2.8 × 4.8, and 3.3	4.4	4.7
Zeolite A	3D	4.2 into α-cage; 2.2 into β-cage	2	7.0
Zeolite L	1D	7.1	6	3.8
Mordenite	2D	2.9 × 5.7	10	2.6
Zeolite omega	1D	7.5	7	3.4
Silicalite-1	3D	5.7-5.8 × 5.1-5.2	50	0.63
Zeolite X	3D	7.4 into supercage; 2.2 into β-cage	2.5	6.4
Zeolite Y	3D	7.4 into supercage; 2.2 into β-cage	4.8	4.4

Note: 1D = one-dimensional, 2D = two-dimensional, 3D = three-dimensional.
Source: Beck [11], Neuzil [18], and Broughton and Gerhold [19].

to alter their acidity and other properties. There is a strong correlation between the total acidity of a zeolite (the sum of both Brönsted and Lewis acids) and the ionic radius of the cation as well as the valence charge of the exchanged cation [21–23]. Exchanged cations with lower ionic radii have higher zeolite acidity. The correlation between zeolite acidity and ionic radius or exchanged cation valence is measured by titration with *n*-butylamine and a Hammett indicator, methylred. Zeolite acidity increases for monovalent exchanged cations from Cs < K < Na < Li and increases from Ba < Sr < Ca < Mg and Ba < Pb < La < Ca < Zn < Co for divalent exchanged cations. For cations of similar ionic radii, divalent cations have higher zeolite acidity than monovalent cations. For example, Ba-zeolite has a higher acidity than K-zeolite, and Ca-zeolite has a higher acidity than Na-zeolite and Li-zeolite. Furthermore, strong acidic Brönsted properties can be observed in hydrated zeolites due to polarization of adsorbed water in the strong electrostatic field between the exchanged cations and the AlO_4 – anions [23]. For hydrated zeolites, a higher moisture content results in higher zeolite acidity.

The importance of zeolite acidic strength with zeolite Y for C8-aromatics (xylenes) systems is illustrated below. In the presence of strong acids, xylene isomers have varying basicity (Table 7.3), with MX being the most basic and PX the least basic

Table 7.3 Base strengths of aromatic hydrocarbons relative to HF.

Compound	Relative basicity at 0.1 molar
Benzene	0.09
Toluene	0.63
p-Xylene	1.0
o-Xylene	1.1
m-Xylene	26
Durene (1,2,4,5-tetramethylbenzene)	140
Isodurene (1,2,3,5-tetramethylbenzene)	16 000

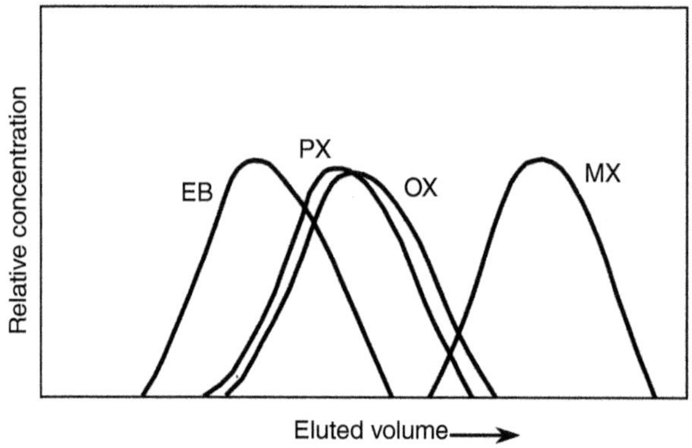

Figure 7.8 Separation of *m*-xylene from C_8-aromatics using NaY adsorbent.

among the C8-aromatics [24]. Based on the basicity of the xylenes, the acidity of Y zeolite can be properly adjusted to selectively adsorb MX or PX. As demonstrated in Figure 7.8, a more acidic zeolite such as NaY will selectively adsorb MX from other C8-aromatics [25, 26]. In contrast, Figure 7.9 shows that a weaker acidic zeolite such as KY will selectively adsorb PX from other C8-aromatics [27, 28]. In both systems, toluene is used as the desorbent. However, acid-based interactions between zeolitic adsorbents and adsorbates do not always correctly predict the trend of adsorbent selectivity. This is illustrated by the adsorptive separation of durene from isodurene. Pulse-test experiments indicated that the adsorbent selectivity for durene/isodurene increases from KX < NaX < LiX, shown in Table 7.4 [29]. Because isodurene is a stronger base than durene (Table 7.3), one would expect that the results for adsorbent selectivity shown in Table 7.4 would be the reverse order (LiX < NaX < KX). This counterintuitive trend may be explained by steric hindrances caused by the differences in ionic radii of the cations, which could play a role in the adsorbent and adsorbate interaction.

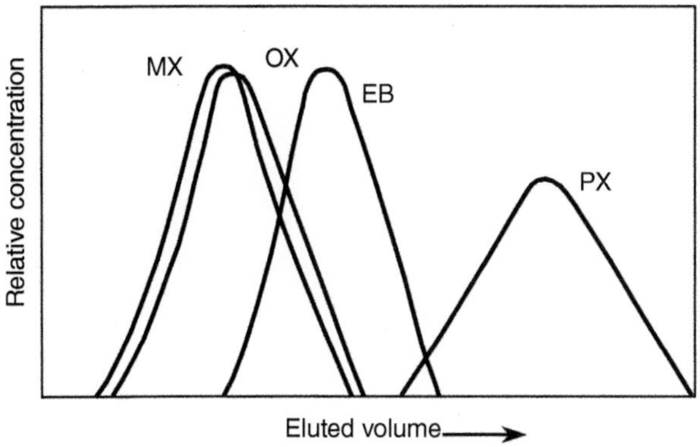

Figure 7.9 Separation of *p*-xylene from C_8-aromatics using KY adsorbent.

Table 7.4 Durene/isodurene selectivity as a function of X-zeolite exchanged cation.

Cation	Durene/isodurene selectivity
K	0.59
Na	0.97
Li	3.50

Zeolite SiO_2/Al_2O_3 Molar Ratio

As mentioned earlier, multiple factors determine zeolite acidity, including the SiO_2/Al_2O_3 molar ratio. Zeolite acidity increases in strength as the molar ratio of SiO_2/Al_2O_3 decreases [22] due to the increase in AlO^{4-} sites, which strengthens the electrostatic field in the zeolite and increases the number of acid sites. However, the wide array of cage and channel arrangements and electrochemical properties that result from various crystalline structures and different SiO_2/Al_2O_3 ratios also affect zeolite acid strength. In certain conditions, a high density of AlO^{4-} in the zeolite framework could lower the acid strength of the adsorbent. The reverse in acid strength can be explained by the dipolar repulsion of the AlO^{4-} groups outweighing the increase in polarizability [21, 30]. For the reasons stated, an increase in zeolite acidity as the molar ratio of SiO_2/Al_2O_3 increases is normally observed. Manipulating both the exchanged cations and the SiO_2/Al_2O_3 ratio offers a great flexibility in tailoring adsorbents for a specific application. However, the more variables that are altered, the more difficult the adsorption behavior becomes to predict.

The adsorptive separation of durene and isodurene is used here to illustrate the effect of zeolite acidity. As shown in Figure 7.10, LiX is more selective toward durene than isodurene [29]. However, LiY is more selective toward isodurene than durene (Figure 7.11). These results are consistent to what is expected since zeolite Y (Si/Al of 2.5) is more acidic than zeolite X (Si/Al of 1.25).

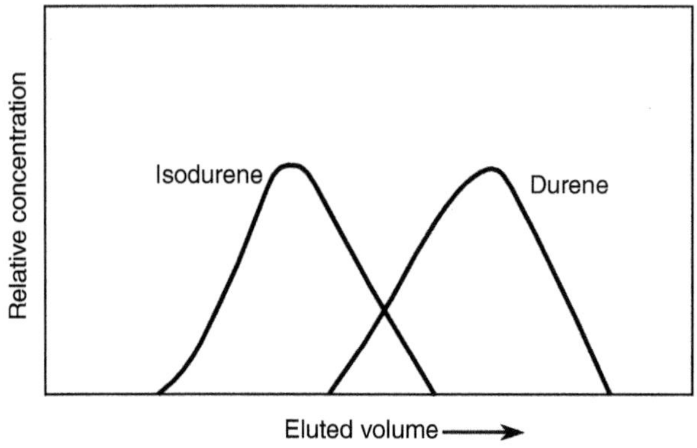

Figure 7.10 Separation of durene and isodurene using LiX adsorbent.

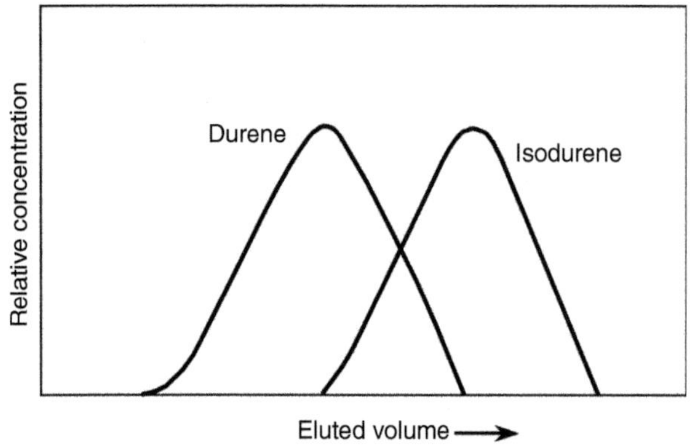

Figure 7.11 Separation of isodurene and durene using LiY adsorbent.

Figure 7.12 shows separation of fructose and glucose using adsorbents CaY, CaX, and KX [31]. The combination of the zeolite framework structure and exchanged cations is instrumental to this intricate mechanism as seen with the separation of fructose and glucose. Ca^{2+} forms complexes with fructose and, thus, CaY shows a high selectivity for fructose over glucose. In contrast, CaX does not exhibit high selectivity for fructose. However, KX zeolite shows a good selectivity for glucose over fructose. No simple explanation can be offered for this elaborate system. In the UOP Sarex^{TM} process, Ca-Y zeolite and, later, Ca-exchanged resin were successfully used as adsorbents that are selective for fructose.

Moisture Content in Zeolite

Adsorbed water molecules on a zeolite adsorbent are polarizable due to the strong electrostatic field between the exchanged cations and alumina framework [23]. Hence, water molecules enhance the acidic properties of the zeolite's Brönsted acids. Adsorbate–adsorbent interactions and, therefore, adsorbent selectivity and adsorbate mass transfer rates are altered due to water polarization. When developing an adsorbent to be used in a commercial adsorptive separation process, the

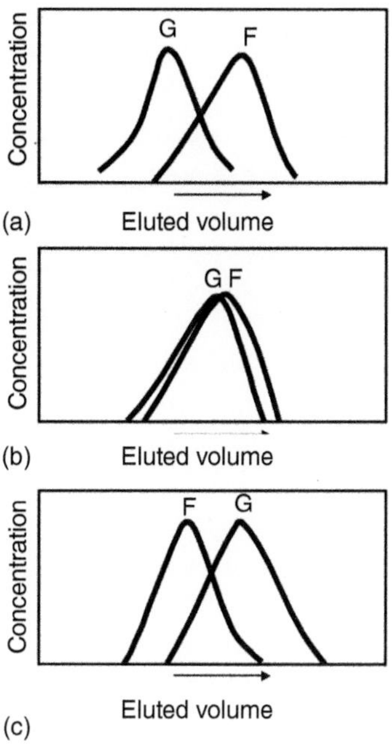

Figure 7.12 Fructose/glucose separation on zeolite adsorbents. (a) Ca-Y adsorbent, (b) Ca-X adsorbent, and (c) K-X adsorbent.

water content of the adsorbent is adjusted to balance adsorbent selectivity and the component mass transfer rate.

Characteristics of the Desorbent

In liquid-phase adsorption, some components of the feed steam are selectively adsorbed or extracted by a solid zeolitic adsorbent. At the same time, other components of the feed stream are rejected by the adsorbent. At equilibrium, the liquid composition within the zeolite pores differs from that of the liquid surrounding the zeolite. In the process, a second liquid component, the desorbent, is also introduced into the system. The function of the desorbent is to desorb and remove any weakly adsorbed impurities and to recover the extracted feed components from the adsorbent. For the desorbent to perform well in the process, a suitable interactive force is needed between the desorbent and the components that are extracted by the adsorbent. If the selectivity of the extracted component is too high, it requires a high desorbent volume to desorb the extracted components from the adsorbent. If the selectivity of the extracted component is too low, the desorbent tends to compete with extracted components for capacity of adsorbent.

A basic guideline for choosing a desorbent is to match the chemical properties of the extracted components with those of the desorbent. Another guideline is an appropriate selection of boiling point differences to allow recovery from the feed components from the desorbent after the adsorption section. The following examples illustrate this concept: (i) for a *n*-paraffin separation, a *n*-paraffin desorbent is preferable, (ii) for an aromatic separation, an aromatic desorbent is preferable, (iii) for a highly polar adsorbate, a desorbent should be selected from a class of alcohol, ketone, acid, or water. Desorbent features are illustrated in Table 7.5

Table 7.5 Separatio of *p*-xylene using KY and BaX adsorbents with various desorbents [38, 39] by pulse tests.

Adsorbent	Desorbent	PX net retention volume (ml)	PX/EB selectivity	PX/MX selectivity	PX/OX selectivity	PX stage time (s)
BaX	Phenyldecane	107	1.84	3.02	2.83	22.2
KY	Phenyldecane	No PX desorption				
KY	Diphenylmethane	10.5	1.68	3.08	1.89	29.7
BaX	Diphenylmethane	0	1	1	1	
BaX	1,3-Diisopropylbenzene	No PX desorption				
BaX	1,4-Diisopropylbenzene	56.4	2.28	3.45	3.34	19.8
KY	1,4-Diisopropylbenzene	No PX desorption				
BaX	1,3,5-Triethylbenzene	77.6	1.3	3.0	2.32	43.9
KY	1,3,5-Triethylbenzene	61.4	2.33	7.58	5.23	40.8
BaX	5-Tertiarybutyl-*m*-xylene	73.2	1.94	3.28	2.87	23.4
KY	5-Tertiarybutyl-*m*-xylene	47.1	2.05	3.72	3.37	54.5

Source: Kulprathipanja [38] and Bland and Davidson [39].

using C_8-aromatic adsorbates with BaX and KY adsorbents. The results in Table 7.5 further emphasize the desorbent characteristic requirement mentioned above. For instance, phenyldecane is a suitable desorbent for PX separation using BaX adsorbent. However, phenyldecane is too weak to desorb PX from KY adsorbents. In contrast, diphenylmethane offers good separation of PX with KY adsorbent, but not with BaX. With BaX adsorbent, PX is separated from other C8-aromatics using 1,4-diisopropylbenzene, but not with other isomers of di-isopropylbenzene, such as 1,3-di-isopropylbenzene.

Operating Temperature

Another critical variable in liquid-phase adsorptive separation is the operating temperature. Liquid-phase adsorption must be operated at a temperature that optimizes selectivity and mass transfer rates. Generally, selectivity increases, and transfer rates decrease at lower temperatures.

Operating Pressure

In general, operating pressure does not affect liquid-phase adsorption. However, enough pressure must be applied to maintain the system in the liquid phase during the entire process.

7.3.2.2 Rate-Selective Adsorption

Although most of the commercial adsorptive separation processes are operated under the selective-equilibrium adsorption mechanism, adsorptive separation may also be based on diffusion rates through a permeable barrier, which are designated as "rate-selective adsorption" processes. In some instances, there may

Table 7.6 Energetics and predictions from modeling molecular diffusion in silicalite.

Organic molecules	Energy barrier (kcal/mol)	Maximum energy (kcal)	Diffusion prediction
n-Decane	2	−31	Fast
2-Methylnonane	8	−29	Moderate
2,6-Dimethyloctane	15	−17	Slow
3,3,5-Trimethylheptane	45	+35	Excluded

be a combination of equilibria as well as rate-selective adsorption. A rate-selective adsorption process yields good separation when the diffusion rates of the feed components through the permeable barrier differ by a wide margin.

Examples of rate-selective adsorption are demonstrated using silicalite adsorbent for separation of C_{10}–C_{14} *n*-paraffins from non-*n*-paraffins [32, 33] and C_{10}–C_{14} mono-methyl-paraffins from non-*n*-paraffins [34–37]. Silicalite is a 10-membered ring zeolite with a pore opening of 5.4 Å × 5.7 Å [18]. In the case of *n*-paraffins/non-*n*-paraffins separation [32, 33], *n*-paraffins enter the pores of silicalite freely, but non-*n*-paraffins such as aromatics, naphthenes, and *iso*-paraffins diffuse into the pores more slowly. However, the diffusion rates of both normal-paraffins and non-*n*-paraffins increase with temperature. So, one would expect to see minimal separation of *n*-paraffins from non-*n*-paraffins at high temperatures but improved separation at lower temperature.

For the rate-selective separation of C_{10}–C_{14} mono-methyl-paraffins from non-*n*-paraffins [34–37], diffusion simulations were carried out using the Solids Diffusion module in the Accelrys Insight II molecular modeling package [36]. The modeling results from the diffusion simulations of four paraffins of varying carbon numbers in silicalite are summarized in Table 7.6.

Note that the predicted diffusivity of the molecules, based on the magnitude of the energy barrier, agrees with the experimental data (see Table 7.6). The retention volumes and hence diffusivities increase in the following order: 3,3,5-trimethylheptane < 2,6-dimethyloctane < 2-methylnonane < *n*-decane. The increase in retention volumes of the paraffins is consistent with the decrease in barriers energies. Both the calculations and pulse-test experiments indicate that the trisubstituted paraffin, 3,3,5-trimethylheptane, is too large to fit in silicalite pores. The disubstituted paraffin, 2,6-dimethyloctane, has limited diffusivity and is mostly excluded from the pores. The monosubstituted paraffin, 2-methylnonane, has good diffusion but is slower than the *n*-paraffin, *n*-decane. These results suggest that normal and monomethyl paraffins can be separated from di-/trisubstituted paraffins, naphthenes, and aromatics using silicalite adsorbent. Laboratory pulse tests have, in fact, demonstrated these separations, as shown in Figure 7.13.

Separation of mono-methylparaffin from depleted kerosene (removal of *n*-paraffins) was also demonstrated in a simulated moving-bed (SMB) pilot plant with >90% mono-methylparaffin purity at >70% recovery [36, 37]. Another example of rate-selective adsorption is the separation of di-isopropylbenzene isomers using a silicalite adsorbent. Figure 7.14 shows the adsorption rates of 1,3-diisopropylbenzene and 1.4-di-isopropylbenzene into silicalite adsorbent. It

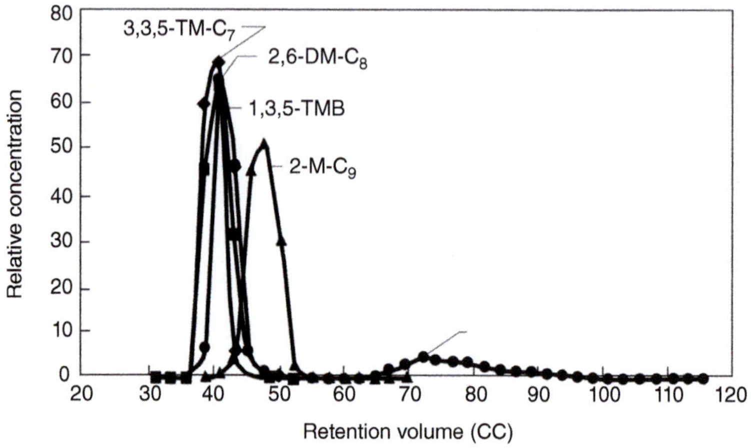

Figure 7.13 Chromatographic separation of mono-methyl paraffin.

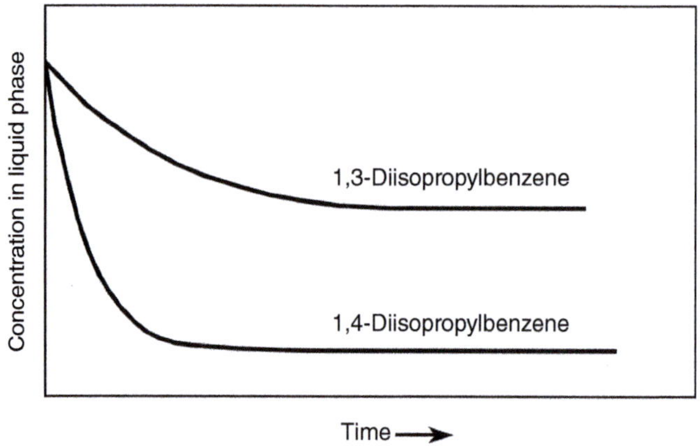

Figure 7.14 Rate-selective adsorption of di-isopropylbenzene isomers on silicalite adsorbent.

illustrates the more rapid adsorption of 1,4-di-isopropylbenzene compared to 1,3-di-isopropylbenzene.

7.3.2.3 Shape-Selective Adsorption

Shape-selective adsorption, also known as molecular sieving, is a process that separates molecules based on inclusion or exclusion from specific zeolite pores. In contrast, the equilibrium- and rate-selective mechanisms are based on adsorbate–adsorbent interactions and molecular diffusion rates through zeolite pores. Thus, for shape-selective adsorption, the theoretical selectivity between molecules that can and cannot enter a specific zeolite pore can be infinite. However, in most cases, adsorbate–adsorbent and/or adsorbate–adsorbate interactions are also involved in the shape-selective mechanism. Figure 7.14 illustrates the separation of n-$C_{5/6}$ and non-n-$C_{5/6}$ in CaA molecular sieves or 5A. The separation mechanism is obvious when the kinetic diameter of the molecules and molecular sieve pore size opening is compared. n-$C_{5/6}$ have kinetic diameters of less than 4.4 Å, which can diffuse freely into the 4.7 Å pores of the CaA molecular sieve, while non-n-$C_{5/6}$ have kinetic diameters of 6.2 Å. A commercial example of shape-selective adsorption is the UOP Molex™ process, which uses a Ca form A zeolite adsorbent as a molecular sieve to separate C_{10}–C_{14} n-paraffins from non-n-paraffins (aromatics, branched, naphthenes).

7.3.2.4 Ion Exchange

Zeolitic adsorbents are composed of many ionic (or potentially ionic) sites. Zeolites are crystalline, hydrated aluminosilicates containing most commonly Na^+, K^+, and H^+ cations. For most zeolites, the aluminosilicate structure is a three-dimensional open framework of Al_2O_4 and SiO_4 tetrahedras linked to each other by oxygen molecules. The framework contains channels and interconnected voids occupied by cations and water molecules. The cations are mobile and can be exchanged with other cations to varying degrees. In ion-exchange separation, cations in liquid are reversibly exchanged with cations on a solid adsorbent. More specifically, cations are interchanged with other cations without changing the structure of zeolites. At equilibrium, electroneutrality in both zeolite and liquid phases is maintained. Liquid separation based on ion exchange is important in industrial applications. Critical factors to consider in developing an ion exchange separation process are adsorbent capacity, selectivity, and kinetics. These factors are described in the following sections.

Ion Exchange Capacity

Ion exchange capacity is a measure of the quantity of cations adsorbed or removed by the zeolitic adsorbent. Theoretical ion exchanged capacities of some common zeolites are calculated and summarized in Table 7.2.

Ion Exchange Selectivity

In general, zeolites have higher ion exchange selectivity for higher-charged cations. For cations having the same valence, the ion exchange selectivity often depends on the hydrated ionic radius. This is seen from the zeolite ion exchange selectivity decreasing with ionic radii [19].

$$Cs^+ > Rb^+ > K^+ > Na^+ > Li^+$$

$$Ba^{++} > Sr^{++} > {}^{++}Ca > {}^{++}Mg$$

Kinetics

Ion exchange involves the formation and breakage of bonds between ions in solution and exchange sites in a zeolitic adsorbent. The reaction equilibrium of the ion exchange process depends most significantly on contact time, operating temperature, and ionic concentration.

7.3.2.5 Reactive Adsorption

In conventional chemical process designs, a chemical reactor is typically sequenced with a downstream separator. The reaction is carried out initially to convert raw materials into value-added products that are then isolated and recovered in the separator. The operating conditions of the reactor and separator can be varied to achieve product yield and purity. However, the variations of operating conditions to achieve optimum performance are subject to prevailing constraints. In many cases, recycle streams are incorporated into the process to reprocess unreacted raw materials or intermediate byproducts back through the reactor and separator to increase overall process yields.

To overcome some of these sequential process constraints, process intensification via reactive separation is an alternative. The reactive separations combine the unit operations of reaction and zeolitic adsorption into a single process operation with simultaneous reaction and separation. In combining sequential processing steps into an integrated process, one can eliminate one or more recycle streams that are associated with optimizing performance of the original sequential process configuration. Besides eliminating some of the recycling streams, the integration may lead to the design of a separation process that cannot be achieved with separate reactor and separator process flow elements. Other advantages of reactive adsorption over the sequential reaction and adsorption include reducing energy and capital costs, overcoming the equilibrium-limited conversion, operating at less severe conditions, and increasing reaction efficiency. More detailed information on reactive adsorption can be found in a book entitled *Reactive Separation Process* [38].

Developments in reactive separation date back to applications of simple chemical treatments, including the use of acidic clays for the removal of olefins in hydrocarbon streams, acid-catalyzed polymerization over clay beds, and the use of solid KOH to remove sulfur from various hydrocarbon streams [39]. More recent developments have followed on the heels of the discovery and development of better-engineered synthetic adsorbents and catalysts for applications in the food, biotechnology, pharmaceutical, chemical, refining, environmental, and nuclear industries. For example, Ag-substituted molecular sieves and ion exchange resins have been used to remove trace iodides by reaction and precipitation from vapor and liquid streams to facilitate safe operation of nuclear reactors [40] and to purify acetic acid produced by methanol carbonylation [41–45]. Many reactive adsorption processes developed to date use traditional fixed-bed and fluidized/moving-bed absorber designs. However, much of the recent development efforts have been focused on improving purification processes by reactive chromatographic methods and SMB technologies.

7.4 Commercial Bulk Liquid-Adsorptive Process

7.4.1 Parex

Recognizing the potential importance of polyester to the synthetic fiber market, UOP looked at many different separation mechanisms and methods to recover purified *para*-xylene from the hydrocarbons that were available in gasoline fractions from refineries. Up to that point, fractional crystallization was being used for purification of *para*-xylene, with the recovery limited by eutectic composition to a little over 60%. The UOP Molex process had already been successfully commercialized for production of high-purity normal paraffins that were used for production of biodegradable detergents. Unfortunately, the molecular dimensions of *para*-xylene and the other feed molecules did not allow molecular sieving to be used for *para*-xylene separation. However, the separation stage efficiencies that could be provided by zeolites were inherently very attractive. When Union Carbide showed that X- and Y-faujasites could be produced economically at high purities, UOP intensified its research on competitive adsorption (where essentially all feed components can access the active sites). Richard Neuzil [46] discovered that a *para*-xylene-selective adsorbent could be made by exchanging barium and potassium onto X- or Y-faujasite, and that it would be possible to use a different aromatic hydrocarbon to desorb the *para*-xylene.

This seminal discovery in 1969 led to the UOP Parex process for recovery of high-purity *para*-xylene from a mixture of C8 hydrocarbons. The first Parex process unit was commissioned in 1970 at URBK in Wesseling, Germany, and was designed to produce approximately 80 000 metric tons per year of 99% *para*-xylene, with recovery that exceeded 90%. This plant used a mixture of diethylbenzene isomers as the desorbent. Thus, the extract and raffinate columns had a fairly simple split to make between the C8 aromatic components of the feed and the C10 aromatic components of the desorbent. These plants are referred to as "heavy-desorbent Parex" because the desorbent has a higher boiling point than the feed components. This original Parex plant was still on stream in 2019. With advances in adsorbent and other PX technologies from UOP, the capacity of this plant, using the same hardware, is now more than 250 000 tons/yr. Today, the UOP Parex process accounts for more than 70% of the world's 50 000 000 tons/yr of *para*-xylene, as shown in Figure 7.15.

The growth in Parex process PX production between 1974 and 2014 was all with heavy-desorbent Parex, with steadily higher-capacity plants being built, and with numerous advances in adsorbent, desorbent, equipment, heat integration, and control systems being introduced by UOP over this time. Single-train Parex plants of up to 3 million tons/d of PX are now offered by UOP. Toray tried to use adsorptive separations for PX with its aromizing process, but that technology suffered from the use of horizontal adsorbent chambers where bed settling and flow maldistribution became problematic. In the 1990s, Axens developed the Eluxyl™ process for PX production, which uses >100 switching valves to accomplish the SMB that a single UOP rotary valve can accomplish. Sinopec developed a similar multiple valve technology in 2010, primarily for deployment within the Sinopec units in China. Neither of these technologies has garnered the acceptance that has been achieved by the UOP Parex process.

In the early days of introducing the Parex process to the industry, some prospective customers voiced concerns about the availability and cost of diethylbenzene as a desorbent. UOP's research had identified an alternative adsorbent formulation that made toluene effective as a desorbent. Four plants of this type were sold. They were

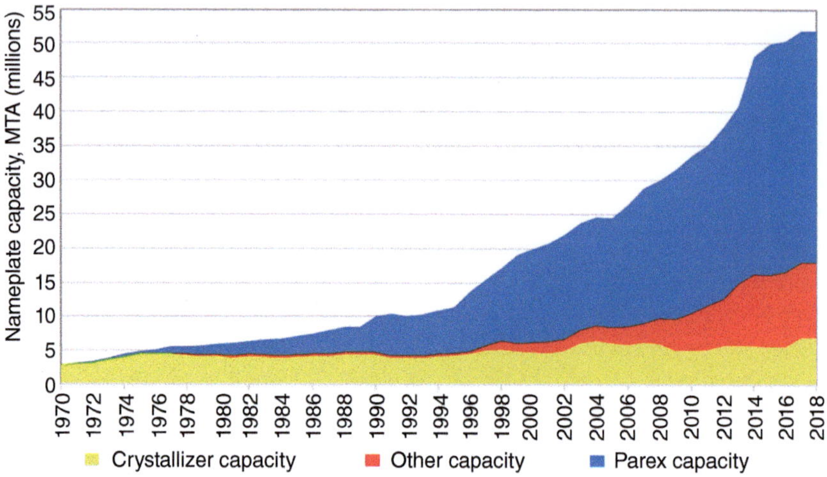

Figure 7.15 The UOP Parex process accounts for more than 70% of the world's 50 000 000 tons/yr of *para*-xylene.

called "light -desorbent Parex" because the desorbent has a lower boiling point than the feed and were licensed. Two of these plants are still operating in this mode, while the other two were revamped to the heavy-desorbent Parex configuration. For these light-desorbent Parex plants, the fractionation in the extract and raffinate columns is more difficult since there is only one carbon number between feed and desorbent, and the desorbent must be boiled overhead. As a result, the utility consumption in the extract and raffinate fractionators was much higher for the light-desorbent compared to the heavy-desorbent Parex plants. For this reason, most customers preferred the heavy-desorbent Parex configuration, particularly since *p*-diethyl benzene is readily available.

In recent years, there has been renewed interest in the energy input per ton of PX. The regions that show the highest growth rates in demand for PX have also tended to be the regions where energy costs are relatively high, for example, fuel gas valued above US$10 or even US$15 per million BTU. Many of these locations also have relatively high costs for electricity. These driving forces have led to significant innovations in PX flowschemes by UOP, with the goal of significantly reducing the fuel fires and the electricity consumed per ton of PX.

These advances have been achieved through a "holistic" view of the flowscheme for converting naphtha to PX. To understand the potential for holistic energy integration (outside the battery limits of the Parex unit), a simplified flowscheme for PX production is shown in Figure 7.16. Naphtha is fed to a CCR Platforming™ process to produce aromatic rings ranging from C6 to C11 in a stream called "reformate." The reformate is fractionated into a C7− and a C8+ fraction. The light fraction can be sent to a Sulfolane™ extractive distillation unit where nonaromatics are rejected and purified benzene and toluene are recovered. The C8+ fraction is sent to a xylene rerun column where the C8 aromatics are taken overhead and sent through the UOP Parex unit, which recovers purified PX. The raffinate from the Parex unit is then sent

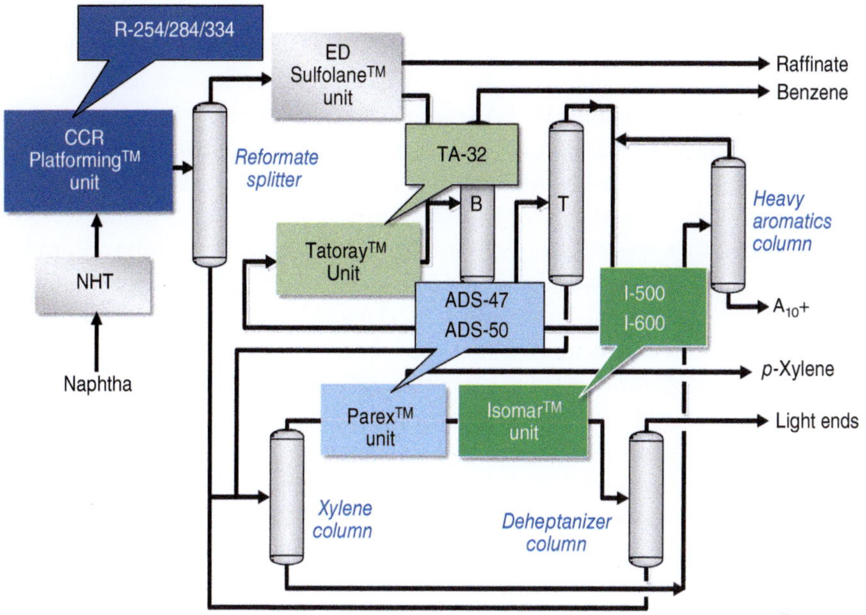

Figure 7.16 The latest-generation UOP Aromatics Technologies product portfolio.

through a UOP Isomar™ unit that re-equilibrates the xylenes, and the isomerate is then sent to the xylene rerun column for removal of heavy byproducts. These heavy aromatic byproducts, along with the toluene and A9+ in the reformate, are sent to a UOP Tatoray™ unit that uses transalkylation reactions to create additional xylenes and benzene. UOP has offered this integrated aromatic technology since the late 1970s with many plants built throughout the world.

Taking this flowscheme as a basis, UOP identified several process intensification changes that could be brought together to provide more than a 30% reduction in energy input per ton of PX. This was done without increasing the capital expenditure of the plant. A key element in this solution was a return to light-desorbent Parex, but with a significantly improved adsorbent, innovative Parex process configuration, and significantly different fractionation overhead and bottom composition limits compared to earlier designs. This new technology has been well accepted by the PX industry, accounting for more than 20 million tons of additional new PX capacity by 2019, a large fraction of the world's demand for *para*-xylene [47].

7.4.2 MX Sorbex™

Of the four C8 aromatic isomers, MX is in highest concentration in most feed streams, amounting to more than 50% of the C8 aromatics. As discussed in Section 7.4.1, MX is sent to the raffinate of the Parex process, and from there it is sent to the Isomar process where it is converted back to additional PX. Despite being the dominant C8 aromatic isomer, the end uses for purified MX are comparatively small relative to those for PX. Whereas PX is primarily used for polyethylene terephthalate

(PET) fiber, resin, and film in relatively large amounts, the dominant use of MX is to produce a co-monomer for PET resin called purified isophthalic acid (PIA) where the co-monomer content is only 2–3% of the resin. Other uses for MX would be for *meta*-xylene diamine (MXDA), which can be combined with adipic acid to produce coatings and gas barriers for containers. The production of high-purity MX was highly influenced by Mitsubishi and Amoco, which used HF/BF3 adduct technology to recover and purify MX for many years.

UOP had known how to purify and recover MX for many years using adsorptive separations [47] and worked steadily to improve the efficiency of this separation. In the 1990s, the applications for PET resin quickly expanded, with sustained growth rates above 10%, for carbonated and noncarbonated beverages. The key to this expansion was the abundance or relatively low-cost MX derived from the UOP MX Sorbex process. Beginning in 1998, UOP has now licensed several of these plants to producers in the United States, Asia, and Europe, even to the two companies that had originally been using HF/BF3 technology.

The MX Sorbex process is very similar to the light-desorbent Parex process in the sense that the product MX is selectively adsorbed and extracted from the mixed xylene feed. Toluene is used as the desorbent, with the desorbent recovered in the overheads of the extract and raffinate columns. Many producers are also in the PX market and can pull a slipstream of feed from their PX production plants to feed their (relatively smaller) MX Sorbex plants. In contrast to the 50 million tons/yr of PX produced in 2018, the worldwide production of high-purity MX is in the range of 1 million tons/yr.

Acknowledgment

The author of the chapter would like to thank James A. Johnson of Honeywell UOP for writing the "Commercial Bulk Liquid-Adsorptive Process," which includes Parex™ and MX Sorbex™ in this chapter.

References

1 Li, L., Xue, B., Chen, J. et al. (2005). Direct synthesis of zeolite coatings on cordierite supports by in situ hydrothermal method. *Applied Catalysis A: General* 292: 312.

2 Sawad, J.A., Alizadeh-Khiavi, S., Roy, S., and Kuznicki, S.M. (2005). High density adsorbent structures. W.I.P.O. Patent WO2005032694.

3 Golden, T.C., Golden, C.M.A., and Battavio, P.J. (2005). Multilayered adsorbent system for gas separations by pressure swing adsorption. US Patent US6893483.

4 Breck, D.W. (1974). *Zeolite Molecular Sieves*. New York: Wiley.

5 Barrer, R.M. (1982). *Hydrothermal Chemistry of Zeolites*. London: Academic Press.

6 van Bekkum, H., Flanigen, E.M., Jacobs, P.A., and Jansen, J.C. (eds.) (2001). *Introduction to Zeolite Science and Practice*, Studies in Surface Science and Catalysis, vol. 137. Amsterdam: Elsevier Science Publishers B.V.

7 Xu, R., Pang, W., Yu, J. et al. (2007). *Chemistry of Zeolites and Related Porous Materials: Synthesis and Structure*. John Wiley & Sons (Asia) Pte. Ltd.

8 Thompson, R.W. (1998). Recent advances in the understanding of zeolite synthesis. In: *Molecular Sieves Science and Technology: Synthesis*, vol. 1 (eds. H.G. Karge and J. Weitkamp), 1–33. Berlin: Spriger-Verlag.

9 Stiles, A.B. (1987). *Catalyst Supports and Supported Catalysts*. Boston: Butterworths.

10 Stiles, A.B. and Koch, T.A. (1995). *Catalyst Manufacture*, 2e. New York: Marcel Dekker.

11 Beck, D.W. (1974). *Zeolite Molecular Sieves: Structure, Chemistry, and Use*. New York: Wiley.

12 Shams, K. and Mirmohammadi, S.J. (2007). Preparation of 5A zeolite monolith granular extrudates using kaolin: investigation of the effect of binder on sieving/adsorption properties using a mixture of linear and branched paraffin hydrocarbons. *Microporous and Mesoporous Materials* 106: 268–277.

13 Jasra, R.V., Tyagi, B., Badheka, Y.M. et al. (2003). Effect of clay binder on sorption and catalytic properties of zeolite pellets. *Industrial and Engineering Chemistry Research* 42: 3263–3272.

14 Kulprathipanja, S. (2010). *Zeolites in Industrial Separation and Catalysis*. New York: Wiley.

15 Serrano, D.P., Sanz, R., Pizarro, P. et al. (2009). Preparation of extruded catalysts based on TS-1 zeolite for their application in propylene epoxidation. *Catalysis Today* 143: 151–157.

16 Gordina, N.E., Prokof'ev, V.Y., and Il'in, A.P. (2005). Extrusion molding of sorbents based on synthesized zeolite. *Glass and Ceramics* 62 (9–10): 282–286.

17 Ruthven, D.M. (1984). *Principles of Adsorption and Adsorption Processes*. USA: Wiley.

18 Neuzil, R.W. (1982). Separation of *m*-xylene. US Patent 4,326,092.

19 Broughton, D. and Gerhold, C. (1961). Continue separation process employing fixed bed of sorbent and moving inlets and outlet. US Patent 2,985,589.

20 Namba, S., Kanai, Y., Shoji, H., and Yashima, T. (1984). Separation of *p*-isomers from disubstituted benzenes by means of shape-selective adsorption on mordenite and ZSM-5 zeolites. *Zeolite* 4: 77–80.

21 Barthomeuf, D. (1996). Basic zeolites: characterization and uses in adsorption and catalysis. *Catalysis Reviews - Science and Engineering* 34: 521–612.

22 Seko, M., Miyake, T., and Inada, K. (1979). Economical *p*-xylene and ethylbenzene separated from mixed xylene. *Industrial & Engineering Chemistry Product Research and Development* 18 (4): 263–268.

23 Ward, J.W. (1968). The nature of active sites on zeolites. III. The alkali and lkaline earth ion-exchanged forms. *Journal of Catalysis* 10: 34–46.

24 Kilpatrick, M. and Luborsky, I.E. (1953). The base strengths of aromatic hydrocarbons relative to hydrofl uoric acid in anhydrous hydrofluoric acid as the solvent. *Journal of the American Chemical Society* 75: 577.

25 Kulprathipanja, S. (1995). Process for the adsorptive separation of metaxylene from aromatic hydrocarbons. US Patent 5,382,747.

26 Kulprathipanja, S. (1983). Separation of bi-alkyl-substituted monocyclic aromatic isomers with pyrolyzed adsorbent. US Patent 4,423,279.

27 Shimura, M., Wakamatsu, S., and Shirato, Y. (1996). Separation of *p*-xylene by using zeolitic adsorbents. Japanese Patent 08,217,700.

28 Zinnen, H.A. (1989). Zeolitic *p*-xylene separation with tetralin heavy desorbent. US Patent 4,886,930.

29 Kulprathipanja, S., Khune, K.K., and Patton, M.S. (1993). Process for separating durene from substituted benzene hydrocarbons. US Patent 5,223,589.

30 Robo, J.A. and Gajda, G.J. (1989-1990). Acid function in zeolites: recent process. *Catalysis Reviews - Science and Engineering* 31 (4): 385–430.

31 Johnson, J.A. and Kulprathipanja, S. (1989). *Proceedings of the International Conference on Recent Development in Petrochemical and Polymer Technologies* (12–16 December). Bangkok, Thailand: Petroleum and Petrochemical College, Chulalongkorn University.

32 Kulprathipanja, S. and Neuzil, R.W. (1984). Low temperature process for separating hydrocarbons. US Patent 4,455,444.

33 Kulprathipanja, S. (2001). Monomethyl paraffin adsorptive separation process. US Patent 6,222,088 B1.

34 Kulprathipanja, S. (2001). Process for monomethyl acyclic hydrocarbon adsorptive separation. US Patent 6,252,127 B1.

35 Sohn, S.W., Kulprathipanja, S., and Rekoske, J.E. (2003). Monomethyl paraffin adsorptive separation process. US Patent 6,670,519 B1.

36 Kulprathipanja, S., Rekoske, R.E., Gatter, M.G., and Sohn, S.W. (2003). *Proceedings of the Third Pacific Basin Conference on Adsorption Science and Technology*, Kyongju, Korea (25–29 May).

37 Kulprathipanja, S. (1991). Adsorptive separation process for the purification of heavy normal paraffins with non-normal hydrocarbon pre-pulse stream. US Patent 4,992,618.

38 Kulprathipanja, S. (2002). *Reactive Separation Process.* New York, USA: Taylor & Francis.

39 Bland, W.F. and Davidson, R.L. (1967). *Petroleum Processing Handbook.* New York, USA: McGraw-Hill.

40 Pence, D.T. and Macek, W.J. (1970). Silver zeolite: iodide adsorption studies. The U.S. Atomic Energy Commission, Idaho Operations Office, Under Contract #AT (10-1)-1230, Nov. 49 Hilton, C.B. (1986) Removal of iodide compounds from nonaqueous organic media. US Patent 4,615,806.

41 Kulprathipanja, S., Spehlmann, B.C., Willis, R.R. et al. (1999). Method for treating an organic liquid contaminated with an iodide compound. US Patent 5,962,735.

42 Kulprathipanja, S., Vora, B.V., and Li, Y. (1999). Method for treating a liquid stream contaminated with an iodide-containing compound using a solid absorbent comprising a metal phthalocyanine. US Patent 6,007,724.

43 Kulprathipanja, S., Lewis, G.J., and Willis, R.R. (2001). Method for treating a liquid stream contaminated with an iodide-containing compound using a cation-exchanged crystalline manganese phosphate. US Patent 6,190,562 B1.

44 Kulprathipanja, S., Sherman, J.D., Napolitano, A., and Markovs, J. (2002). Method for treating a liquid stream contaminated with an iodide-containing compound using a cation-exchanged zeolite. US Patent 6,380,428 B1.

45 Kulprathipanja, S., Vora, B.V., and Leet, W.A. (2003). Combination pre-treatment/adsorption for treating a liquid stream contaminated with an iodine-containing compound. US Patent 6,506,935 B1.

46 Neuzil, R.A. (1971). Aromatic hydrocarbon separation by adsorption. US Patent 3558730 A.

47 Frank (Xin X.) Zhu, James A. Johnson, David W. Ablin, and Gregory A. Ernst (2020). *Efficient Petrochemical Processes – Technology, Design and Operation.* Hoboken NJ: Wiley.

8

Distillation

8.1 Introduction to Distillation

Distillation is the oldest, yet still most commercially important, method of chemical separation and purification. Advances in distillation have been critical to increasing the standard of living for humanity over the past few thousand years [1, 2]. The origins of distillation technologies trace to the third-century BCE in Mesopotamia. Distillation knowledge was initially spread by word of mouth and apprenticeship from its birthplace in the Middle East. Slowly, distillation methods spread to Asia (especially China, Japan, and India) and Europe; these first steps in the development of distillation methods were made largely to enable perfumes, medicines, and alcohols to be isolated.

With the invention of mass printing of scientific information, distillation, and other scientific methods could be shared and taught much more efficiently. These advances in disseminating scientific knowledge were made in China via mechanical wood block prints of the eighth century AD and then Europe with the invention of the mechanical printing press in the fifteenth century AD. The start of the Industrial Revolution in United Kingdom in 1760 enabled step-changes to be made in distillation technologies as the energy required to commercialize these processes could now be harnessed via steam engines. It was around 1800 that Benjamin Thompson, Count Rumford, described some of the benefits of steam-heated distillation, including rapid heating of feeds as well as greater glass vessel life versus direct heating.

The Napoleonic Wars forced continental Europe to develop new methods for the isolation of sugar as the supply of cane sugar from the East and West Indies was cut off by a blockade. One of the scientists tackling this problem was Jean Baptist Cellier-Blumenthal (1768–1840). Blumenthal's key contribution to the forward march of distillation technology was the development of the rectifying, or fractionation, column. Rectification is used today for the fractionation of crude oil and consists of a multistage, continuous distillation process accomplished via counter-current distillation (rectification) in a column. The crude liquid to be separated is fed to the bottom of the column and brought to its boiling point. The vapor produced from boiling moves upwards inside the column, exits at the top, and is condensed. Part of the condensate is removed as the distillate product; the

Modern Petrochemical Technology: Methods, Manufacturing and Applications, First Edition.
Santi Kulprathipanja, James E. Rekoske, Daniel Wei, Robert V. Slone, Trung Pham, and Chunqing Liu.

rest flows back into the column and moves downwards as liquid in the opposite direction of the rising vapor. The upper portion of Blumenthal's column incorporated bubble-cap trays, while the lower part included conical metal caps for contacting vapor and liquid. These advances in distillation not only helped supply the refined sugar Europe needed on the less volatile side of this process but also launched, on the distillate side, the French rum industry.

The coal industry also drove innovations in distillation technology through the nineteenth century. In 1823, Friedlieb Runge discovered phenol and aniline in coal tar and, through further distillation, enabled the use of these tar byproducts in colorants and pharmaceutical product development. In addition, ammonia, benzene, and other aromatic compounds beyond phenol were isolated from the coke gas and were separated by washing, absorption, and distillation. Column and tray designs continued to improve throughout the nineteenth century as a wider range of distilled products were produced. The residue from these processes found use in street paving by the mid-1800s.

The next great leap in distillation technology arrived with the birth of the petroleum industry. While the first generation of crude oil distillation was a simple batch process to produce lamp oil, the outbreak of World War I required larger volumes of oil and fuel to be isolated. With no chemical conversion, only 20–25% of crude oil could be distilled into fuel and oil products. In the 1910s, thermal cracking was utilized to generate lighter oils from the longer-chain hydrocarbons in the crude. In 1916, steam-assisted distillation of petroleum was accomplished, and the 1920s ushered in a switch in gasoline production from batch to continuous distillation. Technologies such as thermal cracking via the Dubbs process [3] allowed much higher yields of gasoline and other smaller-chain hydrocarbons to be derived from crude oil.

These distillation process advances were paired with a greater understanding of the theory and design of distillation. The measurement and prediction of mixture properties, especially vapor–liquid equilibrium (VLE), are crucial to the accurate prediction of separation performance.

For example, Rayleigh pioneered the use of Henry and Raoult's laws to compare test results to data calculations for simple batch distillations. Hausbrand [4] then advanced this work using graphical representations of the VLE properties for non-aqueous and aqueous mixtures. In this manner, the beginnings of VLE theory were established, based on relative volatility (RV) differences in the materials being distilled. Finally, McCabe and Thiel developed their diagram to simplify the correlation of distillation column design with VLE properties and mass balance data [5].

For more than 5 000 years, distillation has served as the primary method for the separation of complex mixtures into refined, purified products. Even today, distillation is the most widely practiced separation technology and is responsible for approximately 50% of capital and operating costs in industrial chemical processing. Distillation accounts for 50% of the annual process energy needs of the chemical and refining industries [6].

Although the refineries of today – as well as those in the future – are far more complex, with higher yields, higher value creation, lower regrettable emissions, and more complex petrochemical production schedules, distillation has maintained its position as the first and most critical separation step within the refinery. In this chapter, we explore the basic principles of crude oil distillation, selection of distillation columns and packing, use of reactive distillation, and explore commercial examples of distillation. Through a basic understanding of the distillation process, we expect the reader will build an appreciation of the criticality of this separation method to the cost and reward of operating a modern petrochemical refinery.

8.2 Principles and Systems

8.2.1 Vapor–Liquid Equilibrium (VLE) Theory

Distillation is a separation process based on VLE properties of a mixture. Whether under atmospheric or vacuum conditions, distillation is based on understanding and taking advantage of the properties of VLE of mixtures. Distillation is fundamentally a separation method in which the variation of equilibrium composition between the vapor phase and liquid phase provides the driving force of the separation. Thermodynamic VLE compositions can be correlated and calculated with the appropriate thermodynamic data and models.

For the successful separation of feed components via distillation, it is required that:

(1) The equilibrium distributions of mixture component compositions in liquid phase must be different from the vapor phase.
(2) The relative volatility of mixture components one to the other must be either greater than or less than one.

To begin, we define the K value of component $\boldsymbol{a}$ as:

$$K_{\mathrm{a}} = \frac{\text{Mole fraction of component } \boldsymbol{a} \text{ in vapor phase}}{\text{Mole fraction of component } \boldsymbol{a} \text{ in liquid phase}} = \frac{y_{\mathrm{a}}}{x_{\mathrm{a}}}$$

In this way, K_{a} represents the tendency of component $\boldsymbol{a}$ to volatilize into the vapor phase and is a function of temperature, pressure, and, to a lesser degree, composition. If K_{a} equals 1, there is an equal amount of component $\boldsymbol{a}$ in the vapor and liquid phases, while $K_{\mathrm{a}} > 1$ means more $\boldsymbol{a}$ in the vapor phase and $K_{\mathrm{a}} < 1$ means more $\boldsymbol{a}$ in the liquid phase [7].

To separate components via distillation, we must understand their relative volatility. The relative volatility of components $\boldsymbol{a}$ and $\boldsymbol{b}$ is defined as $K_{\mathrm{a}}/K_{\mathrm{b}}$. This ratio of volatility is a measure of how easily two components can be separated by distillation. If the ratio is high, component $\boldsymbol{a}$ is much more volatile than component $\boldsymbol{b}$, and the distillation is straightforward. As the ratio ($K_{\mathrm{a}}/K_{\mathrm{b}}$) approaches a value of one, it is more difficult to separate each of the components. In the extreme case of the value of $K_{\mathrm{a}}/K_{\mathrm{b}}$ being exactly one, separation by distillation is not possible other than by azeotropic or reactive distillation methods. It is not possible to have a relative volatility ratio less than one as we always define the ratio as the K value of more volatile component over the K value of the less volatile component.

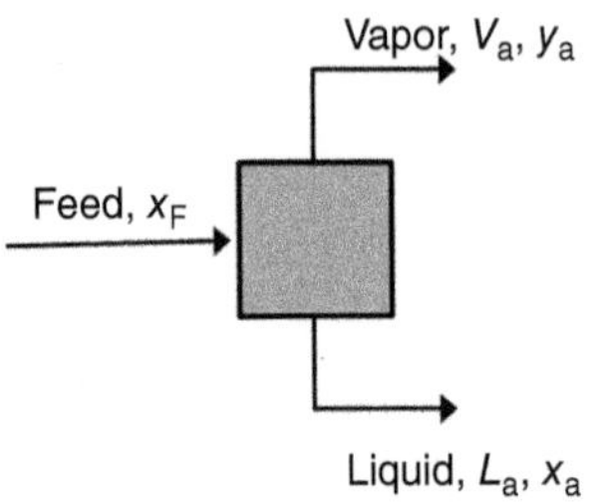

Figure 8.1 Binary separation of components ***a*** and ***b***.

The McCabe–Thiele design method provides a means to model the vapor and liquid composition (mole fraction) of a component of a binary mix within a distillation column. It affords a graphical visualization of distillation principles while also providing a solution to the material balance and equilibrium relationships. This relatively simple approach allows the designer to build on the fundamentals of VLE theory and clarify design basics such as the number of trays required to achieve a given binary component separation as well as where to locate the feed within the tray structure.

Figure 8.1 depicts a binary distillation of a feed of composition x_F into a vapor phase where the mole fraction of the more volatile component, ***a***, is designed as y_a and the mole fraction of ***a*** in the liquid phase is x_a.

Figure 8.2 shows the equilibrium of a single stage of the distillation of this feed (labeled stage 1). Because the system is in equilibrium, there are two conditions that are required to satisfy VLE:

(1) The amount of component ***a*** leaving the stage ($L_a x_a + V_a y_a$) is equal to the amount of vapor and liquid entering the stage from the trays positioned before ($V_{a-1} y_{a-1}$) and after stage 1 ($L_{a+1} x_{a+1}$), respectively.
(2) The amount of component ***a*** in the vapor phase at equilibrium is equal to the K factor times the mole fraction in the liquid phase, $y_a = K_a(x_a)$.

8.2.2 McCabe–Thiele Diagram

The VLE conditions described above allow us to construct the McCabe–Thiele diagram [8]. Shown in Figure 8.3 is a McCabe–Thiele diagram for the distillation of components ***a*** and ***b*** (i.e. a binary mixture).

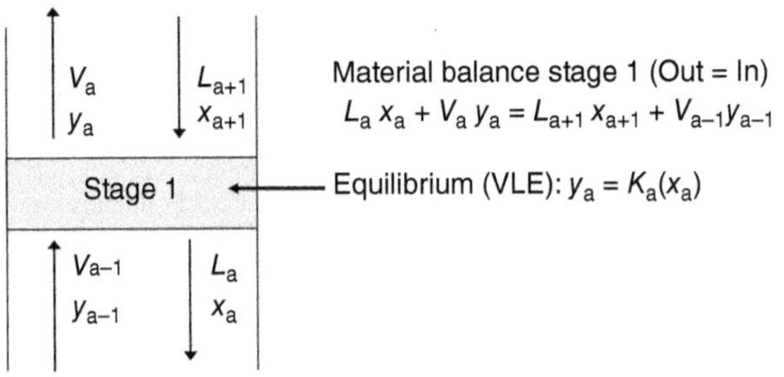

Figure 8.2 Vapor–liquid equilibrium (VLE) concept.

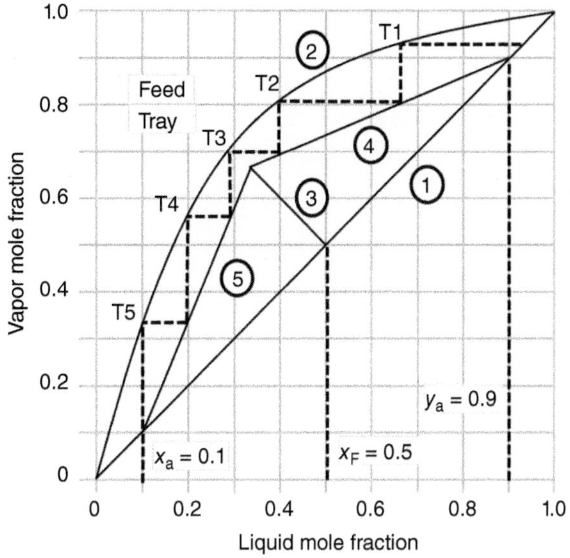

Figure 8.3 McCabe–Thiele diagram.

To assemble this diagram, we:

(1) Label the axis as the mole fraction of the lighter component (***a***) in the vapor phase (y-axis) versus the mole fraction of the same lighter component (***a***) in the liquid phase (x-axis).
(2) Plot **Line 1** with slope 1 and intercept of 0 across the diagram.
(3) Plot **Curve 2**: the VLE curve. This curve is derived from primary thermodynamic reference values. Recall that according to VLE, $y_a = K_a(x_a)$.
(4) Determine the feed composition: In this example, the feed contains 50% component ***a***, that is, $x_F = 0.5$.
(5) Plot **Line 3:** the quality line or q-line, which is shown in Figure 8.3 as a mixture with equal mole of component ***a*** as vapor and liquid in the feed. The line, therefore, has a slope of −1. $M = q/(q-1)$, where q = feed quality (liquid mol fraction of the feed).
(6) Plot **Line 4**: the rectifying line. For a distillate composition of 90% component ***a***, we start this line at 0.9 on the x-axis and intersect Line 1. The slope of this line is $L/(L+D)$, where L = the molar flow rate of reflux and D is the molar flow rate of the distillate product. The line is extended until it intersects Line 3 (the q-line).
(7) Plot **Line 5** (stripping line). For a bottoms composition of 10% component ***a***, we start this line at 0.1 on the x-axis and intersect the point where Lines 1 and 3 meet.
(8) Now, we can plot the tray structure for this distillation by starting at the $x_a = 0.1$ point and creating a stairstep structure, which stops when it intersects either the stripping or rectifying lines.

The picture that emerges is a five-tray distillation structure capable of delivering 90% component ***a*** in the distillate ($y_a = 0.9$) from a feed with 50% component ***a*** ($x_F = 0.5$). Additionally, the McCabe–Thiele diagram allows the feed to be located

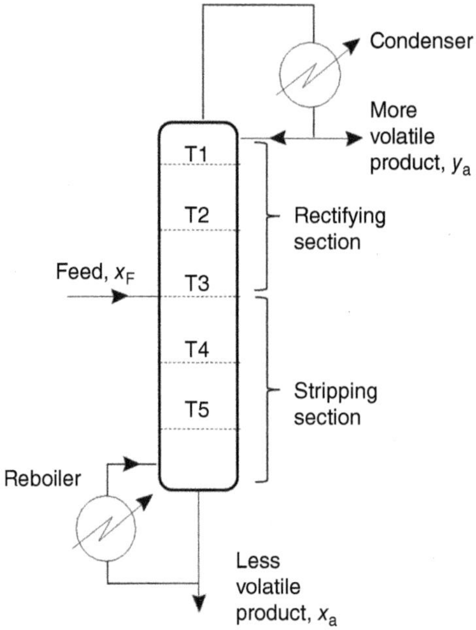

Figure 8.4 Five-tray binary separation distillation column.

within the tray structure. The intersection point of the rectifying line and stripping line falls with the Tray 3 step; therefore, Tray 3 is the best location for the feed in this distillation setup. The resulting column and feed setup is shown in Figure 8.4.

8.2.3 Factors for VLE-Based Distillation Design

The successful design of a distillation using the VLE method involves not only the steps covered above but also more detailed information on several factors, which must be accounted for by the design engineers: relative volatility, activity coefficient, solubility, maximum concentration limit, and surface area.

8.2.3.1 Relative Volatility

When the vapor pressure of a liquid is equal to the pressure exerted by the atmosphere, we can say the liquid has reached its boiling point. As discussed previously, distillation relies largely on differences in the volatility of components of a mixture to separate and purify those materials.

The feed composition influences the boiling point of the feed mixture as the overall feed boiling point is a function of the vapor pressures of each component in the mixture. As the feed mixture is heated, the most volatile components evaporate first and are concentrated in the distillate with the less volatile materials being left behind in the lower range of the distillation column. By knowing the boiling points and relative volatility of each component, the distillation system design engineer determines how difficult a separation will be and how relative volatility needs to be accounted for in this design.

Boiling point is not a static value. Rather, the boiling point of a mixture will vary as temperature, pressure, and composition change across the distillation column. As required by VLE theory, the vapor and liquid within each stage reach equilibrium and no further separation is possible; each stage reaches a unique equilibrium concentration of the mixture components according to their relative volatility. Further purification of the distillate (more volatile components) or bottoms (less volatile components) is possible through the introduction of either more distillation columns or other separation methods such as absorption or filtration.

Mixtures with a relative volatility (RV) value greater than 1.5 are described as a simple distillation and require no special measures to accomplish the desired separation. By contrast, mixtures with relative volatility (RV) values less than 1.2 are described as extractive distillation; these separations can be eased by the addition of heavier components. This method will be described in more detail later in the chapter [9].

8.2.3.2 Activity Coefficient

The separation of mixed feeds through distillation is a function of the relative volatilities of the components, their activity coefficients, the relative ratios of the liquid-phase to vapor-phase flow rates, and the ratio of surface area to liquid within the distillation column. Activity coefficients help account for nonideal behaviors of feed components during distillation such as behavior caused by surface tension (the attraction force of the particles in the surface layer by the bulk of the liquid, which tends to minimize surface area).

As noted above, the presence of more vapor molecules creates a higher surrounding vapor pressure on the mixture within the distillation column. The substance with greatest relative volatility, the highest vapor pressure, will boil more at a given temperature because less energy is required to release molecules from the surface tension. However, the actual amount of a component that boils may differ from ideality, and it is the activity coefficient for the component, which helps correct for this variation from ideal behavior.

8.2.3.3 Solubility

Solubility is a function of temperature and pressure, so the relative solubility of components in the feed as well as throughout the distillation column can make it more or less difficult to separate those components. Differences in solubility can be used, for example, to make possible the separation of materials that have the same relative volatility through the addition of a third component with a higher solubility for one of the components; this method is referred to as azeotropic distillation and is described in more detail later in this chapter.

8.2.3.4 Azeotropes: Maximum Concentration Limit

As purification is carried out, it is common to reach a limit where the vapor and liquid phases include the same concentration of each component. If the liquid in this example is volatilized under the same pressure and temperature, it will yield the same concentration of components ***a*** and ***b***. In this case, it is not possible to further separate these components as they have formed an azeotropic mixture that has hit a constant, maximum concentration limit. One simple example of this maximum concentration behavior is isopropyl alcohol in water; the concentration of isopropyl alcohol cannot rise above 91% as the boiling points for the two components are equal at this maximum concentration limit.

8.2.3.5 Surface Area

As a feed mixture is exposed to an increasing amount of surface area, it is easier to volatilize the feed and separate the components by relative volatility. Equilibrium can be established more quickly and efficiently in systems with larger surface area as the components are vaporized and condensed more rapidly within a given stage of the distillation column. Due to their increased surface area, both random and structured packed columns increase the surface area of a distillation process versus trays alone. A more complete description of column packing materials as well as structured packings is included later in this chapter.

In addition to the five factors above, the distillation system design engineer needs to also account for:

- *Tray hydraulics*: Tray hydraulics and component rheology dictate the flow behavior of a liquid through a tray distillation system. These material properties, along with the tray design properties, will have a significant impact on the ability of the distillation apparatus to create the surface area needed to drive separations.
- *Feed component stability*: The feed boiling point needs to be compared to the individual component's decomposition temperature. Additionally, there is a need to understand any possible reactions that may occur between components within the distillation column as relative volatilities for these components can vary greatly from the feed mixture components.

8.2.4 Operating Modes of Distillation

There are many different distillation strategies possible, and each brings its own unique strengths and weaknesses as well as selection criteria for their use in different separations. In this chapter, we will compare three different modes of distillation:

(a) Batch versus continuous
(b) Tray versus packed column

8.2.4.1 Batch vs. Continuous

Batch Distillation

Batch distillation, used mainly in laboratory and fine chemical separations, involves the charging of the feed to the column in one shot with the distillation then being run with no additional feed being introduced. This mode of operation was the first and only means of distillation available for thousands of years. Today, batch distillation is selected when there are drivers other than cost to consider for a separation process. Some of the advantages of batch distillation include:

(1) *Simplicity*: Multiple components can be separated using only one feed and only one distillation column.
(2) *Flexibility*: One batch distillation column can be used to handle a wide range of feed materials with varying quality and composition.
(3) *Capacity*: Batch distillation is well suited for small volume separations.
(4) *Feed difficulty*: If a feed has high viscosity or high solids levels, it is easier to run a batch distillation than use continuous processing methods.
(5) *Quality control*: Batch distillations can be quality-verified easily based on both feed and product specifications.

Continuous Distillation

Continuous distillation processes complement batch processing in that they are lower cost and are tuned to one specific type of feed. The emphasis for continuous distillations is on stability and the production of a high volume of product stream(s) at consistent quality.

When constructing a continuous distillation strategy, there are several column options to consider:

(1) *Binary column*: Feed contains two components.
(2) *Multicomponent column*: Feed contains more than two components (e.g. crude oil).
(3) *Multiproduct column*: Column has more than two product streams located at different points along the height of the column.
(4) *Extractive distillation*: Where feed appears in the bottom product stream.
(5) *Azeotropic distillation*: Where feed appears at the top product stream.

8.2.4.2 Tray vs. Packed Column

The selection of an appropriate contactor is critical to success in every distillation. There are two general types of contactors available: tray and packed columns.

Tray column: Trays of specified designs (e.g. sieve, valve, and bubble-cap) hold the liquid phase as it flows across the surface and allows for contact between the vapor and liquid phase through openings in the trays to drive enhanced separation. In a crude oil refinery, a 30-tray stack is common in the atmospheric distillation unit column.

Packed column: Packing material (either random or structured) are used to enhance contact between vapor and liquid.

- **Structured packing** provides high surface area with high void fraction to promote efficient vapor–liquid contact.
- **Random packing** is commonly used in absorption, stripping, and fractionation operations in gas, refinery, and chemical plants

The strengths and weaknesses of each mode of column operation are included in Table 8.1.

Table 8.1 Tray columns versus packed columns.

Tray columns	Packed columns
Horizontal flow across the trays	Vertical flow through the packing (counter-current)
Large diameter columns and with high liquid flow rates (>30 $m^3/(m^2\,h)$)	Small-diameter ($D < 0.6$ m) columns with low liquid flow-rate applications (<50 $m^3/(m^2\,h)$)
Metal	Metal, ceramic, and plastic
High liquid residence time	Low liquid residence time (i.e. for heat-sensitive materials)
High-pressure drop (7 mbar per equilibrium stage)	Low-pressure drop (0.1–0.5 mbar per equilibrium stage) as required for vacuum distillation
Easy to clean	Not typically cleaned/reused

8.3 Distillation Columns and Trays

8.3.1 Flow Within the Distillation Column

Now that we have described the basics of distillation theory using the VLE approach and detailed several key factors in distillation strategy and execution, we turn our attention to the structure and function of distillation towers [8]. Figure 8.5 depicts

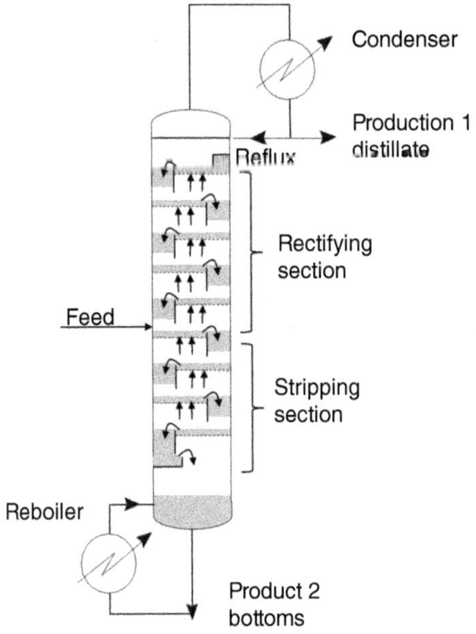

Figure 8.5 Flow within the distillation tower.

a continuous distillation tower containing nine trays with the binary feed mixture brought into contact with Tray 6. Starting at the top of this column, we see that part of the condensate is returned to the column as reflux material to provide liquid flow downward through the column. Likewise, a portion of the liquid bottoms fraction is returned through the reboiler to the column and provides the upward flow of vapor from the bottom of the column. The *stripping section* (below the feed inlet) primarily removes the more volatile components from the liquid, while the *rectifying section* (above the feed inlet) includes an enriched level of volatile components. Once the targeted level of separation is achieved, this binary, continuous distillation results in two products, the *distillate* (more volatile components) and the *bottoms* (less volatile component).

As described earlier, binary systems in which the relative volatility is greater than 1.5 fall in the category of simple distillations and, in some cases, the rectifying section of the column in Figure 8.5 can be removed to yield a simpler *stripping column.*

8.3.2 Multicomponent Distillation

While we have focused primarily on binary (two component) distillations thus far, it is important to note that most commercial feeds (e.g. crude oil) consist of many components and require the use of multicomponent, fractional distillation columns and methods. As shown in Figure 8.6, it is possible to introduce both multiple feeds and withdraw multiple products within a single distillation tower. The function of the distillation column does not fundamentally change with this increase in complexity, and the same key factors described earlier (relative volatility, activity

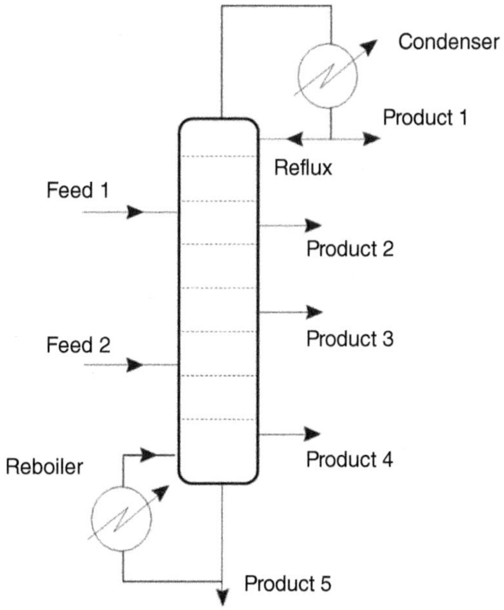

Figure 8.6 Multicomponent continuous distillation column.

coefficient, solubility, maximum concentration limit, and surface area) determine the best approach to distillation of these complex mixtures.

While pure components ***a*** and ***b*** distillate and bottoms products are possible in binary distillations, multicomponent distillations conducted in a single column will yield products, which are, themselves, multicomponent mixtures that are enriched with cuts of differing boiling point ranges. To separate multicomponent feeds into pure component product streams, more complex, multicolumn distillation systems are required.

8.3.3 Tray Types: Sieve, Valve, Bubble-Cap

The distillation of binary and multicomponent feeds begins when the mixed feed is introduced to a contactor within the distillation tower. Both tray and packed-bed contactors are in widespread industrial use. In a tray contactor, the liquid phase flows across the tray as in Figure 8.7, and the vapor moved upward through gaps in the tray (shown in this figure as a simple sieve-style tray). The liquid is retained on the tray by an elevated section called the *exit weir* and is transferred from stage to stage within the distillation towers through vertical channels called *downcomers* and then flows through the *clearance* onto the next stage's tray. While the liquid flows across the tray, the more volatile component(s) are taken into the vapor phase in the *bubbling area*. This vapor can then pass upward from stage to stage through the perforations in the tray. As the vapor moves upward through the above stage's tray openings, the structure of that tray will play a key role in both the cost and performance of the overall distillation tower. There are three main types of cross-flow trays: sieve, valve, and bubble-cap.

There are a few factors that influence the selection of the type of tray to employ:

(1) *Operating range*: The range of vapor and liquid transfer rates over which the tray will efficiently separate components of the feed mix. The operating range is often described in terms of the turndown ratio: the ratio of the highest to lowest flow rates at which the tray will perform its intended separation. Bubble-cap trays have the highest turndown ratio, while sieve trays have the lowest with valve trays in between.

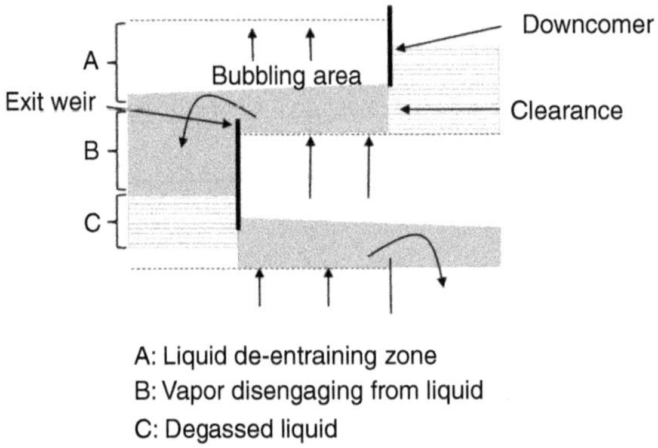

Figure 8.7 Flow pattern in cross-flow plate.

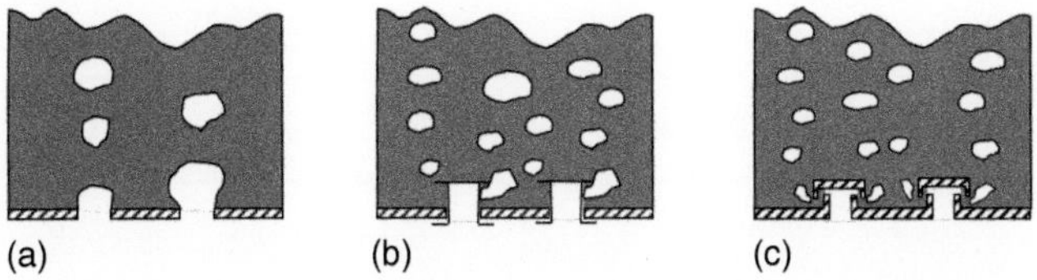

Figure 8.8 Common distillation tray types. (a) Sieve, (b) Valve, and (c) Bubble-cap.

(2) *Operating capacity*: The diameter of the column required to provide a give flow rate is largely invariant across the three types of trays. Sieve trays have a slightly higher capacity, then valve, and finally bubble-cap.
(3) *Cost*: Bubble-cap trays are more complex and, therefore, more expensive than sieve or valve trays.
(4) *Pressure drop*: As might be imagined, sieve trays have the lowest pressure drop followed by valve and then bubble-cap trays. Given the liquid low properties of the feeds involved, vacuum columns need to account more for pressure drop than atmospheric distillation columns.

Shown in Figure 8.8 are the three types of distillation trays we are most interested in (sieve, valve, and bubble-cap). Each of the three types is described below:

(a) *Sieve tray*: This tray is simply a flat plate with 5–6-mm-diameter holes (or slots) punched in it. The liquid flowing across the holes and tray is retained by vapor upflow from below. Due to its simplicity, sieve trays are the least expensive, have the lowest pressure drop, and are most resilient to solids and corrosion (especially when large holes are used). For efficient operation, the velocity of vapor through the sieve tray holes must be sufficiently high to prevent liquid on the tray deck from passing through the perforations to the tray below (weeping). Conversely, if the velocity of the vapor through the sieve tray perforations is too high, liquid from the tray below will transfer to tray above (entrainment) and will cause lower column efficiency. Because of these limitations, the turndown performance for sieve trays is relatively poor at 2 : 1.
(b) *Valve tray*: A valve tray is similar in construction to the simpler sieve tray, but includes a moving valve (e.g. disk), which closes when vapor-flow rates are low to avoid weeping of the liquid to the stage below. For this reason, valve trays are characterized by higher turndown ratings (5 : 1 range) due to improved performance at low vapor-flow rates but are more costly than sieve trays (about 1.2 times more expensive).
(c) *Bubble-cap tray*: A bubble-cap tray has riser or chimney fitted over each opening, and a cap that covers the short riser pipe. The cap is carefully placed to maintain a space between riser and cap, which allows the passage of vapor. This design is the oldest type of cross-flow tray and ensures that liquid is maintained on the tray at all vapor flow rates. While bubble-cap trays bring obvious advantages in turndown performance (10 : 1 and higher), they suffer from high pressure drops, high cost (2× sieve trays), and poor corrosion resistance.

A comparison of these three tray types is included in Table 8.2 below.

Table 8.2 Tray comparison.

Tray type	Capacity	Efficacy	Pressure drop	Entrainment	Turndown	Cost
Sieve	High	High	Medium	Medium	2 : 1	1.0
Valve	High	High	Medium	Medium	4 : 1	1.2–1.5×
Bubble cap	Medium	Medium	High	3× Sieve	15 : 1	2.0–3.0×

8.4 Distillation Packing Materials

While distillation itself dates back thousands of years, packed columns only came into wide use in the 1930s with the introduction of random, but regular, packing materials such as Raschig rings. Shown in Figure 8.9 is a packed column such as that used in distillation or absorption; liquid and vapor flow counter-current within this simple yet effective design.

Packed columns are commonly used for distillation, absorption, stripping, and heat-exchange applications. The design of these systems is a function of their capacity, pressure drop, and resistance requirements (both mechanical and chemical). The packing materials loaded with the column establish the capacity and separation efficiency of these distillation systems and have been steadily increasing in performance over the past several decades to include not only high-performance random packings but also structured packing materials [10].

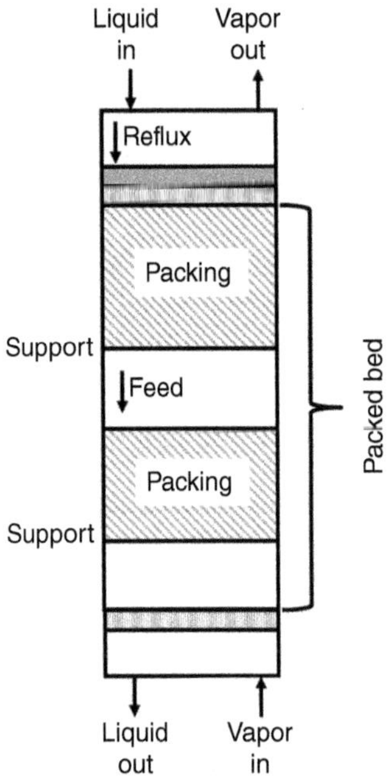

Figure 8.9 Packed distillation column.

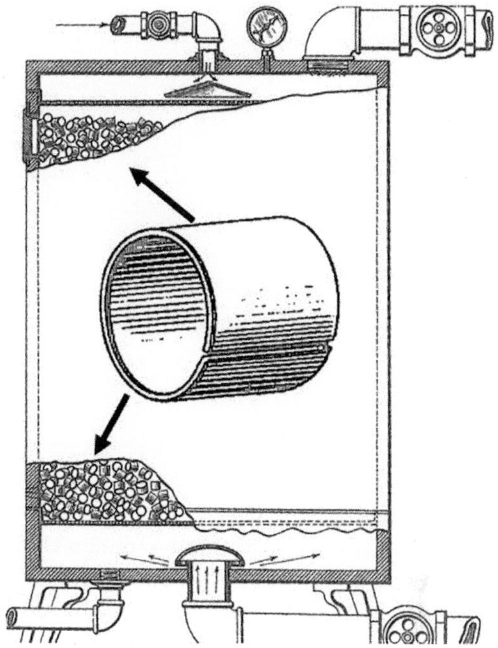

Figure 8.10 Raschig ring column. Source: Raschig [11].

8.4.1 Random Packing

Packed column distillation was introduced in the early nineteenth century to purify alcohol as well as for sulfuric acid absorption. The first packing materials were of relatively low capacity and consisted of glass beads, stones, and coke. Packing material technology advanced greatly with the invention of on-purpose packing materials such as Raschig rings, which were patented in 1915, to improve acid gas absorption processes (Figure 8.10) [11]. These rings were made of glass, porcelain, copper, or iron, were one inch in both diameter and height, and provided improvements in pressure loss as well as mass transfer properties. Engineered random packings were long-lasting and provided intimate vapor–liquid contact while minimizing the holdup of material within the packing and are still used in distillation applications today.

In the words of Fritz Raschig, “By making these cylinders of uniform size it is impossible for them to telescope together or fit one inside of another, but they arrange themselves with only point or line contact with each other and with their axes either horizontal or inclined and with the axes and surfaces of adjacent cylinders, at various angles and inclinations with each other. Since any appreciable surface contact of these cylinders with each other is impossible, and since each cylinder presents a practically unobstructed passage therethrough, it results that pocketing of the liquid is minimized or entirely avoided while the various inclinations and arrangements of the adjacent cylinders are such as to cause the ascending gases or vapors and the descending liquid to pass in a most varied course through the tower, the course being a zig-zag or multivaried course with frequent interruptions and changes in direction. A most intimate and promiscuous intermixing and

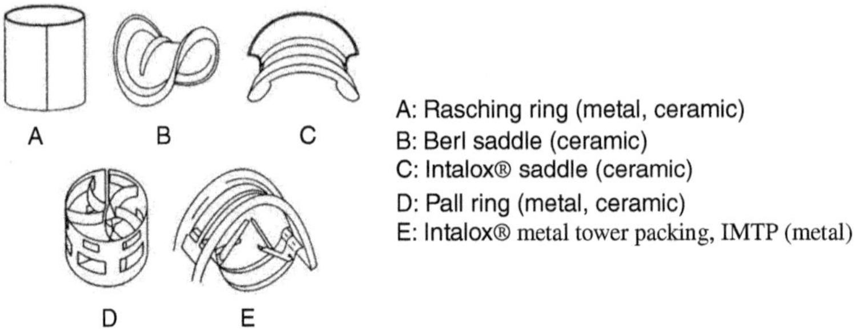

Figure 8.11 Distillation packing elements (random).

intermingling of the ascending gases and vapors and the descending liquid is thus affected" [11].

Raschig rings were just the start of a new, improved suite of distillation options. Shown in Figure 8.11 are a range of packing materials used for packed column distillations and absorption processes today. Pall rings (D) represent a variant of the original Raschig ring cylinder (A) design with strips of material folder into the interior of the cylinder; this change provides more even liquid distribution within the packing phase.

Subsequent generations of packing materials included Berl and Intalox® saddles (B and C), which are typically made of ceramics and exhibit superior wetting characteristics paired with high corrosion resistance. The saddle packings were developed to provide improved liquid distribution versus Raschig rings, and the Intalox saddles were easier to manufacture than the original Berl design. While the saddles themselves are more expensive than Raschig rings, the overall efficiency of Berl and Intalox saddles as well as Pall rings has led to them being favored for most applications today.

Developed in the late 1970s, the Intalox IMTP random packing (E) represents a third-generation packing design, which combines the advantages of the saddle-shaped packing with that of modern high-performance ring-type packings. This shape delivers a lower-pressure drop at the same vapor and liquid loads compared to previous-generation packings as it is one of the most popular random packings in use today.

8.4.2 Structured Packing

The drive to increase separation capability and capacity of distillation and absorption columns led to the development in the 1930s of structured packing materials. Structured packings offer a variety of improvements over trayed systems, including superior separation efficiency, shorter column heights, and more continuous, intimate mixing of vapor and liquid phases throughout the column. Additionally, structured packing option provides higher capacity and lower-pressure drop than random packing, but cost can be a barrier to adopting structured packing.

One of the first structured packing designs was the Stedman column from General Electric for isotope separation [12]. As shown in Figure 8.12, the Stedman packing consisted of wire cloth, which was punched and embossed to form a series

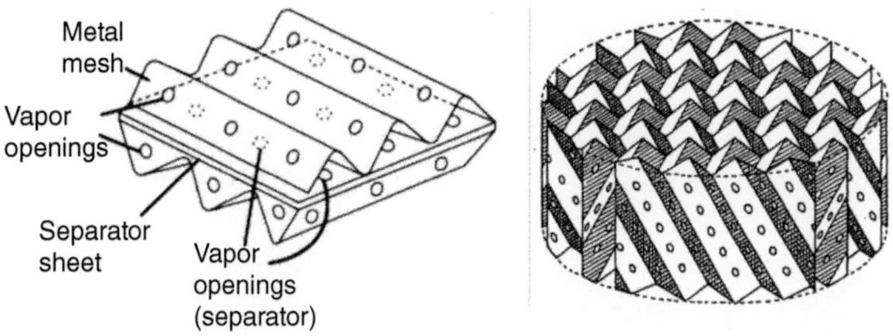

Figure 8.12 Stedman structured packing.

of triangular cells of regular shape and angle. The Stedman column design provided lower holdup, low pressure drop, shorter column height, and low weight. These same advantages drive the selection of structured packing today.

A second generation of structured packing designs and materials was introduced in the 1970s and consisted of corrugated sheet packing, which further raised column capacity while lowering costs and increasing plugging resistance of the system. These second-generation options are now widely adopted as attractive options for column revamps and have in common design elements such as open honeycombed structures with 45° inclined packing angles in combination with perforations to maximize capacity and minimize pressure drop (Figure 8.12). The purpose of the angles and design specifications in a structured packing is to force liquids to take tortuous paths through the column, thereby creating a large surface area for contact between different phases and increasing the separation efficiency.

Some common materials used in the design of structured packing are described below:

Sheet metal: Can tolerate modest levels of fouling and handle a wide range of vapor and liquid rates. Process applications range from low-pressure drop vacuum and atmospheric distillation services to medium high-pressure absorbers.

Gauze: Preferred packing for deep vacuum and low liquid rate applications as it has the lowest pressure drop per theoretical stage, which helps in the purification of temperature sensitive materials such as those found in specialty chemicals and pharmaceuticals.

Grid: Both corrugated metal and stamped blades are grid-type packing and both bring with them a high level of tolerance to fouling as they incorporate open volume to facilitate flow even in the presence of fouling. Corrugated grid packing has a smooth surface to minimize fouling and is constructed of much thicker material than the normal sheet metal packing.

8.5 Commercial Examples

8.5.1 Crude Oil Distillation

The distillation of crude oil is depicted in Figure 8.13 and is performed as a continuous distillation involving both atmospheric and vacuum distillation towers [13].

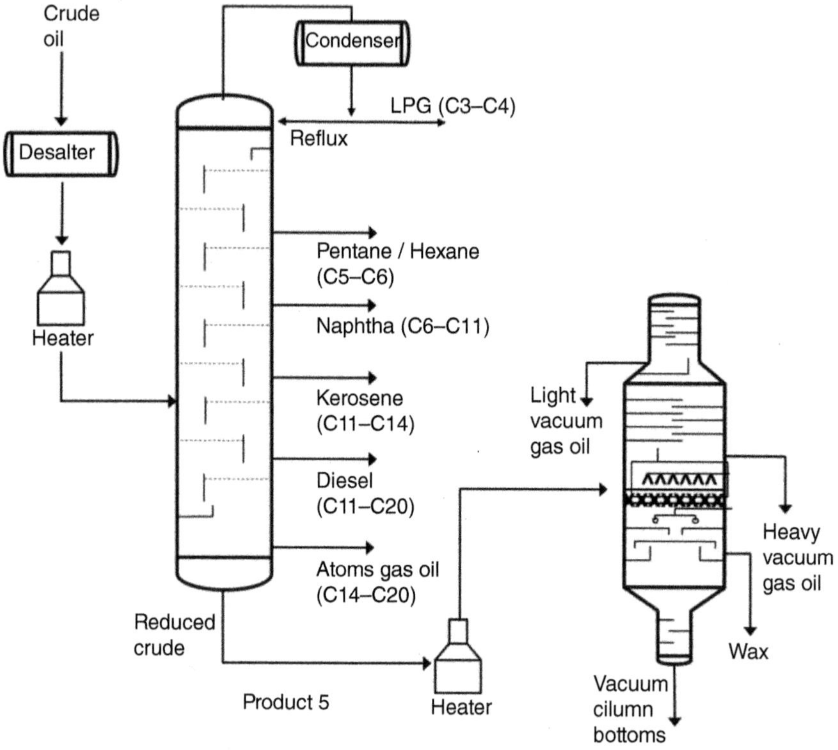

Figure 8.13 Crude oil distillation.

Before distillation can proceed, there is typically a need to clean up the crude oil as crude can include many problematic materials such as sand, inorganic salts, polymeric impurities, and more. In most crudes, NaCl, $MgCl_2$, and $CaSO_4$ form droplets dispersed throughout the crude. If not removed, these impurities can cause issues downstream in the refining process such as abrasion from sand, as well as acidic gases (e.g. HCl) upon heating of the impure crude.

The most common method to desalt crude oil is electrostatic precipitation wherein the crude is first heated to decrease its viscosity. Fresh water is introduced to the heated crude and forms small droplets within the continuous oil phase. This water-in-oil dispersion is then introduced to a pressurized desalter vessel, which includes a high-voltage electrical field that accelerates separation of the salt containing water droplets from the desalted oil. As the separation continues, the water droplets sink within the oil phase and are removed continuously from the bottom of the desalter, along with any collected sediment, for disposal. Desalted oil flows continuously from the top of the pressurized vessel and can then be introduced to the distillation units. Desalting can typically achieve 99% removal rates for common salts such as NaCl, $MgCl_2$, and $CaSO_4$.

Once the desalting process has removed inorganic salts, the resulting crude is heated to approximately 360–400 °C in tube furnaces and fed into the distillation portion of the refinery. Most refineries employ two types of distillation columns in series:

atmospheric distillation and vacuum distillation. A pre-flash column can optionally be used prior to the atmospheric column to remove the most volatile materials and increase system capacity. As their names imply, the atmospheric distillation column is intended to remove the most volatile components, while the vacuum distillation unit allows for an increase in gas oil yield under reduced pressures while avoiding high temperatures that could lead to cracking reactions and degrade the hydrocarbon feed.

To avoid thermal cracking, it is common for vacuum distillation units to run at pressures below 0.05 atm. This reduction in pressure likewise reduces the boiling point of the vacuum distillation components by as much as 100 °C and helps avoid the cracking reactions, which commonly occur if the feed experiences a temperature of 350 °C for extended periods of time [14]. The vacuum distillation unit is fed from the bottoms, or reduced crude, of the atmospheric crude distillation unit (Figure 8.13). Although designs vary greatly in accordance with crude feeds and separation goals, it is common for atmospheric units to utilize tray contactors (e.g. 30 stages) and for vacuum units to employ random or structured packings due to their lower-pressure drop levels.

The major hydrocarbon components of crude oil are:

- *Paraffins (alkanes)*: Straight carbon chains. Formula: C_nH_{2n+2}
- *Isoparaffins (iso alkanes)*: Branched carbon chains. Formula: C_nH_{2n+2}
- *Olefins (alkenes)*: Carbon chains that include at least one double bond. Formula: C_nH_{2n}
- *Naphthenes*: Saturated ring carbon structures with five or six carbon atoms in the ring. Formula: $C_nH_{2n+2} - 2R_n$ (where R_n = # of rings)
- *Aromatics*: Six carbon atoms in aromatic ring with varying levels of alkylation around the ring (e.g. benzene, toluene, xylene, ethylbenzene).

As shown in Figure 8.13, there are many possible product cuts that can be isolated from the atmospheric and vacuum distillation columns. Those highlighted in the figure can be valorized in the following ways:

(1) Liquefied petroleum gas (**LPG**) (**C_3–C_4**) can either be used for home heating/cooking or can be sent to dehydrogenation units such as Oleflex™ from UOP for the production of higher-value petrochemicals such as propylene.
(2) **Pentane–Hexane (C_5–C_6)** can be isomerized (e.g. using Penex™ processing) to increase octane values and is then blended into the gasoline pool.
(3) **Naphtha (C_6–C_{11})** is a versatile feed for either gasoline or petrochemicals production; heavy naphtha is often reformed into aromatics that ultimately find their way into polymer end uses.
(4) **Kerosene (C_{11}–C_{14})** serves as a primary feed for jet fuel and can also be utilized in the manufacture of biodegradable linear alkylbenzene (LAB) detergents.
(5) **Diesel (C_{14}–C_{20})** is a high-value fuel commonly used for transportation.

In similar fashion, the vacuum column products can be converted to higher-value materials using catalytic cracking (e.g. fluid catalytic cracker [FCC]™, Unicracking™) as well as slurry hydrocracking methods (e.g. Uniflex™) to raise the value of the bottoms products from atmospheric distillation in end markets such as olefins and aromatic petrochemicals as well as fuels.

8.5.2 Extractive Distillation

As discussed earlier, the relative volatility difference between components in a feed determines how difficult separation of those components will be. A relative volatility figure >1.5 means simple distillation will be effective. However, once the relative volatility falls below 1.2, extractive distillation becomes a preferred option.

Extractive distillation involves the addition of a miscible, low-volatility solvent, which alters the volatility of the other feed components such that separation becomes easier. The solvent does not form an azeotrope with any of the feed components but can interact within them, for example, through dipole interactions. The net result is that the resulting relative volatility is changed to allow easier separation by distillation.

For example, the separation of isobutane (i-C_4, −12 °C boiling point) and 1-butene (1-$C_4^{=}$, −6 °C boiling point) falls in the extractive range as the relative volatility of these components is 1.16. With the addition of 80% furfural (162 °C boiling point), relative volatility increases to a value of 2.0 (1-butene is soluble in the higher boiling furfural component due to dipole interactions), and the distillation can be carried out in a straightforward manner. Other commercial examples of extractive distillation include propane-propylene (via acrylonitrile), toluene-nonaromatics (via phenol), and butane-butadiene (via furfural).

8.5.3 Azeotropic Distillation

When the relative volatility of different feed components is equal to 1, simple and extractive distillations are no longer possible. Instead, separation must be accomplished via azeotropic distillation. In this strategy, an entrainer (volatile solvent) is added to the feed mixture to form an azeotrope with one of the feed components. As one component interacts more with the volatile solvent, relative volatility of the feed components is increased and thereby makes it easier to accomplish a separation.

As an example, ethanol and water form an azeotrope at 94.6/5.4 wt% (89/11 mol%) ethanol/water (relative volatility = 1). The addition of benzene as solvent/entrainer serves to increase the relative volatility of water (more polar component), while ethanol interacts more strongly with benzene. Therefore, the relative volatility of ethanol/water decreases below a value of 1, and separation can be achieved in the system. Other examples of azeotropic distillations include acetic acid-water (via ethyl acetate), pyridine-water (vis bisphenol), and cumene-phenol (via trisubstituted phosphate).

8.5.4 Dividing Wall Column Distillation

While process intensification has garnered more attention in recent years, distillation approaches, which incorporate the principles of process intensification,

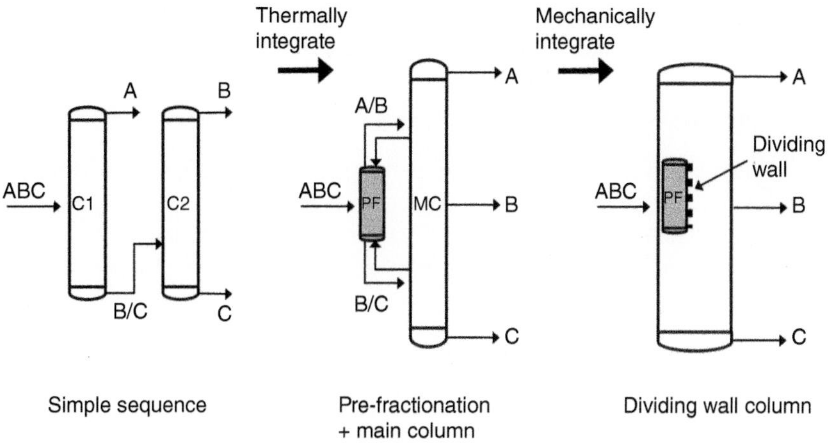

Figure 8.14 Dividing wall column basics.

have been known for many decades. For example, dividing wall columns (DWC) appeared in patent literature as early as 1946 [15]. Deployment of DWC technology took place starting in the mid-1980s as control of these systems advanced greatly with the advance of computer-aided process management tools. The roughly 30% savings in capital investment and operating expenses have led to more than 200 dividing wall column installations globally in 2020.

Shown in Figure 8.14 is a representation of the sequential separation of a three-component feed in two columns alongside the integrated dividing wall column design. In the pre-fractionator section of the dividing wall column, a rough separation is carried out between components A and C. Component A is concentrated at the top of the prefractionator and C at the bottom with component B distributed between the top and the bottom. The pre-fractionator overhead then feeds the upper half of the main fractionator for the binary separation of A and B, while the bottom of the pre-fractionator feeds the lower half of the main column for the separation of B and C.

The main fractionator liquid from the top section refluxes the pre-fractionator top and the main fractionator on the other side of the dividing wall, while the vapor from the bottom section similarly strips the bottom of the pre-fractionator and the main fractionator. The middle component B is withdrawn at the stage where its concentration is at a maximum, avoiding re-mixing losses, which can lead to energy savings of up to 25–30% or more.

There are now many relatively well-known potential applications for dividing wall columns, including straight-run naphtha fractionation upstream or downstream of reformers, FCC naphtha fractionation, C_4 isomer separation, and benzene, toluene, and xylene (BTX) separation. A dividing wall column also is used for UOP's Pacol enhancement process (PEP). In this process, A is pentane, B is benzene, C consists of C7 + olefins, and D comprises C7 + aromatics. To prevent mixing of C and D, a novel trap tray is applied and excessive B is added [16]. The next logical step in DWC performance is to include multiple dividing walls within the DWC column, and designs have been developed to produce six high-purity products from a single column [17].

8.5.5 Reactive Distillation

Reactive distillation is defined as the simultaneous distillation and catalytic chemical reaction within a single process unit and, like the dividing wall column, is another application of process intensification principles. Reactive distillation is classified as a two-phase, counter-current, fixed-bed catalytic process [18, 19]. There are several benefits of reactive distillation, including:

(1) *Lower capital investment*: Catalysis and distillation combined in one process unit.
(2) *Lower energy costs*: Exothermic reaction heat can be used to drive distillation vaporization (lower reboiler heat demand).
(3) *Longer catalyst life*: Especially if the reaction zone is located above the feed.
(4) *Entrainers*: Azeotropes can be broken as reactants or products can serve as entrainers.
(5) *Higher conversion*: Product removal from the reaction phase allows equilibrium to be re-established at higher conversion.

The Ethermax™ process from UOP uses reactive distillation technology and employs the KataMax™ structural packing system from Koch-Glitsch; this packing is specifically designed to expose a solid catalyst to a reacting liquid stream. The feed consists of MeOH or EtOH and hydrocarbon streams containing reactive tertiary olefins such as isoamylene and isobutylene. Ethermax catalyzes the conversion of isobutylene and methanol into a high-octane methyl-*tert*-butyl ether (MTBE) blending agent that contains no benzene or aromatics; the reactive distillation process delivers a high MTBE yield and is currently used in 44 units worldwide.

References

1 Kockmann, N. (2014). History of distillation. In: *Distillation: Fundamentals and Principles* (eds. A. Gorak and E. Sorensen), 1–43. London: Academic Press.
2 Forbes, R.J. (1948). *A Short History of the Art of Distillation*. Leiden: EJ Brill.
3 Dubbs, C.P. (1921). Process of converting hydrocarbons. US Patent 1,392,629, filed 19 March 1919 and issued 4 October 1921.
4 Haubrand, E. (1916). *Die Wirkungsweise der Rektifizier- und Destillier-Apparate mit Hilfe einfacher mathematischer Betrachtungen*, 3e. Berlin: Springer.
5 McCabe, W.L. and Thiele, E.W. (1925). *Graphical design of fractionating columns. Industrial & Engineering Chemistry* 17 (6): 605–611.
6 Gorak, A. and Sorensen, E. (2014). *Distillation: Fundamentals and Principles*. London: Academic Press.
7 Kister, H.Z. (1992). *Distillation Design*. New York: McGraw-Hill.
8 Towler, G.P. and Sinnott, R. (2012). *Chemical Engineering Design*, 2e. Burlington: Elsevier.

9 Gerster, J.A., Mertes, T.S., and Colburn, A.P. (1947). Ternary systems *n*-butane–1-butane–furfural and isobutane–1-butane–furfural. *Industrial & Engineering Chemistry* 39 (6): 797–804.
10 Chilton, T.H. and Colburn, A.P. (1935). Distillation and absorption in packed columns. *Industrial & Engineering Chemistry* 27 (3): 255–226.
11 Raschig, F. (1915). Absorption and reaction tower for acids. US Patent 1,141,266, 26 filed March 1914 and issued 1 June 1915.
12 Stedman, D.F. (1935). Packing for fractionating columns and the like. US Patent 2,047,444, filed 14 January 1935 and issued 14 July 1936.
13 Alfke, G., Irion, W.W., and Neuwirth, O.S. (2007). Oil refining. In: *Ullmann's Encyclopedia of Industrial Chemistry* (ed. James G. Speight), 7e, 208–260. Weinheim: Wiley-VCH.
14 Speight, J.G. (2007). *The Chemistry and Technology of Petroleum*, 4e. New York: CRC Press.
15 Wright, R.O. (1946). Fractionation apparatus. US Patent 2,471,134, filed 17 July 1946 and issued 24 May 1949.
16 Yildirim, O., Kiss, A.A., and Kenig, E.Y. (2011). Dividing wall columns in chemical process industry. *Separation and Purification Technology* 80: 403–417.
17 Piszczek, R., Hergenrother, M., and Wang, Z. (2020, 2020). Introduction to dual-dividing-wall columns. *Chemical Engineering Progress* 4: 22–25.
18 Tower, G.P. and Frey, S. (2002). Reactive distillation. In: *Separation Processes* (ed. S. Kulprathipanja), 18–50. New York: Taylor & Francis.
19 Sakuth, M., Reusch, D., and Janowsky, R. (2007). Reactive distillation. In: *Ullmann's Encyclopedia of Industrial Chemistry* (ed. James G. Speight), 7e, 264–275. Weinheim: Wiley-VCH.

9
Membranes

Membrane technology has been an important enabler for effective and efficient gas, vapor, and liquid separations and a wide range of other potential applications such as for catalytic membrane reactors, membrane distillation, and energy storage. This chapter covers perspectives of membranes from a combination of both academic research and industrial development.

9.1 Principle and Background

A membrane is a selectively permeable barrier between feed and permeate phases that can be used to separate the components of a gas, vapor, or liquid mixture based on their relative permeation rate through the membrane material. Membrane separation is a nonequilibrium process and determined by both the physical and chemical properties of the mixtures to be separated as well as the membrane materials and membrane structures. Membranes are commonly prepared from organic, inorganic, or organic/inorganic composite materials. They are generally not limited by azeotropes or eutectics, less energy intensive, and have lighter weight and smaller footprint than other traditional separation processes such as cryogenic distillation, absorption, and adsorption processes. In addition, membranes have the advantages of modular design, no regeneration requirement, simple and rapid installation, quick start-up, ease of operation, high reliability, low capital and operating cost, ideal for offshore applications such as natural gas purifications on floating, production, storage, and offloading (FPSO) vessels [1].

The state of the art of polymeric membrane-based industrial separation processes has evolved rapidly since the development of asymmetric integrally skinned polymeric membrane by Loeb and Sourirajan [2] and the development of thin-film composite (TFC) membranes by Cadotte [3]. Membranes are competitive or potentially competitive in a variety of separations such as gas, vapor/gas, vapor, aqueous liquid, and organic liquid separations. Membrane technologies have been successfully used for large-scale water purification applications at low cost and minimum environmental impact. Polymeric gas separation membranes are of special interest to

Modern Petrochemical Technology: Methods, Manufacturing and Applications, First Edition.
Santi Kulprathipanja, James E. Rekoske, Daniel Wei, Robert V. Slone, Trung Pham, and Chunqing Liu.

petroleum producers and refiners, chemical companies, and industrial gas suppliers. Several applications of gas separation membranes have achieved commercial success, such as nitrogen enrichment from air, carbon dioxide removal from natural gas and from enhanced oil recovery, separation and recovery of organic vapors, and in hydrogen removal from nitrogen, methane, and argon in ammonia purge gas streams. However, membrane technologies with limitations for organic vapor/vapor or liquid feed separations, such as for olefin/paraffin, aromatic/nonaromatic, and aromatic/aromatic separations, have not been commercially successful so far.

Polymeric membranes often suffer from thermal stability and plasticization or swelling of the organic polymer matrix by the sorbed penetrating molecules such as high concentration of carbon dioxide or olefins, which results in the decrease of membrane selectivity and therefore lower product recovery and/or higher compression cost due to higher recycle rates. Inorganic membranes such as microporous zeolitic membranes can overcome some of the challenges facing polymeric membranes. For example, zeolite membranes have the potential for separations with high efficiency and high productivity under conditions where polymeric membranes cannot survive by taking advantages of their superior thermal and chemical stability, good erosion resistance, and high plasticization resistance to condensable gases or vapors. The main challenges facing inorganic membranes for large-scale industrial applications are their high cost, reproducibility, difficulty of making very thin (<100 nm), grain boundary-free, pinhole-free, and crack-free selective layer.

The significant growth of membrane technologies in refinery, petrochemical, and natural gas markets requires breakthrough development of new membranes, modules, and processes that overcome the challenges facing the current membrane technologies. Innovation in membrane materials, equipment, and process technology integration will play a key role to enable new large-scale, low-energy-intensity membrane-based separations.

9.2 Types and Preparation of Membranes

Three types of membranes, including polymeric, inorganic, and polymeric/inorganic mixed-matrix membranes (MMMs), have been studied extensively in academia and industries for a wide range of potential applications. These membrane applications include gas separations, gas/vapor separations, organic vapor/vapor separations, pervaporation, reverse osmosis (RO), organic liquid RO, forward osmosis (FO), pressure-retarded reverse osmosis (PRO), nanofiltration (NF), organic solvent nanofiltration (OSNF), ultrafiltration (UF), microfiltration (MF), electrodialysis, fuel cell, batteries, membrane contactor, membrane distillation, membrane reactors, etc. Polymeric membranes are the most common membranes for commercial separation applications. For example, almost all of the gas separation and RO commercial membranes are made from polymers. Inorganic microporous zeolite membranes have been applied to commercial-scale pervaporation processes for organic solvent dehydration. Inorganic ceramic membranes have commercial applications in battery, NF, and UF processes.

9.2.1 Polymeric Membranes

9.2.1.1 Fundamentals of Polymeric Membranes

Polymeric membranes are the first membranes commercialized in large industrial scale for separations and have been operated successfully in gas and liquid separations, fuel cells, and some electrochemical processes. Polymers, including both glassy and rubbery polymers, provide a range of properties, including low cost, good permeability, high mechanical stability, and ease of processability that are important for membrane fabrication and applications. For example, cellulose acetate (CA), and cellulose triacetate (CTA) were used to manufacture commercial Separex™ spiral wound membrane for acid gases removal from natural gas [4, 5]. Polysulfone polymer was used to produce PRISM® hollow-fiber membrane for hydrogen separation and air separation [6] and porous polysulfone (PSf) support membrane for the fabrication of some gas separation membranes, RO, and NF membranes. Polyimides and polyaramides were used for the fabrication of gas separation membranes for acid gases removal from natural gas, air separation, hydrogen purification, or helium recovery from natural gas. Silicone rubber polymer was used for the manufacture of rubbery membranes for organic vapor/gas or nitrogen removal from natural gas [7]. The separation layer of most of the current commercial RO and NF membranes was formed from the in situ polymerized polyamides (PAs).

Glassy polymers have stiffer polymer backbones than rubbery polymers and therefore allow smaller molecules such as hydrogen, helium, and water to pass through faster, while larger molecules such as hydrocarbons and salts pass through slower as compared to polymers with less stiff backbones. Gas separation, RO, and pervaporation processes using polymeric membranes are based on a solution-diffusion mechanism [8]. This mechanism involves molecular-scale interactions of the permeating molecule with the polymer. The permeating molecule is first sorbed at the feed side surface of the membrane, then transported by a concentration gradient of the permeating molecule, and finally desorbed at the permeate side surface of the membrane. For a separation process using a polymeric membrane based on the solution-diffusion separation mechanism, the membrane performance in separating a given mixture of gases, vapors, or liquids comprising a more permeable component A and a less permeable component B is determined not only by the intrinsic permeabilities for components A and B (abbreviated as P_A and P_B) and selectivity of A over B (abbreviated as $\alpha_{A/B}$) of the polymeric membrane material, but also by the membrane selective layer thickness and membrane stability. The permeability of component A or B (P_A or P_B), measured in Barrer units (1 Barrer = 10^{-10} cm^3 (STP) cm/cm^2 s (cm Hg)), is the product of the flux of permeating component A or B and the selective skin layer thickness of the membrane, divided by the partial-pressure difference of permeating component A or B across the membrane. The $\alpha_{A/B}$ is the ratio of the permeabilities of permeating component A and B ($\alpha_{A/B} = P_A/P_B$), where P_A is the permeability of the more permeable component A and P_B is the permeability of the less permeable component B. The permeabilities of components A and B through the membrane are determined by solubility coefficient (abbreviated as S) and diffusivity coefficient (abbreviated as D) ($P_A = S_A \times D_A$ and $P_B = S_B \times D_B$).

The solubility coefficient equals the ratio of sorption uptake normalized by some measure of uptake potential, such as partial pressure. The diffusivity coefficient is a measure of the mobility of component A or B passing through the voids between the polymer chains in the polymeric membrane. Therefore, the selectivity of the polymeric membrane for component A over component B ($\alpha_{A/B}$) is the product of the solubility selectivity of component A over component B (abbreviated as S_A/S_B) and diffusivity selectivity of component A over component B (abbreviated as D_A/D_B) ($\alpha_{A/B} = S_A/S_B \times D_A/D_B$). A polymeric membrane can have a high P_A because of high S_A, high D_A, or both, and the membrane can have a high $\alpha_{A/B}$ because of high S_A/S_B, high S_A/S_B, or both. Normally the solubility coefficient increases and the diffusivity coefficient decreases with an increase in the molecular size of the permeating component for a glassy polymeric membrane and vice versa for a rubbery polymeric membrane. Both permeability and selectivity are the intrinsic properties of the polymeric membrane materials and ideally are constant with the change of the thickness of the membrane selective layer, feed pressure, flow rate, and other separation process conditions although they are temperature dependent.

Membrane permeance, measured in gas permeation units (GPUs, 1 GPU = 10^{-6} cm^3 (STP)/cm^2 s (cm Hg)), is the pressure-normalized flux and determined by the intrinsic permeability of the polymer material and the selective skin layer thickness of the membrane. Therefore, a commercially viable polymeric membrane based on a solution-diffusion separation mechanism needs to have a high intrinsic permeability and a thin selective layer to achieve a sufficiently high permeance at the preferred separation conditions. Otherwise, extraordinarily large membrane surface areas are required to achieve high treating capacity. For membrane separation applications, both high permeance and selectivity are desirable because higher permeance decreases the size of the membrane area required to treat a given volume of feed mixture, thereby decreasing the capital cost of the membrane system, and higher selectivity results in a higher-purity product.

Polymeric materials have been studied extensively for the development of membranes for separation applications, particularly for solution-diffusion-mechanism based separations. Many research articles and patents have reported the synthesis of polymeric membrane materials, such as polyimides (PIs), polyethersulfones (PESs), polyethers, polyamides, polybenzoxazoles, polybenzimidazoles (PBIs), and polymers with intrinsic microporosities. Traditional polymeric membrane materials, however, have shown a well-known tradeoff between permeability and selectivity (or so-called polymer upper-bound limit). By comparing the permeation data of over three hundred different polymers, Robeson reported in 1991 and 2008 that the selectivity and permeability of polymeric gas separation membrane materials were inseparably linked to one another, in a relation where selectivity increases as permeability decreases and vice versa [9, 10]. Substantial research effort has been directed to design novel polymer structures or modify existing polymer structures to overcome the permeability and selectivity limits imposed by the polymer upper-bound limit in academia and industrial research labs. Fabrication of defect-free high selectivity membranes with a thin selective skin layer, development of advanced membrane

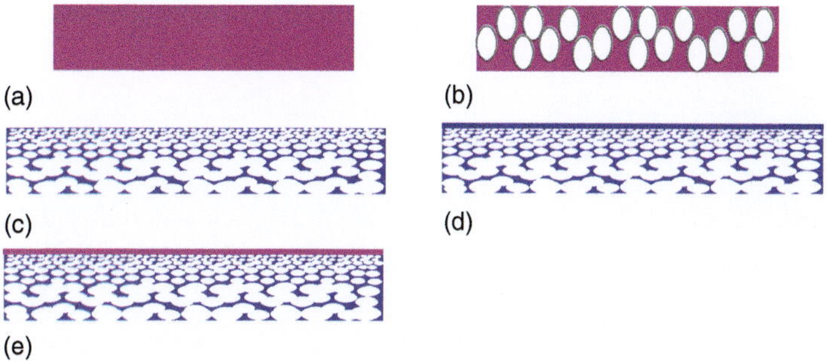

Figure 9.1 Schematic illustration of membrane structures: (a) symmetric nonporous structure; (b) symmetric isotropic porous structure; (c) asymmetric porous structure; (d) asymmetric structure with a thin nonporous selective layer (<100 nm) supported on a porous support layer; and (e) thin-film composite asymmetric structure with a thin nonporous selective layer (<100 nm) supported on a porous support layer. (Photos provided courtesy of Honeywell UOP with a current copyright.)

element/module, and design of novel membrane processes have also played an important role in the advancement of membrane technologies for separations.

9.2.1.2 Preparation of Polymeric Membranes

Polymeric membranes can have symmetric or asymmetric bulk structures and each of them can have a porous or nonporous substructure as illustrated in Figure 9.1, depending on their applications. Polymeric membranes can be fabricated into many different geometries such as flat sheet that can form spiral wound membrane modules, tube, and hollow fiber. Asymmetric polymeric membranes normally have two different structures, including asymmetric integrally skinned membrane structure and TFC membrane structure. Figure 9.2 shows the cross-sectional High-Resolution Scanning Electron Microscopy (HRSEM) images of an asymmetric integrally skinned flat-sheet membrane, a TFC flat-sheet membrane, an asymmetric integrally skinned hollow-fiber membrane, and a TFC hollow-fiber membrane.

Symmetric nonporous dense-film polymeric membranes can be prepared either by dissolving the polymer in an organic solvent followed by solution casting using a casting knife and evaporating the organic solvent or by a melt pressing process using a high-temperature plate under high pressure. Symmetric nonporous dense-film polymeric membranes not only can be used as the proton-conducting membranes for fuel-cell applications, but also can be used to screen membrane materials by measuring the intrinsic properties of membrane materials for gas separations.

Symmetric porous polymeric membranes can be prepared by solution precipitation, irradiation, or physical biaxial stretching of a symmetric nonporous dense-film membrane to form uniform pores. Symmetric porous polymeric membranes such as polypropylene Celgard® and Gore-Tex® membranes prepared by biaxial stretching method can be used for many applications such as separators for battery applications, MF, and membrane contactor applications.

Asymmetric integrally skinned porous polymeric membranes are commonly used for UF, MF, membrane distillation, and membrane contactor applications, and can

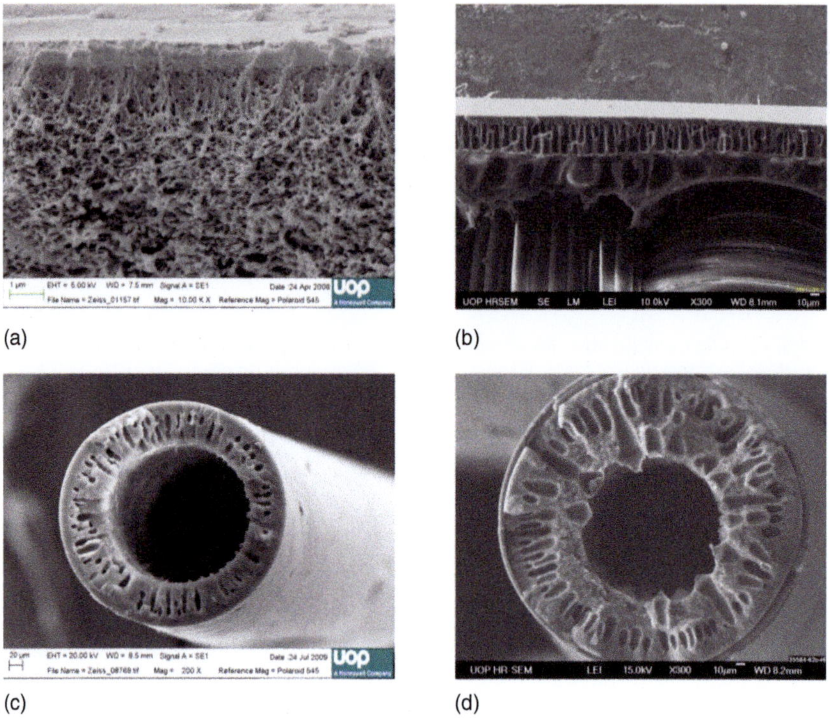

Figure 9.2 Cross-sectional HRSEM images of: (a) asymmetric integrally skinned flat-sheet membrane; (b) TFC flat-sheet membrane; (c) asymmetric integrally skinned hollow-fiber membrane; and (d) TFC hollow-fiber membrane.

be used as the support membranes for the fabrication of RO, NF, and gas separation membranes. This chapter will not cover the preparation of this type of asymmetric integrally skinned porous polymeric membranes in detail.

Asymmetric polymeric membranes for gas separation, RO, NF, and pervaporation applications can have either an asymmetric integrally skinned membrane structure or TFC membrane structure. Such membranes normally consist of a thin, dense, nonporous, selective skin layer of <100 nm and a less-dense void-containing (or porous), nonselective support layer, with pore sizes ranging from large in the support region to very small proximate to the skin layer. The thin, dense, nonporous, selective skin layer provides the membrane resistance and selectivity and the thick, porous support layer provides the membrane with high mechanical strength.

Design and selection of proper polymeric materials with desired properties is a critical initial step for the development of asymmetric polymeric membranes. In addition to the consideration of high permeability and selectivity, the polymer cost, purity, molecular weight, commercial availability or manufacturing possibility, solubility of the polymer in organic solvents for the preparation of polymer solutions, film-forming property and membrane processability, mechanical property, post-treatment possibility, chemical and thermal stabilities, and hydrophilicity/hydrophobicity are all the factors that need to be considered to select the best candidate polymer for the development of high-performance asymmetric polymeric membrane for a particular application.

Preparation of Asymmetric Integrally Skinned Polymeric Membranes

For asymmetric integrally skinned polymeric membranes, the thin, dense, nonporous, selective skin layer on top of the thick, porous support layer is formed simultaneously with the thick, porous support layer by a solution precipitation method from the same polymer solution via a phase-inversion process first developed by Loeb and Sourirajan. Therefore, the thin, dense, nonporous, selective skin layer and the thick, porous support layer are of the same polymer material. For a phase-inversion process, a one-phase polymer solution (or so-called polymer dope) is converted into a two-phase system consisting of a solid (polymer-rich) phase, which forms the membrane structure with either a flat-sheet geometry or a hollow-fiber geometry; and a liquid (polymer-lean) phase, which forms the pores in the final membrane. The advancement of membrane fabrication equipment since the invention of phase-inversion process has enabled better control of membrane fabrication-processing conditions such as tension, humidity, casting speed, and ventilation in the evaporation zone, which results in the development of new membranes from high-performance new polymers.

Preparation of asymmetric integrally skinned flat-sheet membrane via the phase-inversion process comprises many steps, including: (i) design and preparation of a polymer-casting solution: A polymer is dissolved in a mixture of organic solvents to form a homogeneous polymer solution. The composition of the polymer-casting solution is one of the key parameters that determine the membrane property. The polymer-casting solution comprises one polymer or a blend of two or more polymers, organic solvents that can dissolve the polymers such as *N*-methylpyrrolidone (NMP), *N*,*N*-dimethyl acetamide (DMAC), *N*,*N*-dimethylformamide (DMF), 1,3-dioxolane, dimethyl sulfoxide (DMSO), and some other components. Organic nonsolvents that cannot dissolve the polymers such as methanol, ethanol, isopropanol, glycerol, acetone, *n*-octane, citric acid, and lactic acid and other additives such as lithium chloride and lithium nitrate are added to the polymer-casting solution in some cases to achieve the best membrane property; (ii) casting the polymer-casting solution on a highly porous substrate such as a polyester or nylon cloth substrate to form a controlled layer of polymer solution on the surface of the substrate using a casting knife under controlled casting speed, temperature, and humidity; (iii) allowing evaporation of a controlled amount of solvents and nonsolvents from the surface of the layer of polymer solution on the substrate under controlled humidity, temperature, air flow, and time to facilitate the polymer densification and formation of the thin, dense, nonporous, selective skin layer. This step can be avoided for the preparation of asymmetric porous integrally skinned polymeric membranes; (iv) coagulating the layer of polymer solution on the substrate by immersing the partially dried nascent membrane layer together with the substrate into a low-temperature controlled coagulation medium such as a 0–5 °C cold water bath when water is a nonsolvent for the polymer. An asymmetric porous support layer underneath the thin, dense, nonporous, selective skin layer is formed after the organic solvents and nonsolvents diffuse out of the interior of the layer of polymer solution and the nonsolvent liquid in the coagulation medium such as water diffuses into the interior of the layer of polymer solution; (v) washing the nascent asymmetric integrally skinned membrane comprising a thin, dense, nonporous, selective skin layer and a highly porous support layer in a nonsolvent medium such as a water bath to remove the leftover organic solvents, nonsolvents, and additives from the nascent asymmetric integrally skinned membrane; (vi) annealing the nascent asymmetric integrally skinned membrane comprising a thin, dense, nonporous, selective skin layer and a highly porous support layer in an annealing medium with a controlled temperature such as a hot water-annealing tank controlled at 55 °C. The trace amount of leftover organic solvents, nonsolvents, and additives from the coagulating medium will be further removed during this annealing step. In addition, both the thin, dense, nonporous, selective skin layer and the highly porous support layer will be further stabilized with certain degree of shrinkage. The stabilization of the membrane structure is critical to provide the membrane long-term performance stability for high-pressure applications. This annealing step is unnecessary for fabricating certain asymmetric integrally skinned membranes; and finally, (vii) drying the wet asymmetric integrally skinned membrane to remove the coagulation and annealing fluids from the membrane without causing changes to the substructure of the membrane at certain temperature from about 50 to about 100 °C. Some membranes need to be kept wet without drying before use and this drying step can be avoided and replaced with an additional conditioning step by placing the wet membrane in a conditioning medium that may contain water and conditioning agent such as glycerol or polyethylene glycerol. The conditioning agent diffuses into the pores of the membrane to prevent the collapse of the pores during storage. In some cases, some defects on the thin, dense, nonporous, selective skin layer of the membrane caused by gas bubbles, dust, or substrate imperfections cannot be eliminated. These defects can be plugged by coating a thin layer of a highly permeable polymer such as silicone rubber or high-permeability fluoropolymer on the surface of the thin, dense, nonporous, selective skin layer of the membrane.

Preparation of asymmetric integrally skinned polymeric hollow-fiber membranes via the phase-inversion process is similar to that of the asymmetric integrally skinned flat-sheet membranes. One major difference is that a casting knife is used to cast a thin layer of polymer solution, which will then be converted into asymmetric integrally skinned flat-sheet membrane, but an annular spinneret is used to extrude a hollow-fiber polymer solution, which will then be converted into asymmetric integrally skinned hollow-fiber membrane. Another major difference is that a highly porous substrate such as a polyester or nylon cloth, which will provide the membrane mechanical strength, is normally used for the fabrication of asymmetric integrally skinned flat-sheet membranes, but preparation of polymeric hollow-fiber membranes does not need any substrate and the membranes are freestanding membranes. In addition, normally the polymer spinning solution for the preparation of hollow-fiber membranes has higher polymer concentration and therefore higher solution viscosity than the polymer-casting solution for the preparation of flat-sheet membranes. The composition of the polymer-casting solution or spinning solution and the coagulation media are two key parameters that determine the final membrane separation performance.

Overall, the thin, dense, nonporous, selective skin layer of the asymmetric integrally skinned polymeric membranes will not be delaminated easily from the thick, porous support layer because both layers are formed simultaneously from a one-step phase-inversion process. However, fabrication of defect-free, high selectivity asymmetric integrally skinned polymeric membranes sometimes is challenging. The presence of nanopores or defects in the selective skin layer reduces the membrane selectivity. In some cases, the high shrinkage of a high-performance new polymeric membrane on the highly porous substrate during membrane casting and drying process results in unsuccessful fabrication of asymmetric integrally skinned flat-sheet membranes using the phase-inversion process. In some other cases, novel polymeric materials have very limited solubility in common organic solvents that are used for the fabrication of asymmetric integrally skinned polymeric membranes, preventing the development of high-performance membranes.

Preparation of TFC Polymeric Membranes

Since the thin, dense, nonporous, selective skin layer and the thick, porous, non-selective support layer of the asymmetric integrally skinned polymeric membrane are formed simultaneously from the same polymeric material, the cost of a new asymmetric integrally skinned polymeric membrane may be very high when an expensive new polymeric membrane material is used. One approach to reduce the cost of new membranes has been the development of TFC membranes that comprise a thin, dense, nonporous, selective skin layer of a high-cost, high-performance polymer deposited on an asymmetric integrally skinned, low-cost, porous, nonselective support membrane [11]. The asymmetric integrally skinned, low-cost, porous, nonselective support membranes used for the preparation of TFC membranes can be formed from commercially available inexpensive polymers such as CA, CTA, a blend of CA and CTA, polysulfone (PSf), PES, polyetherimide (PEI), polypropylene (PP), or polyether ether ketone (PEEK). The thin, dense, nonporous, selective skin layer on the TFC membranes can be prepared from relatively more expensive, high-performance polymers such as polyimides, polymers with intrinsic microporosities, and fluoropolymers. TFC polymeric membranes have the advantages of low cost and high performance due to the use of inexpensive asymmetric porous support membrane and the use of very small amount of expensive high-performance polymer for the thin, dense, nonporous, selective skin layer. TFC polymeric membrane also provides an option for new polymers that are difficult to be processed as asymmetric integrally skinned membranes. TFC polymeric membranes with a thin, dense, almost nonporous polyamide selective layer on top of an asymmetric integrally skinned porous polymeric membrane are commercially available for RO and NF applications. TFC polymeric membranes with a thin, dense, nonporous selective layer have been used commercially for gas separations.

Most of the TFC polymeric membranes are bilayer membranes prepared from at least two different polymers. For most of the bilayer TFC polymeric flat-sheet membranes and some of the bilayer TFC polymeric hollow-fiber membranes, the thin,

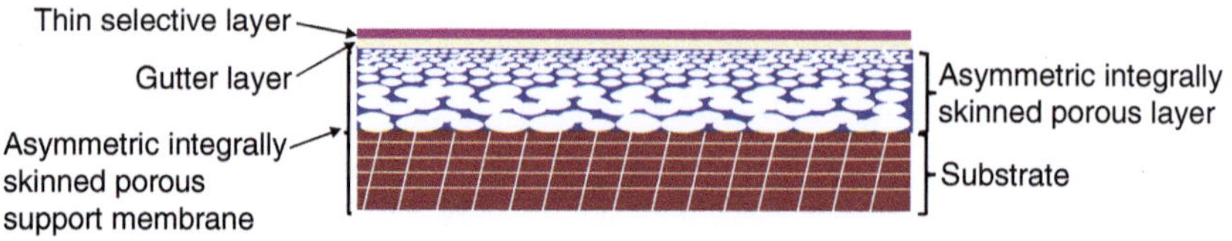

Figure 9.3 Schematic illustration of trilayer TFC polymeric membrane.

dense, nonporous or almost nonporous, selective skin layer and the thick, porous support layer are formed separately in two steps from different polymeric materials. Similar to asymmetric integrally skinned membranes, the thin, dense, nonporous or almost nonporous, selective skin layer is responsible for resistance and selectivity. It is worth noting that the surface pore size control for the asymmetric porous support membrane that will provide the TFC membrane mechanical strength and durability will contribute to the TFC membrane separation performance even though the asymmetric porous support membrane itself has no separation selectivity. The two-step method for the preparation of bilayer TFC polymeric membranes typically includes the fabrication of the asymmetric integrally skinned porous support polymeric membrane via the phase-inversion process followed by adding a thin selective layer formed from a different polymer on the surface of the support membrane by a coating method such as dip coating, spray coating, spin coating, laminating, interfacial polymerization, knife casting, or other methods. Some of the TFC polymeric membranes have trilayers with an intermediate layer (or so-called gutter layer) between the top thin selective skin layer and the bottom thick, porous, nonselective support layer (Figure 9.3) to make a defect-free continuous selective skin layer [12].

Since the thin selective skin layer is formed separately from the thick, porous, nonselective support layer in the TFC polymeric membranes, the thin selective skin layer can be delaminated from the thick, porous, nonselective support layer in some cases, which will result in significantly decreased selectivity for separations. In addition, the porous, nonselective support layer needs to be insoluble in the solvents that will be used for the preparation of the coating solution to form the thin selective skin layer. Otherwise, the porous, nonselective support layer will be damaged during the formation of the thin selective skin layer via the normal coating process. To avoid these issues, TFC polymeric membranes have been prepared via a one-step co-casting [13] or co-extruding [14] method in recent years. TFC polymeric flat-sheet membranes can be fabricated by simultaneously co-casting two different polymeric casting solutions containing the same polymer or different polymers using two casting knives via a one-step phase-inversion process. One solution at the bottom layer forms the low-cost, asymmetric, integrally skinned, porous, nonselective bottom support layer and the other solution forms another asymmetric integrally skinned high-performance selective layer on top of the support layer. TFC polymeric hollow-fiber membranes can be fabricated via a one-step co-extrusion phase-inversion process using two different polymeric spinning solutions and a triple-annular spinneret. The core solution forms the low-cost, asymmetric, integrally skinned, porous, nonselective core layer, and the sheath solution forms the asymmetric integrally skinned high-performance selective sheath layer. The bore fluid normally comprising an organic solvent and water creates the center hollow of the hollow-fiber membrane.

Preparation of TFC polymeric membranes using one-step co-casting or co-extruding approach allows the use of higher-cost, high-performance polymer to form the asymmetric integrally skinned selective layer and the use of inexpensive, commercially available polymer to form the asymmetric, integrally skinned, porous, nonselective support layer without any potential delamination issue and the issue of support membrane damage.

Similar to asymmetric integrally skinned polymeric membranes, TFC polymeric membrane may also comprise a thin coating layer of a highly permeable polymer such as silicone rubber or high-permeability fluoropolymer on the surface of the thin, dense, selective skin layer of the membrane to plug the defects and further improve the membrane selectivity for gas and vapor separation applications.

The development of new polymeric membranes from the starting polymers to the large-scale commercial applications involves many critical steps. Not only polymeric materials, but also membrane engineering need to be studied in detail. The selection or synthesis of the polymeric membrane material for a specific separation application is a key initial step. Membrane engineering, including membrane fabrication, element/module preparation, process design, and simulation, is equally important or even more important than membrane material study. Polymeric casting or spinning solution study allows the successful conversion of a polymeric material to a commercially viable membrane with either flat-sheet or hollow-fiber geometry and each of them can have an asymmetric integrally skinned or TFC structure. The membrane element/module preparation involves not only advanced polymeric membrane, but also unique design of the element/module such as permeate and feed spacer for spiral wound element, as well as the glue that holds membrane leaves or hollow fibers together to prevent leaks. To extract the best value from the membrane, sometimes membrane post-treatment such as chemical or physical cross-linking is used to further enhance the membrane performance. Membrane performance evaluation and technoeconomic analysis allow further membrane optimization. Overall, polymeric material selection, design, and synthesis would be less than about 20% of work in the whole membrane development process.

Consistent, long-term membrane research programs in academia and R&D platforms in industry will continuously advance membrane technologies and lead to breakthrough membranes for separations. These activities should include novel membranes such as molecular design and synthesis of polymers, novel membrane components, improved element/module materials, and application-guided concept discovery, new membrane processes such as physical/computational modeling, process intensification, invention of new membrane-based processes, as well as advanced membrane characterization techniques.

9.2.2 Inorganic Membranes

Inorganic membranes such as microporous zeolite, metal–organic framework (MOF), zeolitic imidazolate frameworks (ZIFs), crystalline organic framework

(COF), carbon molecular sieve, ceramic, silica, metal, and stainless-steel membranes have several advantages over polymeric membranes for a wide range of applications. Inorganic membranes have high thermal stability, high chemical resistance and resistance to harsh environments, high permeability, conductivity, and selectivity for some applications. Therefore, inorganic membranes provide new separation opportunities such as high-temperature catalytic membrane reactor applications. Significant research was reported on four main types of inorganic membranes based on their preparation methods in the early 1990s, including sol–gel membranes such as alumina, titanium, zirconia, and silica membranes; chemical vapor deposition (CVD) membranes such as silica, zirconia, and palladium membranes; pyrolysis membranes such as carbon and carbon molecular sieve membranes; as well as hydrothermal membranes, such as microporous crystalline zeolite membranes. Many different microporous crystalline MOF and ZIF hydrothermal membranes such as MOF-5, ZIF-8, and ZIF-67 have been studied in academia since the report of the first MOF-5 membrane for gas separation in 2009 [15]. This chapter will focus on the discussion of microporous crystalline zeolite, MOF, and ZIF membranes for separations based on molecular sieving separation mechanism.

9.2.2.1 Introduction to Microporous Crystalline Inorganic Membranes

Microporous crystalline inorganic membranes, including traditional zeolitic membranes such as NaA zeolite membrane, SAPO-34 molecular sieve membrane, and the recent MOF and ZIF membranes are composite membranes with a continuous, relatively thin, selective layer of microporous crystalline inorganic material formed by in situ hydrothermal synthesis methods on the surface of a highly porous, thick support membrane. A significant number of microporous crystalline inorganic materials have been studied for the formation of the continuous, relatively thin, selective layer of microporous crystalline inorganic material in the microporous crystalline inorganic membranes since crystalline zeolite membranes were first reported in late 1980s and early 1990s [16, 17]. For example, aluminosilicates, titaniumsilicates, silicates, silicoaluminophosphates (SAPOs), metalloaluminophosphates (MeAPOs), metallosilicoaluminophosphates (MeAPSOs), and their different ion-exchange forms such as NaA, MFI, SAPO-34, SSZ-13, AlPO-18, DD3R, AlPO-17, SAPO-17, AlPO-5, T, NaY, ETS-4, mordenite, as well as MOFs, ZIFs, and COFs such as MOF-5, ZIF-8, ZIF-67, ZIF-7, ZIF-95, MIL-53, NH_2-MIL-53, HKUST-1, and $Zn_2(bim)_4$ have been used for the preparation of the microporous crystalline inorganic membranes. The highly porous, thick support membranes used for the preparation of the microporous crystalline inorganic membranes can be made from either polymeric or inorganic materials such as poly(amide-imide) [18], nylon [19], polyvinylidene fluoride (PVDF) [20], stainless steel [21], ceramic materials such as alumina [22], silica, titania, and glass [23]. The highly porous, thick support membranes can have different geometries such as flat sheet, hollow fiber, disk, tubular, or monolithic multiple channel.

Microporous crystalline inorganic membranes have the potential to overcome some of the challenges facing polymer membranes for a wide range of separation and catalysis applications because of their unique properties, such as uniform micropores, tunable functionalities, superior thermal and chemical stability, good erosion resistance, and high plasticization resistance. For example, SAPO-34 membrane has been synthesized and studied for high-temperature propane dehydrogenation membrane reactor application to significantly enhance the reaction conversion and selectivity by continuously separating H_2 during the reaction process [24]. NaA zeolite membranes have been used commercially for solvent dehydration applications [25].

The microporous crystalline inorganic membranes made from microporous crystalline inorganic materials provide separation properties mainly based on molecular sieving or competitive sorption mechanism. Competitive sorption–based separation takes place for large-pore (≥6 Å) microporous crystalline inorganic membranes when the pore size of the microporous crystalline inorganic material in the continuous, relatively thin, selective layer is much larger than the molecules to be separated. Molecular sieving will play the key role for small-pore (≤4 Å) and some medium-pore (~4–6 Å) microporous crystalline inorganic membranes when the pore size of the microporous crystalline inorganic material is similar or larger than the molecule to be separated in a mixture. Microporous crystalline inorganic membranes need to have a continuous, very thin (<100-nm), grain boundary-free, pinhole-free, and crack-free selective layer formed from a microporous crystalline inorganic material for molecular sieving-based separations with high permeance and high selectivity. However, some minor defects on the continuous, very thin (<100-nm), selective layer formed from the small-pore microporous crystalline inorganic material are acceptable for some applications such as the small-pore microporous crystalline NaA zeolite membranes for solvent dehydration applications via a pervaporation process. The separation for this process is based on preferential adsorption of water from the organic solvent/water mixtures rather than on a real size-exclusion molecular sieving mechanism due to the hydrophilicity of the high framework charge of the membrane; therefore, the membrane with a defective selective layer still can have very high water/solvent selectivity.

9.2.2.2 Preparation of Microporous Crystalline Inorganic Membranes

For potential large-scale industrial separations, such as gas separations and solvent dehydration, or catalysis applications, the microporous crystalline inorganic membranes need to be cost effective, have high permeance, high selectivity, high-pressure resistance for some applications, and high mechanical strength. Proper membrane fabrication methods result in the formation of microporous crystalline inorganic membranes with a continuous, very thin (<100 nm), grain boundary-free, pinhole-free, and crack-free selective layer of a microporous crystalline inorganic material on the surface of a highly porous, relatively smooth, high-surface-area, nonselective support membrane. The highly porous support membrane provides the membrane high enough mechanical strength. The continuous, very thin (<100 nm), grain boundary-free, pinhole-free, and crack-free selective layer, having only intracrystalline micropores and intercrystalline gaps with size smaller than the intracrystalline micropores, provides the membrane high selectivity.

Different from asymmetric integrally skinned polymeric membranes, microporous crystalline inorganic membranes cannot be prepared directly from microporous crystalline inorganic material to form asymmetric integrally skinned membrane structure. A general method for the preparation of the microporous crystalline inorganic membranes is to synthesize a continuous, relatively thin, defect-free, selective layer of microporous crystalline inorganic material on the surface of a highly porous, thick support membrane by an in situ hydrothermal synthesis method using either an amorphous zeolite precursor sol or gel solution or homogeneous MOF or ZIF precursor solutions. Similar to the preparation of TFC polymeric membranes, the highly porous, thick support membrane plays a critical role in the successful preparation of high-performance microporous crystalline inorganic membranes. To make low-cost, reproducible, high permeance, and high selectivity microporous crystalline inorganic membranes with high mechanical strength, the porous support membrane needs to have low cost, high surface area, high thermal and mechanical stability, high chemical and hydrothermal stability, significantly higher permeance than the final membrane (at least 10 times higher), uniform small pores on the surface, high porosity with interconnected pore structure, thermal expansion coefficient similar to that of the microporous crystalline inorganic materials, and high affinity and compatibility with the selective layer of the microporous crystalline inorganic material. Polymeric or inorganic hollow fiber and monolithic multiple channel support membranes are more preferred geometries for the preparation of microporous crystalline inorganic membranes to achieve high membrane area packing density; therefore, smaller membrane area with reduced membrane, engineering, and installation cost is required for large-scale membrane separation processes.

However, there have been numerous challenges for the preparation of cost-effective, reproducible, high permeance, and high selectivity microporous crystalline inorganic membranes using a cost-effective and environmentally benign method. For example, the use of expensive porous support membrane, highly basic, organic structure directing templates, and complicated synthesis procedure, as well as the formation of large amount of synthesis waste such as the large amount of water for membrane washing and poor reproducibility result in high-cost microporous crystalline inorganic membranes. The formation of a very thick selective layer of the microporous crystalline inorganic material, the low permeance from the porous support membrane, and the poor orientation of the microporous inorganic crystals provide the membrane with low permeance. The poor adhesion and the thermal expansion coefficient difference between the selective layer of the microporous crystalline inorganic material and the surface of the porous support membrane, pinholes, grain boundary defects, intra- and intercrystalline cracks result in the membrane with low selectivity.

Significant research efforts have been focused on the preparation of microporous crystalline inorganic membranes with enhanced separation performance from both academia and industry in the past two decades. High-quality NaA zeolite membranes have been commercialized for alcohol dehydration application via a microwave heating method, a secondary growth method, or a ball-milling-assisted secondary growth method using high surface area monolith, porous, ceramic support membranes [25]. Small-pore, high selectivity, microporous DD3R zeolitic membrane was prepared using tubular or high surface area honeycomb-shaped, porous, ceramic support membranes for CO_2/CH_4 gas separation application [26–28]. Small-pore, high selectivity, microporous CHA zeolitic membrane was prepared using tubular, porous, ceramic support membrane for CO_2/CH_4 and other separation applications [29]. Medium-pore, microporous MFI zeolitic membranes were synthesized on a porous, ceramic disk support membrane by microwave heating–assisted secondary growth method without using organic structure directing agents [30].

Several strategies have been evaluated to reduce the cost of the microporous crystalline inorganic membranes such as the use of a cheap, porous support membrane or the use of a low-cost and simple membrane fabrication method that can avoid the formation of extra microporous crystalline inorganic powders and the use of a large amount of water for membrane washing. As an example, low-cost porous polymeric support membranes, such as poly(amide-imide) [18], nylon [19], and PVDF [20], were investigated to replace the expensive inorganic ceramic support membrane to reduce the cost of the microporous crystalline inorganic membranes. Synthesis of the low-cost, continuous, very thin (<100 nm), grain boundary-free, pinhole-free, and crack-free selective layer of the microporous crystalline inorganic material using simplified synthesis procedures is another approach to reduce the total cost of the membrane. Structure-directing template-free synthesis [31], gel-free secondary growth [32], dry-gel synthesis [33], and counter diffusion in situ synthesis [34] have been reported for the preparation microporous crystalline zeolite membranes. Structure-directing template-free synthesis method not only reduces the cost of the membrane, but also allows the formation of a continuous, crack-free selective layer of the microporous crystalline inorganic material without high-temperature calcination step for template removal, which results in high membrane selectivity. Controlling the orientation of the zeolite crystals via oriented secondary growth [35–38], improving the adhesion between the selective layer and the surface of the porous support membrane via counter diffusion method [39], synthesizing grain boundary, pinhole, and crack-free continuous selective layer of the microporous crystalline inorganic material via rapid thermal processing [40, 41], or using high-aspect-ratio microporous crystalline inorganic material [42] has provided high selectivity microporous crystalline inorganic membranes. Controlling the orientation of the zeolite crystals via oriented secondary growth to form vertically oriented microporous zeolite channels with respect to the porous support membrane not only improves the membrane selectivity, but also increases the membrane permeance.

In addition, the approach of using ultra-thin microporous crystalline inorganic nanosheets to reduce the thickness of the continuous selective layer of the microporous crystalline inorganic material and therefore enhance the membrane permeance has attracted significant attention in most recent years. As an example, ultrathin MFI nanosheets of about 3–100 nm in thickness have been used for the preparation of thin MFI nanosheet membranes with a selective MFI layer of about 250 nm to 1 μm in thickness using different membrane fabrication methods such as vacuum filtration deposition of thin sub-100 nm MFI nanosheet suspension on porous inorganic or polymeric support membrane [43, 44], gel-free secondary growth using exfoliated MFI nanosheets of 3.2 nm in thickness as the seeds [32], or direct synthesis of MFI nanosheet layer [45]. As another example, ultra-thin, highly crystalline poly[Zn_2(benzimidazole)$_4$] (Zn_2(bim)$_4$) nanosheets of 1 nm in thickness and large lateral area have been used to fabricate Zn_2(bim)$_4$ nanosheet membrane with a selective Zn_2(bim)$_4$ nanosheet layer of several nm in thickness coated on α-alumina porous support membrane via a hot-drop coating process [46]. However, large-scale production of the cost-effective, reproducible microporous crystalline inorganic membranes with a continuous, thin, defect-free, selective layer of the microporous crystalline inorganic nanosheets is still very challenging. Synthesis of the microporous crystalline inorganic nanosheets or the multistep membrane preparation procedures need to be significantly simplified. The selective layer thickness of the microporous crystalline inorganic nanosheet layer still needs to be reduced to provide high enough membrane permeance. The formation of the nonselective defects on the continuous, thin, selective layer of the microporous crystalline inorganic nanosheets needs to be avoided to achieve high selectivity for the microporous crystalline inorganic nanosheet membranes. Very recently, ultra-thin ZIF-8 membrane with a continuous, very thin, selective ZIF-8 layer of about 17 nm in thickness deposited on low-cost, porous PVDF polymeric hollow-fiber support membrane was fabricated via a gel-vapor deposition method [20]. The membrane showed high permeances and high selectivities for gas separation applications, as well as scaling-up potential due to the use of polymeric hollow-fiber support membrane.

Microporous crystalline inorganic membranes with uniform micropores and tunable functionalities have the potential to achieve "perfect" molecular sieving-based separations featuring super high selectivity and permeance if a very thin (<100 nm), grain boundary-free, pinhole-free, and crack-free selective layer of microporous crystalline inorganic material is formed. Overall, significant advancement in improving the quality, stability, permeance, and selectivity, as well as reducing the cost of the microporous crystalline inorganic zeolite, MOF, and ZIF membranes for gas, vapor, and liquid separations has been made from both academia and industry. However, the commercialization of microporous crystalline inorganic membranes for large-scale industrial application has not been very successful except the successful development of commercial-scale A-type of zeolite membranes for organic solvent dehydration applications [47]. Further advancement in reducing the cost and selective layer defects and thickness, as well as improving the processability and reproducibility is still necessary for microporous crystalline inorganic membranes to replace polymeric membranes for some commercial applications and to be adapted for new commercial applications that are unsuitable for polymeric membranes.

9.2.3 Mixed-matrix Membranes

Both polymeric and inorganic membranes have their unique advantages and some limitations. Up to now, most of the large-scale commercial membranes for gas and liquid separations are polymeric membranes due to their relatively low cost, ease of processability, good mechanical stability, and reasonably good selectivity and permeance. Most of the polymeric membranes, however, have poor contaminant or fouling resistance, low chemical and thermal stability, and low selectivity for some applications. Inorganic membranes such as the microporous crystalline inorganic membranes overcome these limitations of polymeric membrane and have shown much higher selectivities than polymeric membranes for some gas separations and are encouraging for some high-temperature membrane applications. However, they are often much more expensive and more difficult to be fabricated than the commercial polymeric membranes with the current state-of-the-art membrane manufacturing processes, have low area packing density and poor mechanical stability. Breakthrough development of disruptive membranes with the advantages of both polymeric and inorganic membranes will create many new separation opportunities for membrane technologies.

9.2.3.1 Introduction to Mixed-matrix Membranes

MMMs are composite membranes comprising a continuous organic polymeric matrix, dispersed inorganic materials such as microporous zeolites, MOFs, ZIFs, carbon molecular sieves, silica or alumina particles, activated carbons, graphene oxides, and sometimes other liquid additives such as ionic liquids or salts. MMMs are developed to overcome the limitations of polymeric membranes and inorganic membranes and combine the advanced features of both membranes with the potential of being fabricated using polymeric membrane manufacturing process. MMMs have been studied for many potential applications such as gas, vapor, and liquid separations, as well as fuel cell, battery, and catalysis applications.

MMMs comprising dispersed microporous crystalline inorganic materials such as zeolite, MOF, or ZIF materials in a continuous polymeric matrix have been studied most extensively in academia and industry. The separation mechanism of microporous crystalline inorganic material/polymer MMMs is based on the combination of the solution-diffusion separation of polymeric membrane with the molecular sieving separation of microporous crystalline inorganic membrane. Therefore, MMMs have the potential to achieve improved selectivity and enhanced permeance compared to polymeric membranes when microporous crystalline inorganic materials with both higher selectivity and permeability than those of the polymeric membrane materials are incorporated into the MMMs. As illustrated in Figure 9.4, when using a MMM comprising high selectivity and high-permeability microporous crystalline inorganic materials dispersed in a continuous polymer matrix with inferior selectivity and permeability to separate a given mixture of gases, vapors, or liquids comprising a more permeable component A and a less permeable component B, the permeability of component A in the MMM (abbreviated as $P_{A,\,MMM}$) is a combination of the permeability of A in the polymer matrix (abbreviated as $P_{A,\,p}$) and the permeability of A in the dispersed microporous crystalline inorganic materials (abbreviated as $P_{A,\,i}$). $P_{A,\,MMM}$ is greater than $P_{A,\,p}$ because $P_{A,\,i}$ is greater than $P_{A,\,p}$. The less permeable component B has larger size than the micropore size of the dispersed microporous crystalline inorganic materials. Therefore, component B cannot permeate directly through the micropores of the microporous crystalline inorganic materials and needs to permeate through the polymer matrix via a much longer distance than the

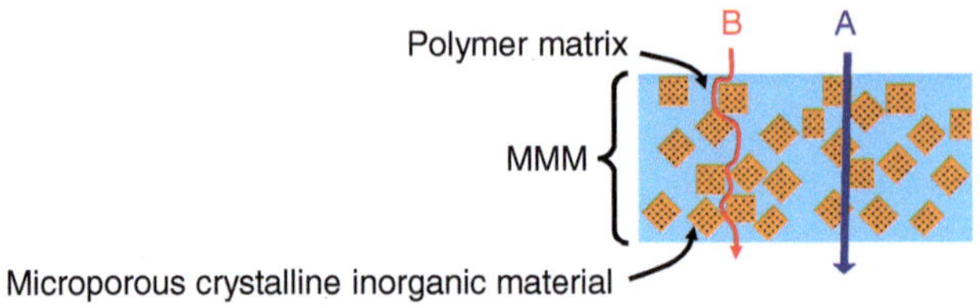

Figure 9.4 Schematic illustration of mixed-matrix dense-film membrane for separating a mixture of A and B.

polymeric material without dispersed microporous crystalline inorganic materials, resulting in lower permeability of component B in the MMM (abbreviated as $P_{B, MMM}$) than that in the polymer matrix $P_{B, p}$. As a result, the selectivity of the MMM for component A over component B (abbreviated as $\alpha_{A/B, MMM}$) is higher than that of the polymeric membrane made from the same polymeric matrix material (abbreviated as $\alpha_{A/B, p}$). Ideally, the improvement of $\alpha_{A/B, MMM}$ and $P_{A, MMM}$ is determined by the amount of high selectivity and high-permeability microporous crystalline inorganic materials dispersed in the continuous polymer matrix and can be predicted by using a Maxwell model [48]. Maxwell model has been used for the proper selection of the polymeric matrix material and the dispersed microporous crystalline inorganic materials with known separation properties for a desired separation.

Some MMMs with good separation property and chemical property match between the dispersed microporous crystalline inorganic materials and the continuous polymer matrix have exhibited separation properties superior to the pristine polymeric membranes. Some of the MMMs have shown gas separation performance well above Robeson's polymer upper-bound limit. As an example, UOP's membrane research group has studied AlPO-14/polymer MMMs for CO_2/CH_4 separation [49, 50]. As shown in Figure 9.5, the CO_2/CH_4 separation performances of AlPO-14/P-1, AlPO-14/P-2, AlPO-14/P-3 dense-film MMMs, comprising dispersed small-pore microporous AlPO-14 molecular sieve particles and P-1, P-2, and P-3 continuous glassy polymer matrices, respectively, are different due to the use of three polymers with different CO_2/CH_4 separation properties even though the same AlPO-14 dispersed particles are used. AlPO-14/P-1 MMM with certain AlPO-14 loading has shown performance exceeded Robeson polymer upper-bound limit for CO_2/CH_4 separation. Most recently, rare-earth face-centered cubic MOF/polyimide and ZIF-71/polyimide mixed-matrix dense films have been prepared for $CO_2/H_2S/CH_4$ and H_2/CO_2 separations, respectively [51, 52].

However, some of the MMMs reported in the literature have shown nonideal MMM performance due to the separation property mismatch or poor adhesion between the polymeric and inorganic phases, poor dispersion of the inorganic materials in the polymer matrix, or not small enough inorganic particles. Significant research studies have been focused on mixed-matrix dense-film membranes. Preparation of commercially viable asymmetric MMMs with integrally skinned or TFC structure has been lack of investigation.

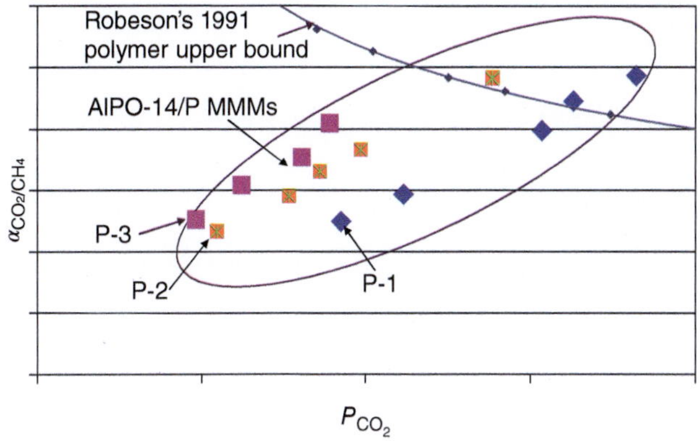

Figure 9.5 CO_2/CH_4 separation performance of AlPO-14/P-1, AlPO-14/P-2, AlPO-14/P-3 dense-film MMMs.

9.2.3.2 Preparation of Mixed-matrix Membranes

A variety of rubbery and glassy polymers have been studied as the continuous polymer matrix and many different microporous crystalline inorganic materials have been used as the dispersed phase for the preparation of MMMs. MMMs can have different structures such as mixed-matrix dense films, asymmetric integrally skinned MMMs, and TFC MMMs. Asymmetric integrally skinned and TFC MMMs can have either flat-sheet or hollow-fiber geometry.

As with polymeric dense films, mixed-matrix dense film has a symmetric structure with a thickness of about 50–100 μm. Mixed-matrix dense films are not commercially attractive for most of the applications, but are useful for measuring the intrinsic separation properties such as selectivity, permeability, and conductivity of the membrane. Most of the mixed-matrix dense films are prepared from a mixed-matrix casting dope comprising inorganic particles, polymer, and organic solvents via a solution casting method. Preparation of mixed-matrix dense films with a thickness of about 50–100 μm is much easier than those of asymmetric integrally skinned MMMs and TFC MMMs. Different membrane preparation approaches have been reported to solve the issues of agglomeration of the inorganic particles in the polymer matrix and poor compatibility and adhesion at the interface of the inorganic particles and the polymer matrix, which will result in MMMs with poor mechanical and processing properties and poor separation performance. As an example, surface modification of the inorganic particles to improve the interfacial adhesion between the two different materials has been studied to improve the separation performance of mixed-matrix dense films [53–55].

Mixed-matrix dense films need to be converted to either asymmetric integrally skinned MMMs or TFC MMMs for large-scale gas, vapor, or liquid separations. Same as asymmetric polymeric membranes, asymmetric integrally skinned MMMs and TFC MMMs have a thin selective layer on a highly porous nonselective support layer as shown in Figure 9.6. Most of the fabrication methods for asymmetric MMMs are similar to those for asymmetric polymeric membranes. Both the thin selective layer and the highly porous nonselective support layer are formed from the

Figure 9.6 Schematic illustration of asymmetric mixed-matrix membrane.

same mixed-matrix material comprising dispersed inorganic particles and the continuous polymer matrix for the asymmetric integrally skinned MMMs (Figure 9.7). TFC MMMs comprise a thin mixed-matrix selective layer on a highly porous nonselective support layer, which can be formed from a polymeric or inorganic material different from the mixed-matrix material in the selective layer. For asymmetric MMMs to achieve improved selectivity and permeance compared to the asymmetric polymeric membranes, the formation of a thin ($<\sim$100 nm), defect-free mixed-matrix selective layer is critical.

Most of the asymmetric MMMs that have been reported so far are asymmetric integrally skinned MMMs with either flat-sheet or hollow-fiber geometry prepared from mixed-matrix casting or spinning dopes via the phase-inversion process [56–59]. The compositions of the mixed-matrix casting or spinning dopes are different from those for mixed-matrix dense films and normally comprise several different organic solvents and nonsolvents as well as some additives to control the mixed-matrix selective layer thickness and the morphology of the MMMs. Preparation of a stable, uniform mixed-matrix casting or spinning dope with the inorganic particles uniformly dispersed in the polymer solution without particle agglomeration is critical for the formation of a defect-free mixed-matrix selective layer, which will provide the MMM with enhanced selectivity. Some of the asymmetric MMMs are TFC MMMs prepared

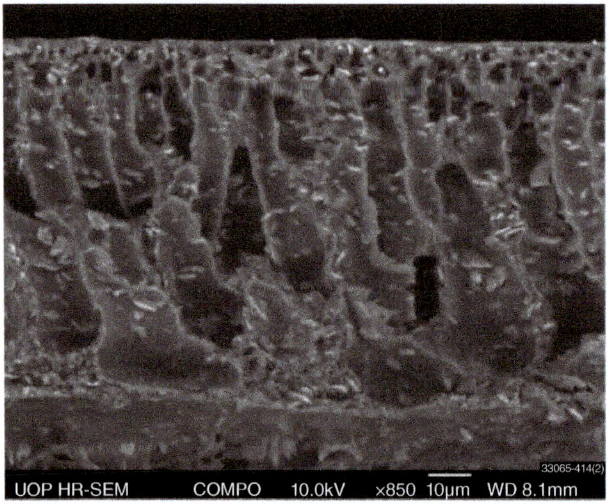

Figure 9.7 Cross-sectional HRSEM image of an asymmetric mixed-matrix flat-sheet membrane. (Photo provided courtesy of Honeywell UOP with a current copyright.)

by dip coating a mixed-matrix solution [60] or interfacial polymerization [61] to form a thin mixed-matrix selective layer on top of a highly porous support membrane.

Different from thick mixed-matrix dense films that can be prepared relatively easily and normally exhibit improved selectivity and permeability compared to the pristine polymeric dense films, almost none of the asymmetric MMMs have shown significantly enhanced selectivity and permeance or have performance reproducibility issue compared to the pristine asymmetric polymeric membranes. One difficulty is the synthesis of nanosized inorganic particles for the preparation of asymmetric MMMs. To make asymmetric MMMs with much better separation performance than asymmetric polymeric membranes, the thickness of the selective mixed-matrix layer needs to be comparable to that of the selective polymeric layer, which normally is ~100 nm or less. The use of large inorganic particles with the particle size in several micrometers range for the preparation of asymmetric MMMs with a selective layer of ~100 nm in thickness results in poor adhesion between the inorganic particles and the polymer matrix and poor mechanical strength that can easily form defects. In addition, the improper selection of the dispersed inorganic material and the continuous polymer material, which results in permeability and selectivity mismatch between the two materials, as well as the formation of interface defects during the phase-inversion process also make defective asymmetric MMMs lack of performance enhancement.

In summary, although MMMs have been extensively studied and significantly advanced for over 30 years, very few MMMs have been commercialized for large-scale industrial applications. Invention of new MMM fabrication techniques to improve the dispersion of the inorganic particles in the polymer matrix, to enhance the interaction, compatibility, and adhesion between the dispersed inorganic materials and the polymer matrix, as well as the discovery of new synthesis methods for inorganic materials particularly the microporous crystalline inorganic material to produce nanosized inorganic particles without agglomeration, or the discovery of new MMM fabrication methods that either allow the use of large inorganic particles or form the inorganic materials in situ during the membrane fabrication process will promote the commercialization of MMMs to advance the membrane-based technologies. As an example, a flexible ZIF-8/PES MMM has been prepared most recently using a new preparation method that allows the formation of the microporous ZIF-8 material after the formation of the asymmetric ZnO/PES composite membrane [62]. The growth of the ZIF-8 particles of <5 μm in diameter is directed by the ZnO seed nanoparticles and restricted to the PES pores generated during the phase-inversion process. The thickness of the effective mixed-matrix selective layer and the defects in the selective layer, however, still need to be reduced to make the membrane commercially viable.

9.3 Membrane Modules

Large-scale industrial membrane separation processes require large membrane areas up to about 240 000 m^2 to meet the unit treating capacity and product recovery.

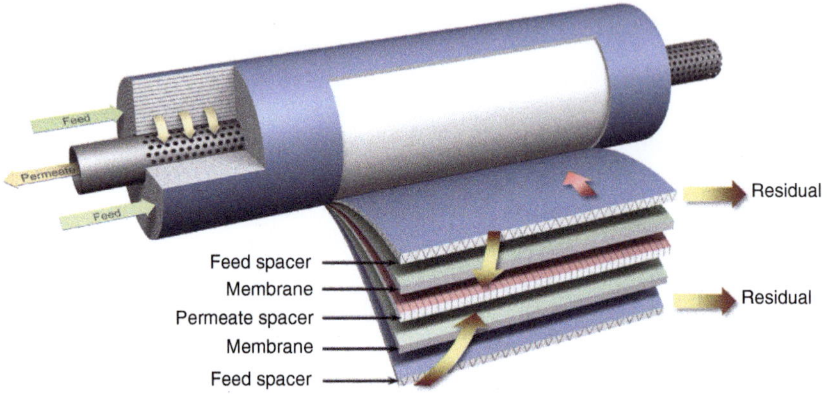

Figure 9.8 Diagram of spiral wound membrane module.

These large membrane areas need to be packaged in cost-effective module configuration with high area packing density. The most common module configurations for polymeric membranes and potentially for MMMs are spiral wound, plate and frame, and hollow fibers. Asymmetric flat-sheet membranes can be incorporated into spiral wound or plate and frame modules and asymmetric hollow fibers can be formed into hollow-fiber modules. The common module configurations for inorganic membranes include tube, hollow fiber, and multichannel monolith.

The areas of the asymmetric polymeric membranes in each spiral wound module with a module diameter of 0.05–0.5 m can be in a range of 10–40 m^2. Spiral wound membrane modules include membrane leaves, plastic sheets, a permeate tube, and glue to hold the leaves together and each membrane leaf includes the asymmetric flat-sheet membrane, permeate spacer, and feed spacer (Figure 9.8). The performance of the spiral wound module is not only determined by the advanced performance of the asymmetric flat-sheet membrane, but also the unique design of feed and permeate spacer that can provide low-pressure drop, as well as the design of the glue that is used to hold the membrane leaves together and prevent module leaks. Spiral wound modules with the features of cross-flow, high-pressure tolerance, high fouling resistance, high reliability, and ease of installation into space-efficient, skid-mounted units have been used for large-scale RO, NF, UF, and gas separation applications.

Asymmetric polymeric hollow-fiber membranes are freestanding membranes and typically have an outside diameter of 100–1000 μm and a wall thickness of 50–200 μm. The areas of the asymmetric polymeric hollow-fiber membranes in each hollow-fiber module can be in a range of 100–300 m^2. In addition, hollow-fiber modules include hollow-fiber membranes and glue to hold the fibers together to prevent module leaks and no feed spacer, permeate spacer, and permeate tube are needed. Therefore, hollow-fiber module with the same diameter and length as those of the spiral wound module can have much larger membrane areas than those in the spiral wound module. Compared to spiral wound module, hollow-fiber module has higher membrane area packing density, but has lower feed-pressure tolerance, higher pressure drop, and more difficulty to repair defects. Hollow-fiber modules have been used commercially for RO, NF, UF, MF, and gas separations. There are two types of hollow-fiber modules for these applications, including shell-side feed modules and bore-side feed modules. Shell-side feed modules are generally used for high-pressure applications up to about 1500 psig feed pressure such as for CO_2/CH_4 separation and H_2 purification with counter-current flow or cross-flow. Fouling on the feed side of the membrane can be a problem with this kind of module design, and pretreatment of the feed stream to remove particulates is required. Bore-side feed modules are generally used for low- and medium-pressure feed streams up to about 200 psig with counter-current flow, such as separation of N_2 or water from air, where good flow control to minimize fouling and concentration polarization on the feed side of the membrane is desired.

Multichannel monolith module is the most preferred module configuration for high-cost inorganic membranes such as the commercial ceramic membranes and NaA zeolite membranes to achieve high membrane area packing density and reduce membrane system cost for large-scale membrane separation processes with high feed treating capacity and high product purity and recovery. The areas of the inorganic membrane in each multichannel monolith module are about 10–15 m^2. Several multichannel monolith zeolite membrane modules have been reported in the literature [63, 64]. NaA zeolite membranes with multichannel monolith module design have been used commercially in isopropanol and bioethanol dehydration plants in China and India [25, 65].

9.4 Membrane Reactors

Membranes have been used for a wide range of gas, vapor, and liquid separations. Membrane separation also has been integrated with a chemical conversion or reaction in two-unit operations and membrane separation process is applied either before or after the reaction process. Membrane reactor concept of combining a chemical conversion process with a membrane separation process in one-unit operation was originally conceived by Alan Michaels 50 years ago [66]. The membrane can have different functions in the membrane reactors, including retention of the catalyst, selective or nonselective addition of the reactants, selective or nonselective removal of the reaction products, or acting as part of the catalyst or catalyst support.

The advantages of the membrane reactor over the reaction/membrane separation two-unit operations are integration of reaction and partial separation into one-unit operation that lowers the Capital expenditure (CapEX), lower energy consumption that lowers the Operating expenses (OPEX), enhancement of thermodynamically limited or product-inhibited reactions that increases product yield or reaction rate, decreases side reactions, or lowers the operating temperature and pressure to avoid coking and reduce catalyst fouling, controlled reaction rates for selective reactions that increases selectivity, as well as mediation and control of the contact between two reactants. The industrial applications of membrane reactors, however, are still facing some challenges due to the technical complexity of the process, difficulty in optimization and design of a suitable reactor, poor sealing of the membrane reactor chamber, high cost of the membrane and membrane module, limited thermal, chemical, and mechanical resistance of the membrane, poisoning or fouling of the membrane, as well as low permeance of the membrane.

Both polymeric membranes and inorganic membranes such as metallic and ceramic membranes have been studied for membrane reactor applications. Despite the advantages of polymeric membrane reactors such as lower cost, commercially mature membrane module design and fabrication, and ease of incorporating homogeneous or heterogeneous catalyst as compared to inorganic membrane reactors, they are only suitable for reactions at relatively low temperatures of less than about 100 °C due to the limited thermal resistance of the polymeric membranes. Some representative reactions for polymeric membrane reactors are selective hydrogenation [67, 68], hydrogenation [69], esterification [70], oxidation [71], and alkylation [72]. Inorganic membrane reactors are suitable for a wide range of high-temperature reactions due to their high thermal stability. Some representative reactions for inorganic membrane reactors are dehydrogenation such as ethane dehydrogenation using Pt-impregnated alumina membrane [73], propane dehydrogenation using SAPO-34 zeolite membrane reactor [24], oxidative coupling of methane using either porous [74] or dense membrane [75], oxidative dehydrogenation [76], hydrogenation [77], steam reforming and methane oxidation [78]. Among these reactions, dehydrogenation reactions for olefin production and steam reforming and methane oxidation for syngas production have been most extensively studied.

For syngas production in a membrane reactor, a mixed ion–conducting membrane based on a dense ceramic inorganic membrane with perovskite structure was used for a four-step O_2 separation process. The first step is the diffusion of O_2 to the surface of the feed side of the membrane. In the second step, O_2 dissociates to oxygen ions and consumes electrons on the surface of the membrane. The third step is the diffusion of the oxygen ions from the feed side to the permeate side of the membrane and electrons moving in the opposite direction. In the last step, oxygen ions recombine or react with CH_4 gas to produce syngas in the presence of a reforming catalyst on the permeate side of the membrane. The driving force for oxygen transport is the chemical activity or oxygen partial-pressure gradient from the membrane feed side to permeate side. The commercialization of this process has not been successful due to several technical challenges such as cost, membrane fabrication and performance, process integration, and membrane reactor scale-up.

For the catalytic dehydrogenation of propane to propylene, UOP has a commercial Oleflex™ process with four reactors used in series. Selective removal of part of the H_2 from the reaction mixture using inorganic ceramic membrane modules has been studied [79]. Some of the requirements for the inorganic ceramic membrane to be commercially viable for this application are low cost, high-temperature stability up to 650 °C, high H_2 over propane and propylene selectivity, and high H_2 permeance at very low pressure of about 10–25 psig.

Despite the potential advantages of membrane reactors as compared to conventional reaction processes, significant advancements in the high temperature and long-term stability, new membrane module design that can provide high membrane area packing density and good sealing at the operating temperature, as well as large-scale membrane module fabrication feasibility are still required for membrane reactors to find their large-scale industrial applications in the petrochemical industry that requires high-temperature reaction processes.

9.5 Commercial Applications of Membranes

Membrane technologies have been used for a wide range of large-scale industrial gas, vapor, liquid, and ion separation processes such as RO, natural gas upgrading, hydrogen recovery, organic vapor/gas and vapor/vapor separations, pervaporation, NF, organic solvent NF, UF, MF, membrane contactor, electrodialysis, fuel cell, and other electrochemical processing. Innovation in membrane materials, equipment, and process technology integration has been instrumental in enabling the widespread use of membranes in the chemical industry. Membrane technologies have been widely adopted in semiconductor, automotive, water, food, pharmaceutical, biotechnology industries, and a wide range of environmental applications. Membrane technology alone or its combination with other separation technologies in integrated systems has been a successful approach for seawater desalination to meet the freshwater demand in many regions of the world at lower cost and minimum environmental impact. Worldwide sales of membrane products and systems have seen a significant growth and increase in applications in the last two decades.

Reverse osmosis (RO): RO membrane technology was the first industrial-scale commercial membrane technology and has been successfully used since 1960s for brackish and seawater desalination and dairy processing. RO is one of the fastest-growing membrane separation technologies. RO is a membrane separation process based on solution-diffusion separation mechanism in which properly pretreated seawater or brackish water is separated by the RO membrane at moderate (200–400 psig for low-salt brackish water) to high (800–1000 psig for high-salt seawater) pressures. The RO membrane rejects most solute ions and molecules with high rejection rate, while allowing water to pass through with high water flux to produce high-purity water up to 99.9% purity. Different from gas separation and pervaporation membranes that have a thin, nonporous, selective skin layer, RO membranes have a thin, relatively porous, selective skin layer with $\leq$1 nm small pores. The first-generation RO membrane was CA asymmetric integrally skinned membrane prepared from low-cost CA polymer based on Loeb-Sourirajan phase-inversion process and the membrane was packaged in spiral wound module for desalination process. The second-generation RO membrane was cross-linked polyamide (PA) TFC flat-sheet membrane in spiral wound modules and the thin selective layer of PA was formed by interfacial polymerization on the surface of a highly porous support membrane. CA and aromatic PA hollow-fiber RO membranes have also been developed in later years. CA RO membranes have much higher chlorine and fouling tolerance than PA RO membranes, but have lower solvent resistance and narrower pH range for operation. The PA polymer on the polyamide TFC membrane has been modified to further reduce the energy consumption and improve membrane stability and fouling resistance. Several advanced RO membranes have been commercialized by Dow FILMTEC and Nitto Denko.

Gas separations: Gas separation membrane was first commercialized for large-scale industrial hydrogen separation process by Permea (now part of Air Products) in 1980 and PRISM silicone rubber–coated polysulfone hollow-fiber membrane was used in this process [80]. PRISM membrane has also been used for air separation. Dried CA/CTA spiral wound membrane technology developed by Separex (now part of Honeywell UOP) was first used commercially for acid gas removal from natural gas in a small membrane unit in 1981. Cynara (now part of Schlumberger) introduced the first dried CTA hollow-fiber membrane system also for acid gas removal from natural gas in early 1980s. Generon produced the first N_2/air membrane separation system in 1982. Since then, gas separation membranes have been used for a wide variety of applications and N_2/air separation is the largest application for gas separation membranes. Ube Industries commercialized polyimide hollow-fiber membranes for nitrogen separation from air, H_2 purification, and He recovery. Air Liquide MEDAL commercialized polyimide and other hollow-fiber membranes for air separation, H_2 purification, and acid gas removal from natural gas. Membrane Technology and Research, Inc. (MTR) has several commercial TFC spiral wound membranes for vapor/gas separation, N_2/CH_4, and other applications. Typical polymers for the development of commercial gas separation membranes include glassy and rubbery polymers such as CA, CTA, polysulfone, polyimide, polyimide-aramide, polycarbonates, polyphenylene oxide, and silicone rubber.

Large-scale H_2 is produced by steam methane reforming. Refineries are large-volume producers and consumers of H_2. The demand for H_2 recovery in refinery for existing and new hydrotreating and hydrocracking units has grown dramatically. Refinery gases and various other hydrocarbon streams normally contain H_2, light hydrocarbons (C_1–C_5), and light heteroatom-containing gases. Improving the recovery of H_2 from refinery gases could significantly improve the economics of the H_2-valued refinery processes. The first large scale membrane system for H_2 recovery is to recover H_2 from ammonia purge stream using Prism™ polysulfone hollow-fiber membrane. Since then, membrane technology has been extended to other H_2 recovery applications such as adjustment of H_2/CH_4 or H_2/CO ratios for syngas production. UOP supplies Polysep™ hollow-fiber membrane systems for H_2 separation from a variety of feed sources. H_2 recovery from refinery streams to meet the increased demand of H_2 is an emerging field for gas separation membranes for hydrotreating, hydroprocessing, or hydrodesulfurization processes.

Natural gas is expected to become the second-largest source of energy in the world by 2035. With growing demand, the industry needs innovative solutions to efficiently process gas reserves and capture valuable natural gas liquids. CO_2 and H_2S acid gases in raw natural gas, in combination with water, are highly corrosive and can rapidly destroy natural gas pipelines and equipment. In addition, H_2S is toxic and at relatively modest levels can be life threatening. CO_2 reduces the heating value of natural gas stream and wastes pipeline capacity. Amine absorption is the widely used technology for CO_2 and H_2S removal from natural gas. Amine plants, however, suffer from problems such as high capital cost, complicated operation, and expensive maintenance. Membrane separation is an alternative technology for this natural gas upgrading application. Both spiral wound and hollow-fiber polymeric membranes have been adopted commercially for this separation. Some subquality natural gas reserves have high N_2 content, which needs to be removed to meet natural gas pipeline specifications. Cryogenic distillation and pressure swing adsorption (PSA) technologies are currently used for this separation. MTR has commercialized NitroSep™ membrane, which is more permeable to methane, ethane, and other hydrocarbons than to N_2, for separating N_2 from natural gas [7].

UOP's Separex membrane technology has been an important enabler for effective and efficient separation of contaminants such as CO_2 and H_2S from natural gas for onshore and offshore applications. Separex membrane technology was first commercialized in small membrane units of <10 MMSCFD (million standard cubic feet per day) treating capacities in early 1980s and it was used when the remote locations of natural gas reserves required a simple and reliable separation technology. Separex membrane systems became larger and larger with the gas-treating capacity increased to 1000 MMSCFD in 2009 (Figure 9.9) and have been extended from onshore applications to offshore platforms and most recently offshore FPSO vessels that require lower-weight and smaller-footprint membrane systems (Table 9.1). Figure 9.10 shows a single-lift Separex module that offers small footprint and low weight for FPSO vessels. The simplest Separex membrane-processing scheme is a one-stage flow scheme. For the one-stage flow scheme (Figure 9.11), a natural gas feed is separated into a hydrocarbon-rich, low CO_2 and H_2S residual product stream and a CO_2 and H_2S-rich permeate stream. High hydrocarbon recovery (>95%) can be achieved using a two-stage (Figure 9.12) or multistage flow scheme where the low-pressure, first-stage permeate is compressed and processed in a second-stage membrane. For example, a two-stage Separex membrane system was used for a large offshore platform in Malaysia (Figure 9.13) to process 680 MMSCFD of natural gas. The system reduced CO_2 from 44 to 8 mol%. Another Separex membrane-processing scheme is a two-step flow scheme (Figure 9.14) where the residue from the first membrane is sent to the second membrane to further remove CO_2 and H_2S and the permeate from the second membrane is compressed and sent back to the first membrane feed. The two-step membrane system has better hydrocarbon recovery than that of the one-stage system, but is worse than that of the two-stage system.

Separex commercial membrane systems include not only the membrane unit, but also the membrane pretreatment unit. Proper membrane pretreatment design is critical to the performance of the membrane systems. Many of Separex membrane pretreatment systems include UOP MemGuard™ pretreatment system (Figure 9.15), which is a regenerative system that can provide an absolute cutoff of heavy hydrocarbons and can remove water, mercury, and other contaminants along with the heavy hydrocarbons. MemGuard pretreatment systems have been used for both onshore and offshore membrane systems. The invention of new

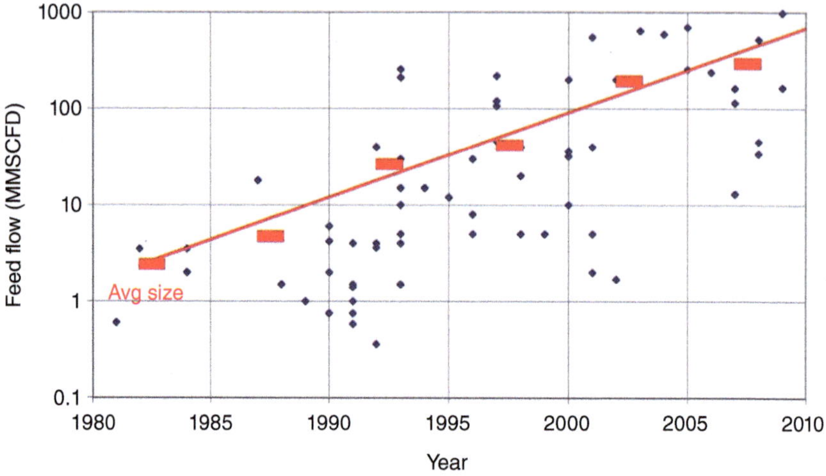

Figure 9.9 Separex commercial membrane systems for natural gas upgrading with significantly increased gas treating capacities from 1981 to 2009.

Table 9.1 Separex commercial membrane systems for offshore CO_2 removal from natural gas applications.

Location	Treating capacity (MMSCFD)	CO_2 removal (%)	Stages
Thailand	32	50 to 20	1
Thailand	480	35 to 23	1
Thailand	580	40 to 23	1
Malaysia	680	44 to 8	2
FPSO Thailand	11	28 to 3	1
FPSO Brazil (14 units)	176–240	10–40 to 3–5	1–2

membrane materials and new membrane fabrication methods, as well as the advances in membrane fabrication and engineering capabilities have enabled significant improvement in the performance of Separex membranes, providing reduced size and weight of the membrane system, as well as significantly increased hydrocarbon recovery.

Vapor/gas and vapor/vapor separations: Extending gas separation membrane technology to large-volume organic vapor/gas and vapor/vapor separations such as olefin recovery from polyolefin manufacturing plant and olefin/paraffin separations requires membranes with not only high selectivity and permeance, but also high contaminant resistance and hydrocarbon plasticization resistance, as well as long-term performance stability in the presence of organic vapors. Most of the gas separation membranes are not suitable for these vapor/gas and vapor/vapor separations. As with membranes for gas separations, membranes for vapor/gas separations with a thin, nonporous, dense, selective layer on the surface of a

Figure 9.10 Single-lift Separex module for FPSO applications. (Photo provided courtesy of Honeywell UOP, with a current copyright.)

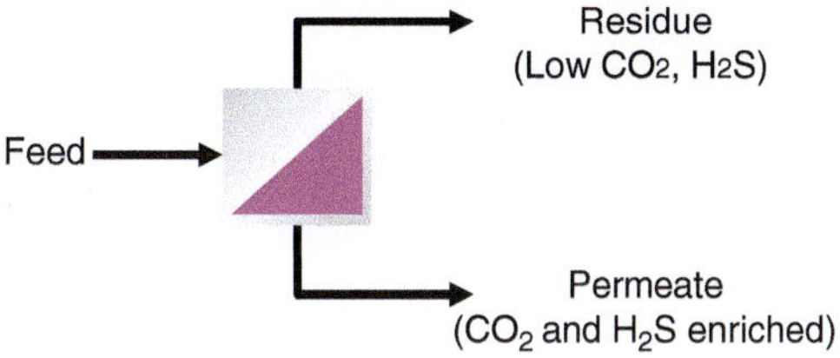

Figure 9.11 One-stage membrane flow scheme.

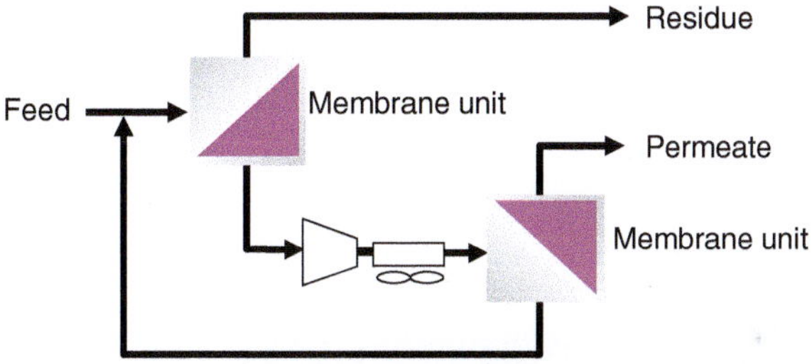

Figure 9.12 Two-stage membrane flow scheme.

Figure 9.13 Separex two-stage membrane system in Malaysia. (Photo provided courtesy of Honeywell UOP, with a current copyright.)

highly porous support layer are also based on solution-diffusion separation mechanism. Rubbery polymers such as silicone rubber with high solubility selectivity for vapor over N_2 or other gases have been used for the development of commercial membranes for vapor/gas separations. Vapor/gas separation membranes are commercially available from MTR, Dalian Eurofilm Industrial Ltd., and BORSIG Membrane Technology GmbH.

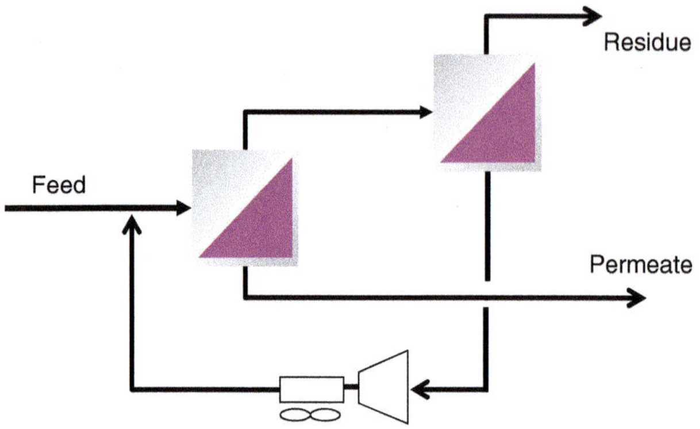

Figure 9.14 Two-step membrane flow scheme.

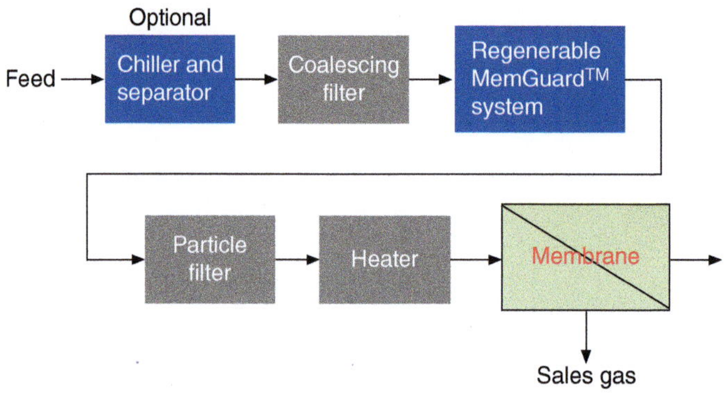

Figure 9.15 Separex membrane system, including MemGuard system and membrane unit.

There is still no commercially available membrane for vapor/vapor separations such as olefin/paraffin and paraffin isomer separations for petrochemical and refining industries. Expensive and energy-intensive cryogenic distillation separation technology has been used for many years for the separation of olefin and paraffin components. Large capital expense and high energy cost have created incentives for extensive research in developing low energy-intensive membrane technology for this separation. Facilitated transport membranes (FTMs), polymeric membranes, MMMs, and inorganic membranes have been studied in academia and industry. FTMs, comprising metal ions as the complexing agent for olefins, have very high olefin/paraffin separation selectivity and high potential for commercialization. FTMs for olefin/paraffin separation are based on facilitated transport mechanism. Olefins form reversible metal cation complexes such as silver(I) complexes with the metal cations via the pi bonds on the surface of the high-pressure feed side of the FTM, whereas no interaction occurs between the metal cations and the paraffins (Figure 9.16). The reversible metal cation/olefin complexes will transport through the selective layer of the FTM via complex association/dissociation in the presence of water vapor and finally the olefins will be dissociated from the reversible metal cation/olefin complexes and released at

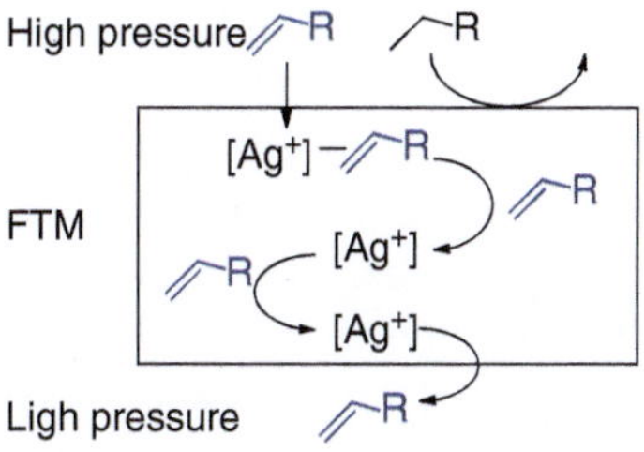

Figure 9.16 Facilitated transport mechanism of facilitated transport membrane (FTM) for olefin/paraffin separation.

the low-pressure permeate side of the FTM. FTMs have been prepared in lab scale and some of them have been evaluated in demo unit for propylene/propane separation [81–83]. Membrane performance stability needs to be further improved and the issue of complexing agent poisoning by contaminants needs to be solved for commercial applications of FTMs for olefin/paraffin separations.

Pervaporation: Pervaporation is a membrane separation process to separate liquid mixtures that combines evaporation of the volatile liquid component with permeation of the vapor-phase component through a pervaporation membrane. Vacuum is normally applied to the permeate side of the membrane to facilitate the separation and recover the permeate. Both polymeric and inorganic membranes have been used for commercial-scale pervaporation processes to separate azeotropic mixtures such as alcohol and organic solvent dehydration, but sales of pervaporation systems are much lower than other membrane separation processes. Pervaporation using polymeric membranes is based on solution-diffusion separation mechanism same as that for polymeric gas separation membranes. As with polymeric gas separation membranes, polymeric pervaporation membranes prepared from polymers such as polyvinyl alcohol or silicone rubber have a thin, nonporous, selective skin layer on the surface of a highly porous support layer. Pervaporation using small-pore microporous inorganic zeolite membranes such as NaA membrane is based on preferential adsorption and/or molecular sieving separation mechanism and the membranes have very high selectivity. Most of the pervaporation membranes are commercially available from GKSS (polymeric membrane), MTR (polymeric membrane), Jiangsu Nine Heaven (NaA), Mitsui-BNRI ZeoSepA (NaA), Mitsubishi Konker™ (NaA), and Mitsubishi ZEBREX™ (high Si/Al ratio CHA).

Nanofiltration (NF): NF membrane is a loose RO membrane with an average pore size of 1–2 nm in the thin selective skin layer and has been widely used for water softening and treatment of high color and high organic content feed water. NF separation is based on sorption-sieving separation mechanism. NF is superior to both RO and UF for the treatment of bleaching effluents from pulp and paper manufacturing plants due to its higher water flux than RO and higher rejection of bleaching agent than UF. NF membrane can be operated at lower pressure than RO membrane and has high rejection for divalent ions (>95%) and organics, but has lower rejection for NaCl (10–50%) than RO membrane. Most of the commercially available NF membranes are PA TFC membranes offered by Dow FILMTEC, hydranautics, and several other companies. OSNF membranes are also loose RO type of membranes with an average pore size of 1–2 nm in the thin selective skin layer and have high resistance to organic solvents. The most common polymers for OSNF membranes are polyimides. OSNF-based MAX-DEMAX process developed by W. R. Grace and ExxonMobil has been used commercially for crude oil dewaxing. OSNF is still an emerging membrane technology and there have been very limited large-scale commercial sales for OSNF systems.

Ultrafiltration (UF): UF membrane is a porous membrane with an average pore size of 1–100 nm in the surface skin layer and has been widely used for surface water treatment, RO pretreatment, colloidal silica and iron removal from water, color reduction, oil/water separation, milk processing, fruit juice clarification, automobile paint recovery, biomolecule separation, as well as therapeutic proteins, DNA, and RNA purification and filtration. UF membrane separation is based on not only the pore size of the membrane, but also the shape and size of the molecules to be rejected, as well as the solute/membrane interactions. The most common UF membranes are prepared from CA, PSf, PES, polyacrylonitrile (PAN), PVDF, PEI, or ceramic inorganic materials. Most recently, several approaches have been studied to enhance the flux, reduce fouling, and improve solvent resistance of UF membranes such as polymer chain modification, use of solvent-resistant polymers such as PEEK and PBI, addition of hydrophilic polymer, blending of two different polymers, modification of membrane surface, and coating of ceramic membrane with silica or zirconia.

Microfiltration (MF): MF membrane is another porous membrane with an average pore size of 0.1–10 μm in the surface skin layer and has been widely used for bioengineering to remove microorganisms from the fermentation products or remove yeast from alcoholic beverages, removing particulates, colloidal materials, and complexed heavy metals from liquid suspensions, and cell harvesting. MF membrane separation is more complicated than simple sieving due to the pore size reduction by the particles or macromolecules to be separated. The most common MF membranes are prepared from CTA, PSf, PVDF, PP, polyethylene (PE), polycarbonate, polyester, PI, polytetrafluoroethylene (PTFE), nylon 6, nylon 6,6, or ceramic inorganic materials.

UF and MF membranes are commercially available from a significant number of companies such as Hydranautics, Suez, Hyflux, Pall, Tianjin MOTIMO Membrane Technology, Litree Purifying Technology, Beijing Scinor Membrane Technology, Beijing OriginWater Technology, Mitsubishi, and Jiangsu Jiuwu Hi-Tech.

Membrane contactor: Membrane contactor is a membrane-based device that is used to achieve gas/liquid mass transfer by allowing a gaseous phase and a liquid phase to come into direct contact with each other without dispersion of one phase within another. In a membrane contactor, the membrane separation is integrated with a conventional phase contacting operation such as absorption or extraction to exploit the benefits of both technologies. The key features for membrane contactors are nonsensitivity to flooding, channeling, or back-mixing, modular design, much higher surface area per unit volume as compared to conventional columns or reactors, and use of vacuum and inert gas to create the driving force for mass transfer. Common membranes used for membrane contactors are UF or MF hollow-fiber membranes prepared from hydrophobic PP, PVDF, or PTFE polymers. Membrane contactors have been commercialized for the removal of gases from water such as stripping dissolved oxygen from ultra-high-purity water for semiconductor industry, CO_2 removal in beer production, carbonation of beverage, and blood oxygenation.

Ion-exchange membrane applications: Ion-exchange membranes (IEMs) are electrically conductive, semipermeable membranes that are permeable to either positively or negatively charged ions. IEMs are made of hydrophobic organic polymeric or inorganic ceramic materials with charged ionic side groups and movable counterions. IEMs can be classified into cation exchange membranes with negatively charged fixed ions such as sulfonic or carboxyl groups and anion exchange membranes with positively charged fixed ions such as quaternary ammonium groups. A most recent review discussed about the materials, preparation, and applications of IEM in detail [84]. IEMs have been used in electrodialysis, diffusion dialysis, continuous deionization, or bipolar membrane electrodialysis in commercial scale for seawater desalination, industrial wastewater treatment of highly scaling waters, food and beverage production, and other industrial wastewaters. IEMs have also been studied for energy storage and transformation device applications such as reverse electrodialysis, fuel cell, and redox flow batteries. Some representative companies that offer commercial IEMs for fuel cell or redox flow battery applications are Asahi Glass (Flemion), Asahi Chemical Industry, Tokoyana Soda (NEOSEPTA-F), Kashima-Kita Electric Power, Sumitomo, and Evonik.

Since the discovery of the phase-inversion process to make asymmetric integrally skinned polymeric membrane by Loeb-Sourirajan, membrane technology has been an important enabler for effective and efficient separations and is widely used in gas, water, and chemical industries. Innovation in membrane materials with long-term stability, chemical resistance, and specific to applications, fabrication methods, module design, new process design, and process technology integration has been instrumental to the widespread use of membrane technology, and it will continue to enable membrane growth for new applications such as low-energy intensive olefin/paraffin separations, aromatic/nonaromatic separations, aromatic isomers separations, and paraffin isomer separations.

References

1 (2012, January). Honeywell UOP technology is used to clean natural gas on FPSO vessels. *Membrane Technology*: 5.

2 Loeb, S. and Sourirajan, S. (1964). High flow porous membranes for separating water from saline solutions. US Patent 3,133,132. Issued: 12 May 1964.

3 Cadotte, J.E. (1981). Interfacially synthesized reverse osmosis membrane. US Patent 4,277,344. Issued: 7 July 1981.

4 Tang, M., King, W.M., Wensley, C.G. (1989). Air dried cellulose acetate membranes. US Patent 4,855,048. Issued: 8 August 1989.

5 Chiou, J. (2002). Epoxysilicone coated membranes. US Patent 6368382. Issued: 9 April 2002.

6 Henis, J.M.S. (1994). *Polymeric Gas Separation Membranes* (eds. D.R. Paul and Y.P. Yampolskii), 441. Boca Raton, FL: CRC Press.

7 Bernardo, P., Drioli, E., and Golemme, G. (2009). Membrane gas separation: a review/state of the art. *Industrial and Engineering Chemistry Research* 48: 4638.

8 Matsuura, T. (1994). *Synthetic Membranes and Membrane Separation Processes*, Chapter 2. Boca Raton, FL: CRC Press.

9 Robeson, L.M. (1991). Correlation of separation factor versus permeability for polymeric membranes. *Journal of Membrane Science* 62: 165.

10 Robeson, L.M. (2008). The upper bound revisited. *Journal of Membrane Science* 320: 390.

11 Riley, R.L., Hightower, G.R., and Lyons, C.R. (1973). Thin-film composite membrane for single-stage seawater desalination by reverse osmosis. *Applied Polymer Symposia* 22: 255.

12 Chiou, J. (1994). Composite gas separation membrane having a gutter layer comprising a crosslinked polar phenyl-containing-organopolysiloxane and method for making the same. US Patent 5,286,280. Issued: 15 February 1994.

13 Kools, W. (2001). Process of forming multilayered structures. WO 01/89673.

14 Yates, S. F., McGuirl, M. C., Tonev, T. G., et al. (2012). Photo-crosslinked gas selective membranes as part of thin film composite hollow fiber membranes. US Patent 8,337,598. Issued: 25 December 2012.

15 Liu, Y.Y., Ng, Z.F., Khan, E.A. et al. (2009). Synthesis of continuous MOF-5 membranes on porous alpha-alumina substrates. *Microporous and Mesoporous Materials* 118: 296.

16 Suzuki, H. (1987). Composite membrane having a surface layer of an ultrathin film of cage-shaped zeolite and process for production thereof. US Patent 4,699,892. Issued: 13 October 1987.

17 Haag, W.O. and Tsikoyiannis, J.G. (1991). Membrane composed of a pure molecular sieve. US Patent 5,019,263. Issued: 28 May 1991.

18 Brown, A.J., Brunelli, N.A., Eum, K. et al. (2014). Interfacial microfluidic processing of metal-organic framework hollow fiber membranes. *Science* 345: 72.

19 Yao, J., Dong, D., Li, D. et al. (2011). Contra-diffusion synthesis of ZIF-8 films on a polymer substrate. *Chemical Communications* 47: 2559.

20 Li, W., Su, P., Li, Z. et al. (2017). Ultrathin metal–organic framework membrane production by gel–vapour deposition. *Nature Communications* 8: 406.

21 Martínez Galeano, Y., Cornaglia, L., and Tarditi, A.M. (2016). NaA zeolite membranes synthesized on top of APTES-modified porous stainless steel substrates. *Journal of Membrane Science* 12: 93.

22 Ma, J., Shao, J., Wang, Z., and Yan, Y. (2014). Preparation of zeolite NaA membranes on macroporous alumina supports by secondary growth of gel layers. *Industrial and Engineering Chemistry Research* 65: 6121.

23 Dong, W.Y. and Long, Y.C. (2000). Preparation of an MFI-type zeolite membrane on a porous glass disc by substrate self-transformation. *Chemical Communications* 12: 1067.

24 Kim, S.J., Liu, Y., Moore, J.S. et al. (2016). Thin hydrogen-selective SAPO-34 zeolite membranes for enhanced conversion and selectivity in propane dehydrogenation membrane reactors. *Chemistry of Materials* 28: 4397.

25 Rangnekar, N., Mittal, N., Elyassi, B. et al. (2015). Zeolite membranes-a review and comparison with MOFs. *Chemical Society Reviews* 44: 7128.

26 Nakayama, K., Suzuki, K., Yoshida, M., et al. (2006). Method for preparing DDR type zeolite membrane, DDR type zeolite membrane, and composite DDR type zeolite membrane, and method for preparation thereof. US Patent 7,014,680. Issued: 21 March 2006.

27 Uchikawa, T., Yajima, K., Nonaka, H., and Tomita, T. (2013). Method for production of DDR type zeolite membrane. US Patent 8,377,838. Issued: 19 February 2013.

28 Teranishi, M., Miyahara, M., Ichikawa, M., and Suzuki, H. (2016). Porous body and honeycomb-shaped ceramic separation-membrane structure. US Patent 9,321,016. Issued: 26 April 2016.

29 Imasaka, S., Masaya, I., Asegawa, Y., et al. (2016). Zeolite membrane, production method therefor, and separation method using same. WO Patent Application 2016/006564. Published on 14 January 2016.

30 Tang, Z., Kim, S.J., Gu, X., and Dong, J. (2009). Microwave synthesis of MFI-type zeolite membranes by seeded secondary growth without the use of organic structure directing agents. *Microporous and Mesoporous Materials* 118: 224.

31 Shi, H. (2015). Organic template-free synthesis of SAPO-34 molecular sieve membranes for CO_2–CH_4 separation. *RSC Advances* 5: 38330.

32 Agrawal, K.V., Topuz, B., Pham, T.C.T. et al. (2015). Oriented MFI membranes by gel-less secondary growth of sub-100 nm MFI-nanosheet seed layers. *Advanced Materials* 27: 3243.

33 Truter, L.A., Ordomsky, V.V., Nijhuis, T.A., and Schouten, J.C. (2012). Preparation of ZSM-5 zeolite coatings within capillary microchannels. *Journal of Materials Chemistry* 22: 15976.

34 Kwon, H.T. and Jeong, H.K. (2013). In situ synthesis of thin zeolitic-imidazolate framework ZIF-8 membranes exhibiting exceptionally high propylene/propane separation *J. American Chemical Society* 135: 10763.

35 Bein, T. (1996). Synthesis and applications of molecular sieve layers and membranes. *Chemistry of Materials* 8: 1636.

36 Lai, Z., Bonilla, G., Diaz, I. et al. (2003). Microstructural optimization of a zeolite membrane for organic vapor separation. *Science* 300: 456.

37 Pham, T.C.T., Nguyen, T.H., and Yoon, K.B. (2013). Gel-free secondary growth of uniformly oriented silica MFI zeolite films and application for xylene separation. *Angewandte Chemie, International Edition* 52: 8693.

38 Pham, T.C.T., Kim, H.S., and Yoon, K.B. (2011). Growth of uniformly oriented silica MFI and BEA zeolite films on substrates. *Science* 334: 1533.

39 Kwon, H.T., Jeong, H.K., Lee, A.S. et al. (2015). Heteroepitaxially grown zeolitic imidazolate framework membranes with unprecedented propylene/propane separation performances. *Journal of the American Chemical Society* 137: 12304.

40 Choi, J., Jeong, H.K., Snyder, M.A. et al. (2009). Grain boundary elimination in a zeolite membrane by rapid thermal processing. *Science* 325: 590.

41 Schillo, M.C., Park, I.S., Chiu, W.V., and Verweij, H. (2010). Rapid thermal processing of inorganic membranes. *Journal of Membrane Science* 362: 127.

42 Huang, Y., Wang, L., Song, Z. et al. (2015). Growth of high-quality, thickness-reduced zeolite membranes towards N_2/CH_4 separation using high-aspect-ratio seeds. *Angewandte Chemie, International Edition* 54: 10843.

43 Varoon, K., Zhang, X., Elyassi, B. et al. (2011). Dispersible exfoliated zeolite nanosheets and their application as a selective membrane. *Science* 334: 72.

44 Zhang, H., Xiao, Q., Guo, X. et al. (2016). Open-pore two-dimensional MFI zeolite nanosheets for the fabrication of hydrocarbon-isomer-selective membranes on porous polymer supports. *Angewandte Chemie, International Edition* 55: 7184.

45 Jeon, M.Y., Kim, D., Kumar, P. et al. (2017). Ultra-selective high-flux membranes from directly synthesized zeolite nanosheets. *Nature* 543: 690.

46 Peng, Y., Li, Y., Ban, Y. et al. (2014). Metal-organic framework nanosheets as building blocks for molecular sieving membranes. *Science* 346: 1356.

47 Morigami, Y., Kondo, M., Abe, J. et al. (2001). The first large-scale pervaporation plant using tubular-type module with zeolite NaA membrane. *Separation and Purification Technology* 25: 251.

48 Mahajan, R. and Koros, W.J. (2000). Factors controlling successful formation of mixed-matrix gas separation materials. *Industrial and Engineering Chemistry Research* 39: 2692.

49 Liu, C., Chiou, J.J., Wilson, S.T. (2009). Cross-linkable and cross-linked mixed matrix membranes and methods of making the same. US Patent 7,485,173. Issued: 3 February 2009.

50 Liu, C., Tang, M. W., Wilson, S. T., Lesch, D. A. (2010). Method of making high performance mixed matrix membranes using suspensions containing polymers and polymers stabilized molecular sieves. US Patent 7,815,712. Issued: 19 October 2010.

51 Liu, G., Chernikova, V., Liu, Y. et al. (2018). Mixed matrix formulations with MOF molecular sieving for key energy-intensive separations. *Nature Materials* 17: 283.

52 Japip, S., Liao, K.S., and Chung, T.S. (2017). Molecularly tuned free volume of vapor cross-linked 6FDA-durene/ZIF-71 MMMs for H_2/CO_2 separation at 150 °C. *Advanced Materials* 29: 1603833.

53 Miller, S.J., Kuperman, A., Vu, D.Q. (2007). Mixed matrix membranes with small pore molecular sieves and methods for making and using these membranes. US Patent 7,166,146 B2. Issued: 23 January 2007.

54 Kulkarni, S.S., David, H.J., Corbin, D.R., Patel, A.N. (2003). Gas separation membranes with organosilicone-treated molecular sieve. US Patent 6,508,860. Issued: 21 January 2003.

55 Marand, E., Pechar, T.W., Tsapatsis, M. (2006). Mixed matrix membranes. US Patent 7,109,140 B2. Issued: 19 September 2006.

56 Kulkarni, S.S. and Hasse, D.J. (2005). Novel polyimide based mixed matrix membranes. US Patent Application 2005/0268782 A1. Published on 8 December 2005.

57 Ekiner, O.M. and Kulkarni, S.S. (2003). Process for making mixed matrix hollow fiber membranes for gas separation. US Patent 6,663,805. Issued: 16 December 2003.

58 Jiang, L.Y., Chung, T.S., and Kulprathipanja, S. (2006). An investigation to revitalize the separation performance of hollow fibers with a thin mixed matrix composite skin for gas separation. *Journal of Membrane Science* 276: 113.

59 Ismail, A.F., Kusworo, T.D., and Mustafa, A. (2008). Enhanced gas permeation performance of polyethersulfone mixed matrix hollow fiber membranes using novel Dynasylan Ameo silane agent. *Journal of Membrane Science* 319: 306.

60 Jia, M., Peinemann, K.V., and Behling, R.D. (1992). Preparation and characterization of thin-film zeolite-PDMS composite membranes. *Journal of Membrane Science* 73: 119.

61 Jeong, B.-H., Hoeka, E.M.V., Yan, Y. et al. (2007). Interfacial polymerization of thin film nanocomposites: a new concept for reverse osmosis membranes. *Journal of Membrane Science* 294: 1.

62 Hess, S.C., Grass, R.N., and Stark, W.J. (2016). MOF channels within porous polymer film: flexible, self-Supporting ZIF-8 poly(ether sulfone) composite membrane. *Chemistry of Materials* 28: 7638.

63 Liu, W., Post, P., Williams, J. L., et al. (2007). Multi-channel cross-flow porous device. US Patent 7,169,213. Issued: 30 January 2007.

64 Fekety, C. R., Kinney, L. D., Liu, W., Song, Z. (2009). Zeolite membrane structures and methods of making zeolite membrane structures. WO Patent Application 2009005745A1. Issued: 8 January 2009.

65 Caro, J. and Noack, M. (2009). Zeolite membranes – status and prospective. In: *Advances in Nanoporous Materials*, vol. 1 (ed. S. Ernst), 1. Elsevier.

66 Michaels, A.S. (1968). Separation techniques for the CPI (chemical process industry). *Chemical Engineering Progress* 64: 31.

67 Liu, C., Xu, Y., Liao, S. et al. (1997). Selective hydrogenation of propadiene and propyne in propene with catalytic polymeric hollow-fiber reactor. *Journal of Membrane Science* 137: 139.

68 Liu, C., Xu, Y., Liao, S., and Yu, D. (1998). Mono- and bimetallic catalytic hollow fiber reactors for the selective hydrogenation of butadiene in 1-butene. *Applied Catalysis A: General* 172: 23.

69 Brandao, L., Fritsch, D., Mendes, A.M., and Madeira, L.M. (2007). Propylene hydrogenation in a continuous polymeric catalytic membrane reactor. *Industrial and Engineering Chemistry Research* 46: 5278.

70 Kita, H., Tanaka, K., Okamoto, K.I., and Yamamoto, M. (1987). The Esterification of oleic acid with ethanol accompanied by membrane separation. *Chemistry Letters* 16: 2053.

71 Solovieva, A.B., Belkina, N.V., and Vorobiev, A.V. (1996). Catalytic process of alcohol oxidation with target product pervaporation. *Journal of Membrane Science* 110: 253.

72 Binning, R.C. and Kelly, J.T. (1959). Hydrocarbon conversion with dialytic separation of the catalyst from the hydrocarbon products. US Patent 2,913,507. Issued: 17 November 1959.

73 Champagnie, A.M., Tsotsis, T.T., Minet, R.G., and Webster, A.I. (1990). A high temperature catalytic membrane reactor for ethane dehydrogenation. *Chemical Engineering Science* 45: 2423.

74 Lafarga, D., Santamaria, J., and Menendez, M. (1994). Methane oxidative coupling using porous ceramic membrane reactors - I. reactor development. *Chemical Engineering Science* 49: 2005.

75 Lu, Y., Dixon, A.G., Moser, W.R. et al. (2000). Oxidative coupling of methane using oxygen-permeable dense membrane reactors. *Catalysis Today* 56: 297.

76 Capinelli, G., Carosini, E., Cavani, F. et al. (1996). Comparison of the catalytic performance of V_2O_5/γ-Al_2O_3 in the oxidehydrogenation of propane to propylene in different reactor configurations: (i) packed-bed reactor, (ii) monolith-like reactor and (iii) catalytic membrane reactor. *Chemical Engineering Science* 51: 1817.

77 Torres, M., Sanchez, J., Dalmon, J.A. et al. (1994). Modeling and simulation of a three-phase catalytic membrane reactor for nitrobenzene hydrogenation. *Industrial and Engineering Chemistry Research* 33: 2421.

78 Udovich, C.A. (1998). Ceramic membrane reactor for the conversion of natural gas to syngas. In: *Natural Gas Conversion V, Studies in Surface Science and Catalysis*, vol. 119 (eds. A. Parmaliana, D. Sanfilippo, F. Frusteri, et al.), 417.

79 van Veen, H.M., Bracht, M., Hamoen, E., and Alderliesten, P.T. (1996). *Fundamentals of Inorganic Membrane Science and Technology* (eds. A.J. Burggraaf and L. Cot), 641.

80 Baker, R.W. (2002). Future directions of membrane gas separation technology. *Industrial and Engineering Chemistry Research* 41: 1393.

81 Pinnau, I., Tpy, L.G., and Casillas, C. (1997). Olefin separation membrane and process. US Patent 5,670,051. Issued: 23 September 1997.

82 Herrera, P.S., Feng, X., Payzant, J.D., and Kim, J.-H. (2008). Process for the separation of olefins from paraffins using membranes. US Patent 7,361,800. Issued: 22 April 2008.

83 Feiring, A.E., Laareri, J., Majumdar, S., and Shsnggusn N. (2018). Membrane separation of olefin and paraffin mixtures. US Patent 10,029,248. Issued: 24 July 2018.

84 Ran, J., Wu, L., He, Y. et al. (2017). Ion exchange membranes: new developments and applications. *Journal of Membrane Science* 522: 267.

10

Absorption

10.1 Introduction to Absorption

Absorption is a method of separation commonly used in the oil and gas and chemical industries for removal of acid contaminants such as H_2S (hydrogen sulfide) and CO_2 (carbon dioxide) from hydrocarbons; this separation strategy is often employed in the purification, transport, and storage of natural gas. There are two primary classes of absorption: chemical absorption and physical absorption. Chemical absorption involves a liquid absorbent, typically an aqueous solution of amines, reacting with an acidic, non-hydrocarbon contaminant (e.g. CO_2 or H_2S). In physical absorption, there is no reaction between the liquid absorbent and CO_2, and the method of desorption tends to be through pressure or temperature swing depending on the absorption liquid employed [1]. Through the chemical or physical removal of H_2S and CO_2, natural gas producers are able to better protect their pipelines and production infrastructure from damage due to corrosion.

With a stronger focus over time on reducing the CO_2 emissions associated with energy generation and usage, natural gas has risen as a percentage of overall energy consumption. Table 10.1 compares the CO_2 emissions (normalized to natural gas) for a variety of hydrocarbon fuels. While coal emits nearly double the level of CO_2 per unit energy produced, gasoline, heating oil, and diesel are also significantly less efficient than natural gas from a CO_2 emissions' perspective. Traditionally, natural gas extraction was achieved through the drilling of vertical wells into natural gas–bearing formations. In conventional natural gas deposits, the hydrocarbon gas can flow easily up through the wells to the surface.

In North America, natural gas now can be produced efficiently from shale and other types of sedimentary rock formations by forcing water, chemicals, and sand (or other particles) down the well under high pressure. This process, called hydraulic fracturing, or fracking, breaks up the rock formation, releases the natural gas from the rock, and allows the natural gas to flow upward to the surface. At the top of the well on the surface, natural gas can then be transported via pipelines and sent to natural gas–processing plants.

The combination of low CO_2 emissions relative to other hydrocarbons as well as advances in extraction technologies such as hydraulic fracturing have led to a rapid growth in the importance of natural gas as a source of energy both in North America

Modern Petrochemical Technology: Methods, Manufacturing and Applications, First Edition.
Santi Kulprathipanja, James E. Rekoske, Daniel Wei, Robert V. Slone, Trung Pham, and Chunqing Liu.

Table 10.1 Normalized CO_2 emissions per unit of energy for various hydrocarbon fuels.

Fuel	CO_2 emissions
Coal (anthracite)	1.95
Coal (bituminous)	1.76
Coal (lignite)	1.84
Coal (subbituminous)	1.83
Diesel fuel and heating oil	1.38
Gasoline (without ethanol)	1.34
Propane	1.19
Natural gas	1.00

Source: US Energy Information Administration [2].

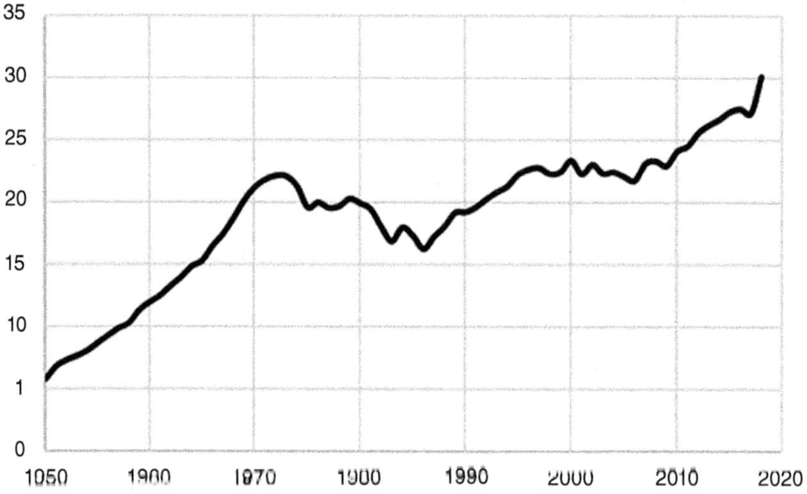

Figure 10.1 US Natural Gas Consumption (1950–2018) in trillion cubic feet. Source: US Energy Information Administration [3].

as well as around the world (Figure 10.1). Natural gas production has risen steadily over the past 50 years. As a result, the importance of absorption methods to the oil and gas and chemical industries has been, likewise, increasing steadily. Proven natural gas reserves are around 190 trillion m^3 globally, but total reserves are much larger than this figure [4]. The Middle East and Russia hold the largest total natural gas reserves, while the largest current national producer and consumer of natural gas is the United States.

The main constituent of natural gas is methane (Table 10.2), but it can contain a range of other higher hydrocarbons. While these hydrocarbons represent the main attraction of natural gas extraction due to their energy value through combustion, there also are many non-hydrocarbon materials present in natural gas such as H_2S

Table 10.2 Composition of natural gas (US Energy Information Administration).

Component	Molar fraction
Methane (CH_4)	0.75–0.99
Ethane (C_2H_6)	0.01–0.15
Propane (C_3H_8)	0.01–0.10
Butane (C_4H_{10})	0.00–0.02
Nitrogen (N_2)	0.00–0.15
Carbon dioxide (CO_2)	0.00–0.30
Hydrogen sulfide (H_2S)	0.00–0.30
Helium (He)	0.00–0.05

Source: Hammer et al. [5]. © 2006, John Wiley & Sons.

and CO_2, which can interfere with the transport and storage of natural gas and also can damage expensive energy infrastructure if not properly treated via methods such as chemical and physical absorption. In addition, the presence of CO_2 can lead to a reduction in energy value of the natural gas as there is no energy value unlocked with the attempted combustion of this contaminant.

Given its importance in the energy market, the application of gas separation methods to the removal of acidic contaminants such as H_2S and CO_2 from natural gas will be the focus of this chapter. It is important to note that many of these same chemical and physical absorption methods also are effective in the treatment of synthesis gas ($CO + H_2$), as both H_2S and CO_2 require sweetening for the same reasons as natural gas.

Gas absorption can be described as the removal of one or more pollutants from a contaminated gas stream via intimate contact of that contaminated gas with a liquid (absorber) that enables the pollutants to dissolve within the absorption liquid. The principal factors dictating absorption performance are the solubility of the contaminants in the absorbing liquid as well as the ability of the absorption units to efficiently mix gas and absorption liquid through the creation of a large amount of interfacial surface area.

10.2 Acid Gas Removal

As described previously, natural gas has been growing in importance over the last few decades and brings with it concerns regarding how to properly treat natural gas for transport and storage. H_2S and CO_2 are often found in natural gas streams and bring specific risks and challenges. H_2S (hydrogen sulfide) is a highly toxic gas that is corrosive to carbon steels and, if allowed to remain in natural gas, will oxidize into SO_2, which, when emitted, can harm the environment. CO_2 is, likewise, corrosive to equipment and reduces the energy value of gas when combusted as part of the natural gas stream.

The corrosion chemistry of H_2S and CO_2 is described in the equations below. Mild grades of steel (e.g. carbon steel) often are pitted through the action of acid gas contaminants. Removal of these materials via chemical or physical absorption helps preserve energy infrastructure, including wells and pipelines. In addition to H_2S and CO_2 levels, corrosion also is influenced by temperature, water chemistry (esp. pH), flow velocity, oil or water wetting, and composition and the surface condition of the steel.

CO_2 corrosion reactions:

$$CO_2(g) \rightarrow CO_2(aq) \tag{10.1}$$

$$CO_2(aq) + H_2O \rightarrow H_2CO_3 \tag{10.2}$$

$$H_2CO_3 \rightarrow H^+ + HCO_3^- \tag{10.3}$$

$$HCO_3^- \rightarrow H^+ + CO_3^{2-} \tag{10.4}$$

$$Fe^0(s) + CO_2 + H_2O \rightarrow FeCO_3 + H_2 \tag{10.5}$$

H_2S corrosion reactions:

$$H_2S(aq) \rightleftharpoons HS^-(aq) + H^+(aq) \tag{10.6}$$

$$HS^-(aq) \rightleftharpoons H^+(aq) + S_2^-(aq) \tag{10.7}$$

$$Fe^0(s) + S_2^-(aq) \rightleftharpoons FeS(s) \tag{10.8}$$

Conventional processes for removing acid gases typically involve their countercurrent absorption from the contaminated natural gas or syngas stream using a regenerative solvent in an absorber column. This approach of contacting a gas stream to a liquid absorber for the removal of acid gases is common across a wide range of process industries, including natural gas production, refining, and chemical production. Because each of these industries requires different degrees of acid gas removal, the absorption solvent employed will vary significantly.

Shown in Table 10.3 are some common solvents (both chemical and physical) employed in acid gas absorption/separation. The criteria for solvent selection in chemical and physical absorption will be covered in more detail later in this chapter. For now, please note that primary, secondary, and tertiary amines all find use in chemical absorption processes. In addition, non-amine systems are used (e.g. potassium carbonate in the removal of CO_2 via the Benfield™ process).

Prior to selecting either a chemical of physical absorption approach, the strengths and weaknesses of each need to be considered. While the loading of a physical solvent is proportional to the partial pressure of a component in natural gas, the loading of a chemical solvent is determined by the stoichiometric reaction between a component in natural gas and chemical solvent (i.e. it is a weak function of pressure). At low partial pressure of the component (generally low concentration), a chemical

Table 10.3 Common solvents for acid gas absorption.

Chemical	Physical
MEA (methanolamine)	Selexol™ (dimethyl ethers of polyethylene glycol)
DGA (diglycolamine)	Sulfolane™ (tetrahydrothiophenedioxide)
DEA (diethanolamine)	Rectisol® Methanol
DIPA (diisopropylamine)	Purisol® *N*-methyl-2-pyrrolidone
MDEA (methyldiethanolamine)	
TEA (triethanolamine)	
Potassium carbonate	

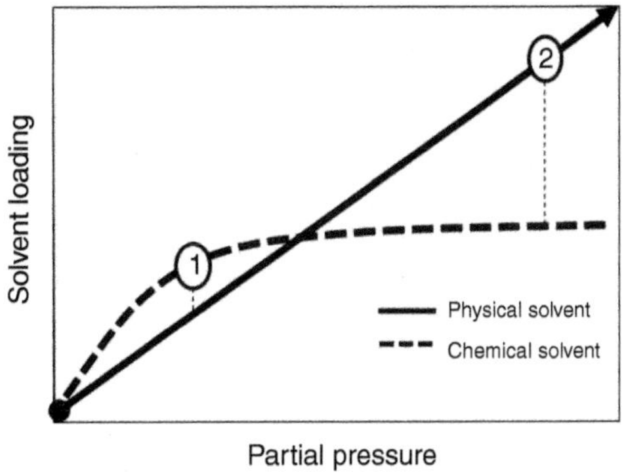

Figure 10.2 Physical vs. chemical absorption.

absorption system generally is preferred (such as condition 1 in Figure 10.2). At high partial pressure of the component to be removed (as in condition 2 in Figure 10.2), physical solvent has an advantage due to its high loading capability.

In selecting a solvent, the primary consideration is establishing how low the levels of H_2S and CO_2 need to be, post absorption processing. A secondary consideration is the impact of solvent selection on capital (i.e. equipment installation) and ongoing operating expenses. Solvent selection will affect capital expenses for the absorption unit and the design of the regeneration system. It is estimated that 50–70% of the initial capital investment for an amine-based acid gas absorption unit is associated with the magnitude of the solvent circulation rate, and 10–20% of the initial capex depends on the regeneration energy required for the selected amine solvent system [6]. In addition, about 70% of operating costs, excluding labor, result from the regeneration process.

10.3 Chemical Solvent

As demonstrated in 10.2, when impurity concentrations are low (e.g. 1–10% H_2S and CO_2), chemical absorption is favored over physical absorption. The ideal solvent

Table 10.4 Qualitative characteristics of the amines for the removal of H_2S and CO_2.

Amine type	Primary (MEA)	Secondary (DEA)	Tertiary (TEA, MDEA)
Enthalpy of reaction	High	Medium	Low
Enthalpy of evaporation	High	Medium	Low
Reaction rate	High	Medium	Low
Corrosivity	High	Medium	Low
Loading	Low	Medium	High
Concentration (wt%)	<32%	25–30%	35–55%

Source: Yildirim et al. [7]. © 2012, Elsevier.

for this process is capable of a fast, reversible chemical reaction with the gaseous impurity. When impurity levels are even lower, solvents that react irreversibly are employed, but such irreversible systems bring drawbacks such as solid waste treatment and added cost. Chemical solvents include aqueous solutions of primary, secondary, and tertiary amines (see Table 10.4). We will separately consider inorganic, alkaline salts such as potassium carbonate (K_2CO_3) in our summary of the Benfield process below.

Through acid-base reactions, aqueous solutions of these basic compounds capture and remove acid gases and other impurities by forming weak chemical bonds with dissolved acid gases (H_2S and CO_2) in the absorber. The bonds are broken by heat in the regenerator to release the acid gases and regenerate the solvent for reuse.

10.3.1 Amines Process

The first industrial use of amines for chemical absorption separation of acid gas from hydrocarbon was patented by R.R. Bottoms of the Girdler Corporation, Louisville, Kentucky, in 1930; claim 1 from this patent is included below [8]. As indicated in the claim, the need for intimate mixing of liquid amine with hydrocarbon gas for effective absorption processes has been known since the beginning of industrial chemical absorption.

> *1. The process of separating an acidic gas from gaseous mixtures, which includes effecting intimate contact of a gaseous mixture with an absorbent agent in liquid form including an amine selected from the group consisting of aliphatic and cycloparaffin amines, and which is free from carboxyl or carbonyl groups, and which has a boiling point not substantially below 100 °C.* – Bottoms [8]

The basic design of a chemical absorption unit is included in Figure 10.3. The flow of gas upward and liquid amine downward encourages the sort of intimate mixing required of a high efficiency chemical (or physical) absorption process. The packing material within the column bed plays a role in creating the mixing needed to efficiently remove acid gas impurities from natural gas and other hydrocarbon streams.

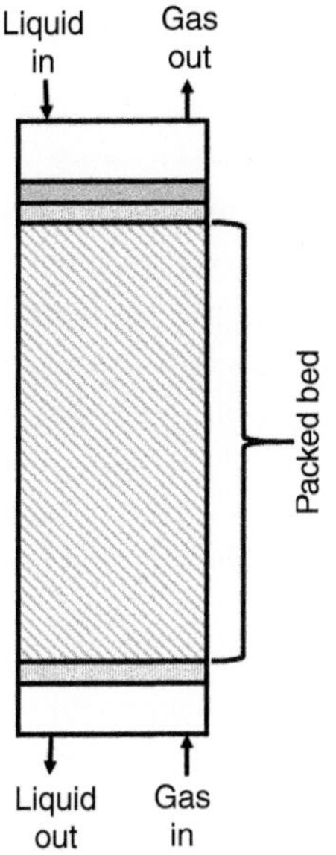

Figure 10.3 Packed absorption column.

The first commercial chemical absorption systems used triethanolamine (TEA) for acid gas removal. While TEA has been largely replaced due to its low capacity, low reactivity, and poor stability, amines have remained the absorption solvent of choice to the present day. The selection of an amine solvent is centered on four key factors:

(1) Required purity of cleaned gas stream for further processing or end use
(2) Composition of feed gas
(3) Utility requirements, process costs
(4) Corrosion and solvent degradation/solvent losses.

A comparison of these selection criteria is included in Table 10.4. Choosing the best amine solvent for a specific absorption end use will affect all aspects of performance and payback, including absorption unit safety, capital requirements, unit life, and operating cost. In many cases, simply changing the amine solvent can unlock large gains in efficiency.

The concentration of amine in the solution is an important operating parameter. High amine concentrations will reduce the amine solvent circulation rate and thereby reduce the operating and capital cost of the gas plant. Two major factors prevent high concentration of amine:

(1) Corrosion at high amine concentration.
(2) Temperature rise in the absorption column due to chemical reaction: high-concentration amine (low circulation) will cause high-temperature rise, which will reduce the equilibrium loading.

We will now dive a bit deeper into the strengths of each main category of amine solvent for chemical absorption.

10.3.1.1 Monoethanolamine (Primary Amine)

Monoethanolamine (MEA, $H_2NCH_2CH_2OH$) is a common solvent for chemical absorption [9]. It is often used at a concentration of 10–20 wt% in water (i.e. below 32%). While MEA itself is not considered to be highly corrosive, degradation products of MEA are extremely corrosive. Specifically, MEA will react with oxidizing agents such as COS, CS_2, SO_2, SO_3, and oxygen to form soluble products, which must then be removed from the chemical absorption unit to avoid serious corrosion damage. Degradation or deactivation of MEA also lowers the effective amine concentration, but regeneration of the MEA is possible to avoid the expense and complexity of having to add fresh MEA.

As a result, acid gas loading levels are typically limited to 0.3–0.35 mol acid gas per mole of amine for carbon steel equipment. In stainless steel, it is possible to increase loading levels up to 0.7–0.9 mol acid gas per mole of amine without harming corrosion performance.

MEA is a primary amine with a high pH, which enables MEA solutions to produce a sweetened gas product containing less than 4 ppm (by volume) of acid gas at very low H_2S partial pressures. For MEA systems, nearly all the CO_2 will be absorbed to produce gas that meets the H_2S ppm specification. Because the enthalpy of reaction for MEA with CO_2 is high, a feed gas containing high concentrations of CO_2 will cause either extremely high reboiler duty or poor acid gas stripping. For all amines, the enthalpy of reaction is a function of loading and other conditions.

10.3.1.2 Diethanolamine (Secondary Amine)

Diethanolamine (DEA, $HN(CH_2CH_2OH)_2$) is often used in the 25–30 wt% range. Like MEA, the total acid gas loading for DEA is also limited to 0.3–0.35 mol acid gas per mole of amine for carbon steel surfaces. For stainless steel, the DEA loading can be raised to 1 mol acid gas per mole of amine. The degradation products of DEA are much less concerning than MEA degradation products. Unlike MEA, DEA is not reclaimable in most units because it decomposes below its boiling point at ambient pressure.

Since DEA is a secondary alkanolamine, its affinity for H_2S and CO_2 is lower and may not be capable of producing pipeline specification natural gas for some low-pressure streams. As gas pressure is reduced, more stripping steam must be used or a split-flow design needs to be employed. DEA is selective toward the removal of H_2S and will permit a significant level of CO_2 to remain in the product gas stream.

10.3.1.3 Methyldiethanolamine (Tertiary Amine)

Methyldiethanolamine (MDEA, $H_3CN(CH_2CH_2OH)_2$) is used in the range of 35–55 wt%. Lower weight % solutions of MDEA are often used in low-pressure, high-selectivity applications such as the cleanup of tail gas in Shell Claus off-gas treatment (SCOT) units. Acid gas loadings as high as 0.7–0.8 mol acid gas per mole of amine in carbon steel equipment are possible with MDEA due to their relatively

Table 10.5 Four typical AGFS (amine guard formulated solvent) flow schemes.

Scheme	Feed gas quality	Desired product gas quality	Typical application	Reboiler duty (MBtu/lbmol CO_2 removed)
Flash only	Very high acid gas (>12%)	>2% CO_2	Pipeline NG	8–10
Conventional	Low acid gas (<7%)	50 ppm CO_2	LNG	45–60
1-stage	High acid gas (7–12%)	50–1000 ppm CO_2	LNG, NGL	32–40
2-stage	Very high acid gas (>12%)	500 ppm CO_2	Ammonia	12–18

low corrosivity. Oxidation of MDEA forms corrosive acids, which can result in the buildup of iron sulfide in the absorption system. [10].

MDEA offers several unique advantages over primary and secondary amines, including lower vapor pressure, lower heat of reaction, higher resistance to degradation, lower corrosivity, and higher selectivity toward H_2S in the presence of CO_2; being a weaker base, MDEA reacts much more quickly with H_2S than CO_2. Its lower heat of reaction allows MDEA to be used in pressure swing units for bulk CO_2 removal. In a pressure swing plant, the rich amine is flashed at or around atmospheric pressure and little or no heat is required for stripping.

Once the correct amine for chemical absorption has been selected, the design of the flow scheme must be considered. There are four basic flow scheme options for an amine guard formulated solvent (AGFS) chemical absorption unit (Table 10.5).

The first scheme (flash) is a simple flow scheme and is relatively inexpensive, has a low energy requirement, is ideal for bulk removal of CO_2 and partial removal of H_2S to generate pipeline quality natural gas. The second flow scheme (conventional, depicted in Figure 10.4) is suitable for both LNG and natural gas liquids (NGL); it can achieve CO_2 levels below 50 ppm, H_2S levels below 4 ppm, and requires less solvent than the flash system. The third option (1-stage) is capable of the same CO_2 and H_2S removal as a conventional unit but at lower energy usage. Finally, the 2-stage system is suitable for LNG and ammonia and provides the flexibility to tradeoff solvent flow rate for thermal regenerator duty (at much lower regeneration duty requirements than the 1-stage system).

10.3.2 Benfield™ Process

Originally developed at the U.S. Bureau of Mines in the early 1950s, the hot potassium carbonate (HPC) process was modified and optimized by Benson and Field [11, 12] and, subsequently, Honeywell UOP. The chemical solvents available for acid gas removal in the 1950s were water, monoethanolamine (MEA) and other amines. These solvents require operating temperatures below 125 °F (50 °C) to prevent solvent degradation. Water could be used to remove CO_2, but it is very inefficient and

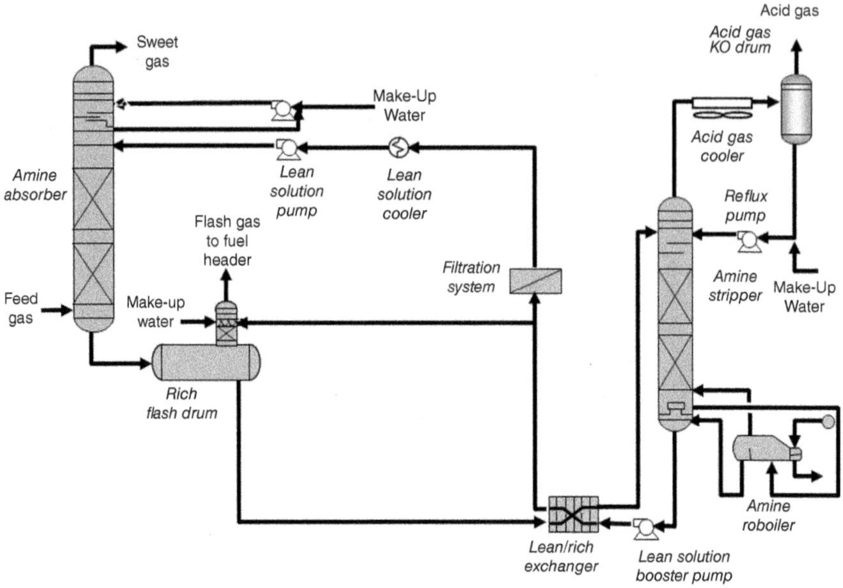

Figure 10.4 Honeywell UOP conventional chemical solvent – amine process (suitable for LNG and NGL).

does not provide acceptable treated gas purity. Caustic solutions, either NaOH or KOH, could remove the acid gases very effectively, but could not be regenerated. Using a solution of KOH to first pick up CO_2 would generate potassium carbonate in solution, which could absorb still more acid gases. Thus, the "hot potassium carbonate" process was born.

The UOP Benfield Process is a thermally regenerated cyclical solvent process using an activated, inhibited hot potassium carbonate solution to remove CO_2, H_2S, and other acid gas components. More than 700 Benfield units have been put into commercial service. In addition to its wide use in ammonia and hydrogen production, the Benfield process has been applied in 50+ natural gas plants and in direct iron ore reduction plants.

The absorption of CO_2 into an aqueous potassium carbonate (K_2CO_3) system means there are two reaction paths available. Either hydroxide ions react with CO_2 to directly form bicarbonate ions:

$$CO_2(aq) + OH^-(aq) \rightleftharpoons HCO_3^-(aq)$$

or water can hydrate CO_2 to form carbonic acid, which subsequently deprotonates to form bicarbonate:

$$CO_2(aq) + 2\,H_2O(l) \rightleftharpoons HCO_3^-(aq) + H_3O^+(aq)$$

The resulting drop in pH from either reaction is then buffered by the carbonate-bicarbonate equilibrium, thus giving the overall reaction as:

$$CO_2(g) + CO_3^{2-}(aq) + H_2O(l) \rightleftharpoons 2\,HCO_3^-(aq)$$

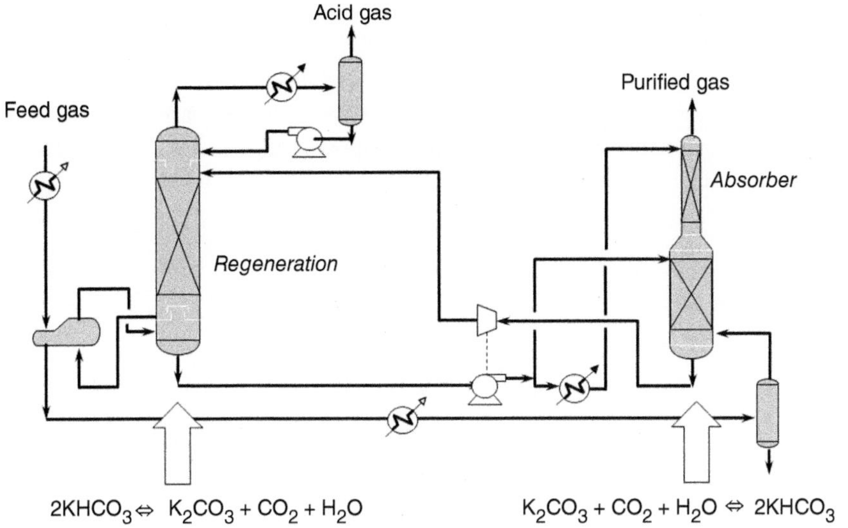

Figure 10.5 Benfield™ process: chemical absorption based on K_2CO_3 (conventional flow scheme).

Potassium carbonate chemical absorption processes possess a number of advantages over the amine-based absorption liquid processes, including the ability for absorption to occur at higher temperatures, which makes the regeneration process more efficient and economical [13]. Potassium carbonate also has a low cost, is less toxic, nonvolatile, and less prone to degradation effects that plague amines at high temperatures and in the presence of oxygen and other minor gas components such as SO_x and NO_x.

A typical Benfield chemical solvent composition is:

Potassium carbonate (K_2CO_3), in water (30%)
Activators – DEA and ACT-1
Corrosion inhibitor – V_2O_5

Shown in Figure 10.5 is the conventional flow scheme for the Benfield process, including the thermal regeneration system. With over 700 units in commercial operation, the Benfield process offers efficiency coupled with low capital investment. This method of separation achieves very low CO_2 concentrations (<50 ppm) in treated gas as well as low H_2S levels (<4 ppm)

10.4 Physical Solvent

As shown in Figure 10.2, physical solvent absorption systems offer the advantage of operating at high loading. Indeed, solvent loading increases linearly with the partial pressure of the contaminant to be removed. Therefore, when the component to be removed (e.g. CO_2 or H_2S) is present at a high partial pressure, physical absorption systems are favored over chemical absorption systems. Additionally, contaminants can often be removed from physical solvents simply by reducing pressure on the physical solvent and with a minimum of heat applied. Finally, physical solvents tend to be less corrosive than their chemical absorption amine-based counterparts. Due to these advantages, physical solvents have grown in popularity for the removal of contaminants in coal gasification applications as well as synthesis gas ($CO + H_2$) treatment.

10.4.1 Solvent Selection

Solvent selection remains a key performance determinant for physical absorption systems, and physical absorption system span a range of chemistries ranging from propylene carbonate and glycol-based solvents to alcohols, alcohols, phosphates, and pyrrolidones (Table 10.6). Cost and performance factors such as solvent volatility, contaminant solubility, and process conditions such as pressure and temperature often guide the choice of solvent in physical absorption systems.

Table 10.7 shows the relative solubility of typical gas stream components in three commercial solvents for physical absorption systems: Selexol™ (DMPEG), Rectisol™ (methanol), and Fluor (PC). The data have been normalized based on CO_2 solubility. The relative strength of each of these three physical absorption solvents is detailed below. [15]

10.4.1.1 DMEPG (Dimethyl Ether of Polyethylene Glycol, ($CH_3O(C_2H_4O)_nCH_3$))

DMEPG is used in UOP's Selexol process and is a mixture of dimethyl ethers of polyethylene glycol where *n* falls in the range of two to nine. DMPEG is often used to physically absorb H_2S, CO_2, and mercaptans from gas streams. As can be seen from the data in Table 10.7, DMEPG can be used for selective H_2S removal, which requires stripping, vacuum stripping, or a reboiler. Depending on downstream unit configuration, the H_2S-rich feed can be sent onward to the Claus unit. It also is possible to use

Table 10.6 Physical solvent processes.

Process name	Solvent	Process licensor
SELEXOL™	Dimethyl either of polyethylene glycol (DMPEG)	UOP/DOW
Fluor solvent	Propylene carbonate (PC)	Fluor Daniel
Sepasolv MPE	Methyl isopropyl ether of polyethylene glycol (MPE)	BASF
Purisol	*N*-methyl-2-pyrrolidone (NMP)	Lurgi
Rectisol	Methanol	Linde AG and Lurgi
Ifpexol	Methanol	IFP
Estasolvan	Tributyl phosphate (TBP)	IFP/Uhde
Methylcyanoacetate	Methylcyanoacetate	Unocal

Table 10.7 Relative solubility of chemical species at 25 °C (CO_2 = 1).

Component	Selexol (DMPEG)	Rectisol (Methanol)	Fluor (Propylene carbonate)
Carbon dioxide, CO_2	1	1	1
Hydrogen sulfide, H_2S	8.82	3.29	7.06
Hydrogen, H_2	0.013	0.0054	0.0078
Methane, CH_4	0.066	0.051	0.038
Ethane, C_2H_6	0.42	0.42	0.17
Propane, C_3H_8	1.01	2.35	0.51
n-Butane, C_4H_{10}	2.37	—	1.75
n-Pentane, C_5H_{12}	5.46	—	5
n-Hexane, C_6H_{14}	11	—	13.5
n-Heptane, C_7H_{16}	23.7	—	29.2
Carbonyl sulfide, COS	2.3	7.06	3.29
Water, H_2O	730	—	300

Source: Stewart [14]. © 2014, Elsevier.

DMEPG for selective H_2S removal with deep CO_2 removal typically accomplished via a two-stage process with two absorption and regeneration columns. The first column selectively removes H_2S, while CO_2 is removed in the second absorber. The second-stage solvent can be regenerated with air or nitrogen for deep CO_2 removal, or using a series of flashes if bulk CO_2 removal is required.

Compared to the other two solvents (methanol and propylene carbonate), DMEPG is of higher viscosity. This difference in rheology reduces mass transfer rates and tray efficiencies and increases packing (or tray number) requirements. These differences will become even more important when operating at reduced temperatures. DMEPG service temperatures range from −18 °C up to 175 °C.

10.4.1.2 MeOH (Methanol, H_3COH)

Methanol is used in the Rectisol process from Linde and was the first commercial physical absorption process based on an organic solvent. Methanol often is used in synthesis gas applications (especially purification of syngas derived from the gasification of heavy oil and coal). As the vapor pressure of methanol is relatively high, low-temperature operations that require refrigeration are employed to prevent high solvent losses. The need for refrigeration tends to drive the cost of operating a Rectisol unit higher, and the solvent flow rate often is reduced to counteract this cost disadvantage. Water washing of effluent streams often is used to recover the physical solvent in methanol absorption systems. Methanol service temperatures in Rectisol processes range from −62 °C up to −40 °C.

Like DMEPG, methanol offers a high degree of selectivity for H_2S over CO_2, combined with the ability to remove COS. While the need for methanol refrigeration can bring added cost, acid gas solubility in physical solvents increases significantly as the temperature decreases, so low-temperature operation may be of interest. Low temperature also reduces solvent losses by lowering the vapor pressure of the methanol in the product streams. On balance, if H_2S is to be removed from a gas with CO_2 remaining in the treated gas, Selexol tends to be a preferred option versus Methanol/Rectisol.

10.4.1.3 PC (Propylene Carbonate, $C_4H_6O_3$)

The Fluor solvent process uses propylene carbonate, a cyclic carbonate ester, since the late 1950s. Propylene carbonate is favored when H_2S levels are low and CO_2 removal is the focus of the physical absorption taking place. As seen in Table 10.7, the solubility of light hydrocarbons (methane through propane) is lower than DMEPG and methanol; this lower solubility results in lower recycle gas compression requirements and lower hydrocarbon loss in the CO_2 vent gas stream.

The vapor pressure of propylene carbonate is higher than DMEPG; however, solvent losses tend to be low. DMEPG is, by contrast, more selective than PC for H_2S removal from gases containing CO2. PC cannot be used for selective H_2S treating because high temperatures degrade the solvent. For this reason, Selexol is preferred for selective H_2S removal. The operating temperature for PC ranges from −18 °C up to 65 °C.

10.4.2 Flow Schemes

There are basically three sections to physical absorption unit:

(1) *Absorber*: High pressure and/or low temperatures
(2) *Regeneration unit*: Low pressure and/or higher temperature stripper
(3) *Intermediate section*: for heat exchange and solvent concentration. This section is tailored to achieve a certain acid gas quality, suitable for the sulfur recovery unit.

For each project, a specific design is required to achieve the desired product quality and the optimum acid gas quality at the same time. There are three general flow schemes for physical absorption, which differ primarily in their regeneration approach (Table 10.8):

A flash regeneration system involved the enriched solvent being regenerated solely through pressure let-down; there is no thermal regeneration involved. This method is best suited for bulk CO_2 removal to meet pipeline specs (<3% CO_2) and is not applicable for meeting low CO_2 (e.g. treated gas) specifications or for high sulfur

Table 10.8 Three typical physical absorption flow schemes.

Scheme	Desired product gas quality	Typical application
Flash Regen	<3% CO_2	Bulk CO_2 removal
Thermal Regen	<1 ppm S	Sulfur removal only (no CO_2)
Flash and thermal	<1% CO_2; <1 ppm S	Sulfur and CO_2 removal

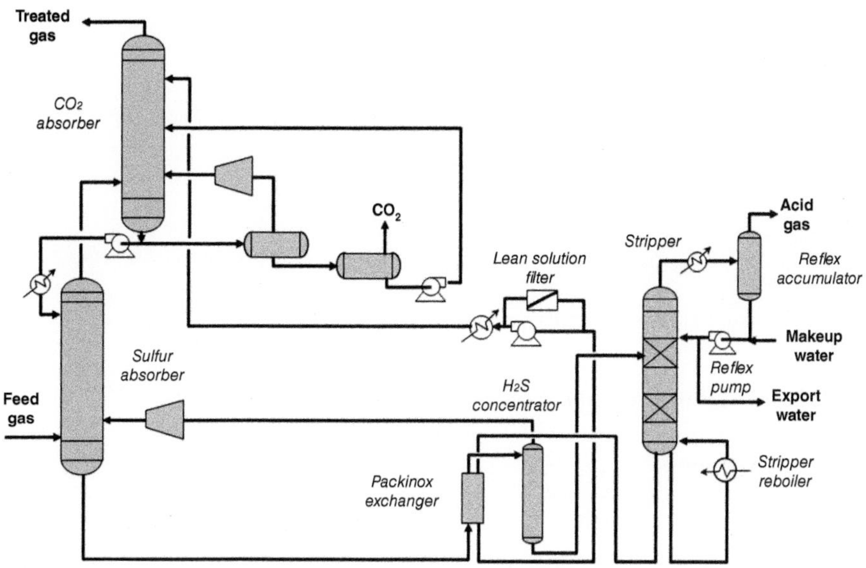

Figure 10.6 Physical solvent process flow scheme 3: Thermal and Flash Regeneration.

(H_2S) feeds. Because heat energy is not used, flash regen is a low-energy approach to physical absorption.

Thermal regeneration has been applied to sulfur removal (not CO_2) and allows a low (<1 ppm) sulfur specification to be realized. This method works best with high thermal stability solvents like DMEPG; it is not used with propylene carbonate and that solvent would degrade under high-temperature regeneration. The use of thermal regen does, of course, add energy cost to the physical absorption process, so the performance needs of a process need to be balanced with cost considerations.

The third physical absorption scheme includes both thermal and flash distillation and can accomplish sulfur and CO_2 removal. While more complex, a primary advantage of this approach is that the product acid gas levels can be adjusted to meet specific customer requirements. This flow scheme is shown in Figure 10.6 and is representative of the process design for Selexol physical absorption unit, which will now be described in more detail.

10.4.3 Selexol™ Process

Selexol from UOP is a commercial physical absorption process consisting of a mixture of dimethyl ethers of propylene glycol (DMEPG), which is selective for both CO_2 and H_2S and can also remove COS and RSH contaminants as well. The product streams from Selexol can be made essentially sulfur free with CO_2 levels tunable by need. For this reason, a wide range of end uses have been identified for Selexol, including natural gas treatment, synthesis gas purification, acid gas removal with simultaneous dehydration, and landfill gas treating. Selexol is not well suited for applications that involve low acid gas partial pressures (use chemical absorption methods below 50 psi) and hydrocarbon-rich streams.

Table 10.9 Selexol™ relative selectivity.

	Methane	Hydrogen
H_2S	135	680
CH_3SH	340	1700
CO_2	9	45
CO	0.43	2.2
COS	35	175

Table 10.10 Selexol™ advantage summary.

Selexol characteristic	Advantage
Low vapor Pr	Low solvent loss
Non fouling	Clean service
High on-stream efficiency	High availability
Low CW requirements	Lower OPEX
Low power requirements	Lower OPEX
100% organic and non-toxic	Safe operation

The selectivity profile of Selexol is summarized in Table 10.9. The high selectivity of DMEPG of H_2S over CO_2 allows efficiency recovery of CO_2 in the flash drum (and minimizes CO_2 in acid gas) and saves energy and money in the solvent recovery process. Likewise, the selectivity of Selexol for CO_2 over H_2 minimizes downstream H_2 purification requirements.

Table 10.10 summarizes a few of the key advantages of Selexol versus other physical absorption methods. While most of these advantages speak to lower operating costs (e.g. less solvent loss, higher availability, and low energy/power requirements), other advantages such as lower toxicity, lower corrosion (no formation of heat-stable salts), minimal process effluent, and protection of downstream equipment through the capture of metal carbonyls add to the benefits of Selexol versus other absorption options.

Shown in Figure 10.6 is a standard flow scheme for a Selexol unit tasked with the removal of H_2S, COS, CO_2, HCN, and metal carbonyls. Given the low corrosivity of DMEPG, a Selexol unit is made mostly of carbon steel.

The first column (sulfur absorber) removes most of the H_2S (and a limited amount of CO_2) from the feed syngas, which then flows to the second column (CO_2 absorber), which removes most of the CO_2. The rich solvent leaving the CO_2 absorption column is flashed in drums, from which relatively pure CO_2 is recovered. The solvent in the sulfur absorber column must be stripped in a column with a reboiler to remove the high H_2S content gases.

Table 10.11 Selexol™ commercial experience summary.

Applications	Number of units	Feed gas composition, %	
		CO_2	H_2S
Natural gas			
Bulk CO_2 removal	12	28–65	0–0.01
Selective H_2S removal	12	2–90	0.7–1.0
Synthesis and fuel gases			
Steam reforming	6	14–19	—
Partial oxidation	3	33–35	0.02–0.09
Coal gasification	12	33–42	0.7–1.0
Landfill gases			
Bulk CO_2 removal	9	~50	—

Table 10.11 provides some perspective on the number and variety of Selexol units in operation. As can be seen, Selexol's combination of selectivity, energy efficiency, and low corrosivity allows it to be used across a range of gas treatment end uses.

10.5 Additional Commercial Uses of Absorption

10.5.1 Glycol Dehydration

In natural gas systems, water vapor must be removed to a low level to avoid corrosion problems and pipeline blockage concerns caused by the formation of hydrates. To avoid free water from forming within the pipeline, the temperature cannot fall below the water dewpoint. The acid gas removal methods described previously in this chapter can result in water-saturated gas, which requires dehydration before the treated natural gas can be placed into the pipeline. Product specifications require the maximum quantity of water in the gas to be 4–7 lbm/MMcf. To meet this specification, glycol dehydration using diethyleneglycol (DEG) or triethyleneglycol (TEG) is used. [16]

Glycol dehydration systems are efficient in the removal of water from natural gas streams, and are also effective in the removal of volatile organic compounds (VOCs) such as benzene, toluene, ethylbenzene, and xylene (BTEX). The absorption solvents used in this process, DEG and TEG, have a chemical affinity for water and remove water from the natural gas stream (i.e. DEG serves as a liquid desiccant). The glycol solution is intimately mixed with the wet gas stream in a glycol contactor (Figure 10.7). While in the contactor, the glycol solution will absorb water from the wet gas. The heavier water-bearing glycol moves to the bottom of the contactor where it is removed. The natural gas, having been stripped

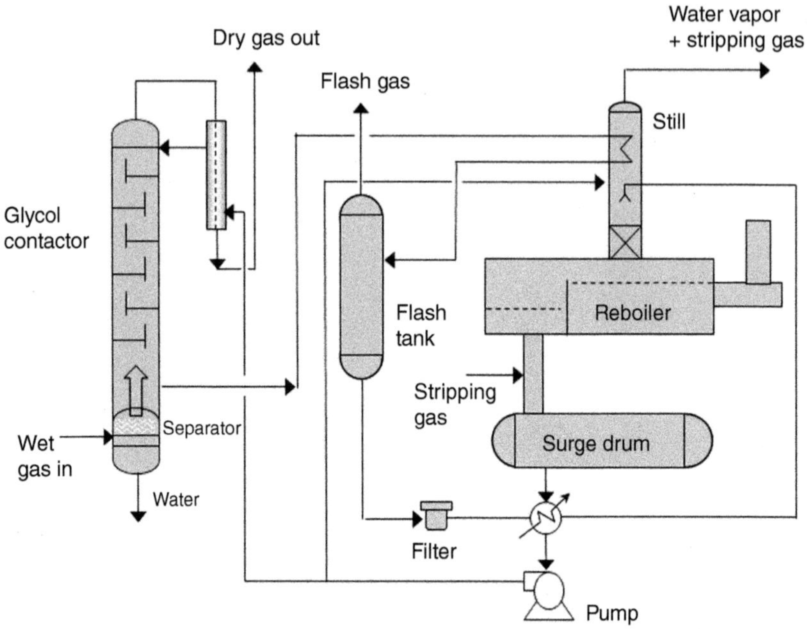

Figure 10.7 Glycol dehydration unit flow scheme. Source: Bentley [17]. © 1991, American Institute of Chemical Engineers.

of most of its water content, then is transported out of the top of the contactor dehydrator.

The glycol solution now contains the water removed from the natural gas and is processed in a reboiler designed to remove only the water from the solution. The boiling-point differential between water (100 °C) and glycol (204 °C) makes it relatively easy to remove water from the glycol solution, allowing it be reused in the dehydration process.

In addition to water from the wet natural gas stream, the glycol solution may carry with it small amounts of methane and other compounds found in the wet gas. Previously, this methane was simply vented out of the boiler. In addition to losing a portion of the natural gas that was extracted, this venting contributes to greenhouse gas emissions. To decrease the amount of methane and other compounds that are lost, flash tank separator-condensers are employed to remove these compounds before the glycol solution reaches the reboiler. The flash tank separator reduces the pressure of the glycol solution stream, which allows the methane and other hydrocarbons to vaporize (flash). The glycol solution then travels to the boiler, which may also be fitted with air or water-cooled condensers to capture any remaining organic compounds in the glycol solution. In total, absorption systems are capable of recovering 90–99% of the methane that would otherwise be flared into the atmosphere. The regeneration (stripping) of the glycol is limited by temperature as both DEG and TEG decompose at or below their respective boiling points.

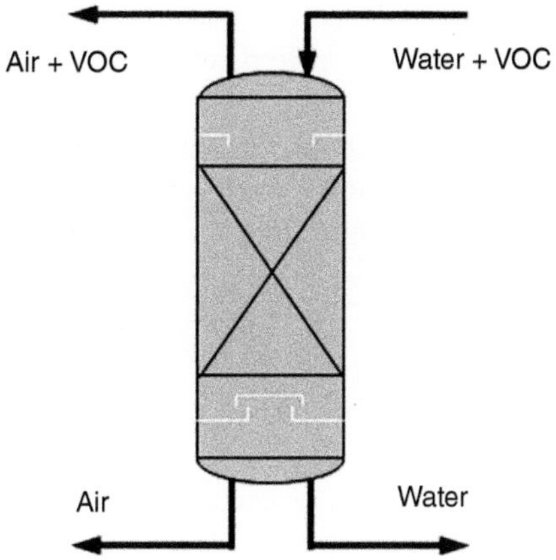

Figure 10.8 Air stripping of VOCs from water.

10.5.2 Organics from Air (Air Stripping)

Air-stripping columns commonly are used for the removal of VOCs from water (Figure 10.8). By forcing air upward through a descending mist of contaminated water that then flows downward through a packed-bed column, VOCs are evaporated from the water as air passing through the column absorbs the contaminated vapors. The air then can be removed for proper treatment of the VOC emissions (e.g. carbon absorption or thermal oxidation).

The two most common types of air strippers are packed-columns and sieve-tray air strippers, though diffused-aeration tanks often can act as an effective means of air stripping as well. The design of air stripping columns is governed by differences in volatility between the VOCs in the contaminated water. This method of VOC abatement can achieve 99+% removal of VOCs. Periodically, the bed-packing material must be removed and regenerated with an acid wash.

References

1 Abu-Zahra, M.R.M., Sodiq, A., and Feron, P.H.M. (2016). Commercial liquid absorbent based PCC processes. *Absorption-Based Post-combustion Capture of Carbon Dioxide* (ed. Paul H.M. Feron), 757–778. Cambridge: Woodhead Publishing.

2 US Energy Information Administration (2020). How much carbon dioxide is produced when different fuels are burned?, https://www.eia.gov/tools/faqs/faq.php?id=73&t=11 ()

3 US Energy Information Administration (2019). Natural gas. Monthly Energy Review (ed. Michael Kopalek), Table 4.3, November 2019, https://www.eia.gov/totalenergy/data/monthly/.

4 Breeze, P. (2016). *The Natural Gas Resource, Gas-Turbine Power Generation*, 9–19. Cambridge, MA: Academic Press.

5 Hammer, G., Lübcke, T., Kettner, R. et al. (2006). Natural Gas. In: *Ullmann's Encyclopedia of Industrial Chemistry*. Weinheim: Wiley-VCH.

6 Asmara, Y. (2018). The roles of H_2S gas in behavior of carbon steel corrosion in oil and gas environment: a review. *Jurnal Teknik Mesin* 7 (1): 37–43.

7 Yildirim, O., Kiss, A.A., Hüser, N. et al. (2012). Reactive absorption in chemical process industry: a review on current activities. *Chemical Engineering Journal* 231: 371–391.

8 Bottoms, R.R. (1930) Process for separating acidic gases. US Patent 1,783,901, filed 7 October 1930 and issued 2 December 1930.

9 Abu-Zahra, M.R.M., Schneiders, L.H.J., Niederer, J.P.M. et al. (2007). CO_2 capture from power plants: Part I. A parametric study of the technical performance based on monoethanolamine. *International Journal of Greenhouse Gas Control* 1: 37–46.

10 McCann, N., Maeder, M., and Attalla, M. (2008). Simulation of enthalpy and capacity of CO_2 absorption by aqueous amine systems. *Industrial & Engineering Chemistry Research* 47 (6): 2002–2009.

11 Benson, H.E., Field, J.H., and Haynes, W.P. (1956). Improved process for CO_2 absorption used hot carbonate solutions. *Chemical Engineering Progress* 52 (10): 433–439.

12 Tosh, J.S., Field, J.H., Benson, H.E., and Haynes, W.P. (1959). *Equilibrium Study of the System Potassium Carbonate, Potassium Bicarbonate, Carbon Dioxide, and Water* (ed. U.S.Bureau of. Mines), 5484. Pittsburgh.

13 Smith, H., Nicholas, N.J., and Stevens, G.W. (2016). Inorganic salt solutions for post-combustion capture. In: *Absorption-Based Post-Combustion Capture of Carbon Dioxide* (ed. P.H.M. Feron), pp. 145–166. Cambridge: Woodhead Publishing.

14 Stewart, M.I. Jr., (2014). Gas sweetening. In: *Surface Production Operations Volume 2: Design of Gas-Handling Systems and Facilities*, 3e, vol. 2 (ed. M.I. Stewart Jr.,), 433–539. Houston: Gulf Professional Publishing.

15 Burr, Barry & Lyddon, Lili. (2008). A comparison of physical solvents for acid gas removal. *GPA Annual Convention Proceeding*, January 2008.

16 Speight, J.G. (2014). *The Chemistry and Technology of Petroleum*, 5e. Boca Raton, FL: CRC Press.

17 Bentley, M.T. (1991). The basics of operating glycol dehydration units. In: *1991 AIChE Spring Annual Conference*. TX: Houston 1–11 April 1991.

11

Extraction Technology

11.1 Introduction

Though there are many extraction processes, this chapter will focus on liquid–liquid extraction (LLE) and supercritical fluid extraction (SCFE). Critical factors that control and affect extraction performance, including solvent selection and type of extractors, will be described. Selecting the right solvent and type of extractors are key to achieving high purity and efficiency in the extraction process. After introducing the fundamentals of extraction, the commercial extraction examples of hydrometallurgy and hydrocarbon will be introduced.

11.2 Extraction Processes

11.2.1 Liquid–Liquid Extraction (LLE)

LLE is a method to separate components based on their relative solubility in two different immiscible liquids such as water and isooctane. LLE is a technology used to extract one or more components from one liquid (feed) phase into a second liquid (solvent) phase. The transfer of the components from the feed to the solvent is controlled by the solubility of each component in the two separate phases. The two resulting phases from the extraction process are the enriched extract phase and the depleted raffinate phase. After extraction, the solvent can be regenerated using another separation technology such as distillation.

Figure 11.1 shows the four streams involved in the LLE process. Heavy Feed A + B flows down from the top of extractor. Light solvent flows up the extractor. Since Feed A is more soluble in the solvent, Feed A will be extracted by the solvent and enter the extract separator to recover Product A. The solvent is then recycled back to the extractor. The unextracted Feed B flows down the extractor with some solvent and enters the raffinate separator to recover Product B. The solvent is then recycled back to the extractor.

The extraction process is commonly operated at ambient temperature and near-atmospheric pressure. For LLE to work, the component being removed from the feed phase must preferentially distribute itself into the solvent phase from the

Modern Petrochemical Technology: Methods, Manufacturing and Applications, First Edition.
Santi Kulprathipanja, James E. Rekoske, Daniel Wei, Robert V. Slone, Trung Pham, and Chunqing Liu.

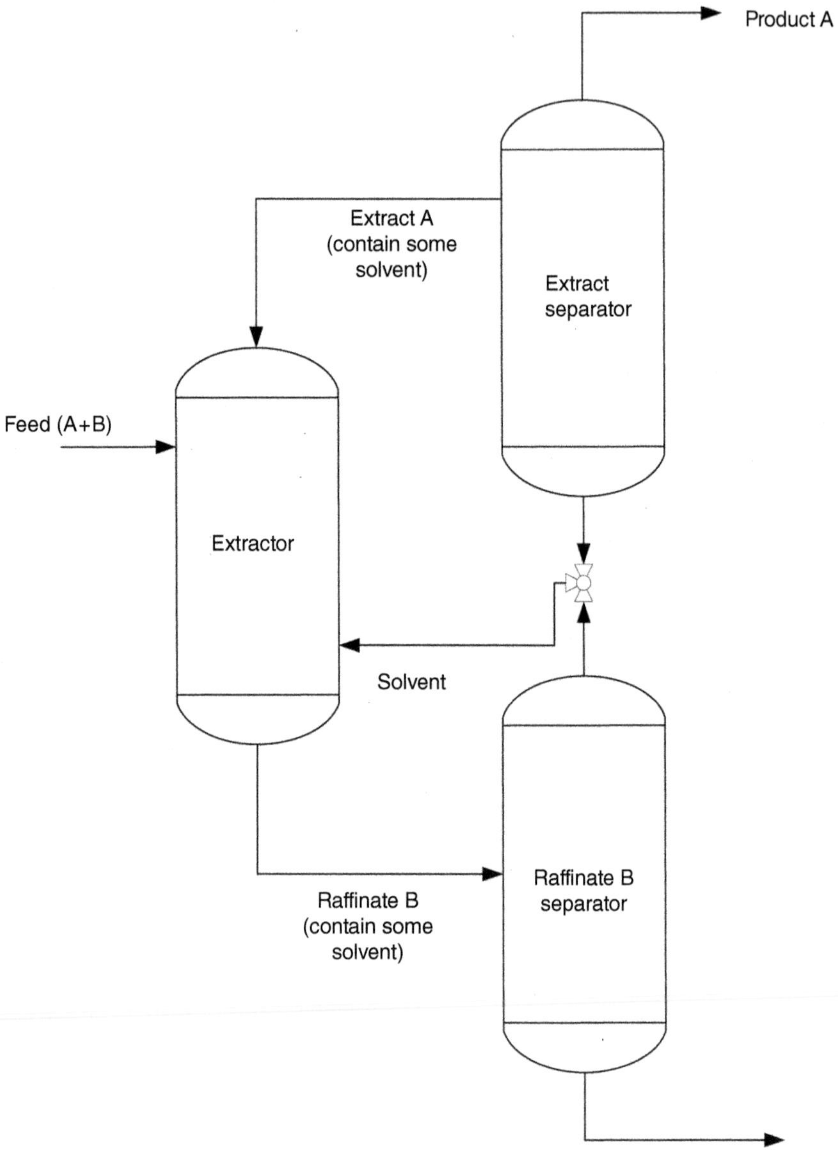

Figure 11.1 Liquid–liquid extraction and solvent recovery system.

bulk feed. In addition, the solvent phase must be substantially immiscible from the feed phase. Extraction efficiency can be enhanced by maximizing dispersion of solvent phase into the feed phase. Extraction columns normally contain trays, packing, or mechanical agitators to enhance the solvent dispersion. The function of trays or packing is also to enhance solvent drop formation in the disperse phase. Decreasing size of solvent drop increases mass transfer efficiency of feed components to dissolve into the solvent phase. Mass transfer fundamentals in extraction are like those in distillation.

11.2.2 Supercritical Fluid Extraction (SCFE)

SCFE is a method that uses a supercritical fluid such as carbon dioxide as the extraction solvent. SCFE can be used when the components to be separated have overlapping boiling points and distillation cannot be used. SCFE is also used to recover thermally sensitive components. Carbon dioxide is a nontoxic solvent that is often used in the food and pharmaceutical industries. In environmentally sensitive areas, SCFE is used as substitutes for organic chemicals.

One application of SCFE involves replacing chlorofluorocarbon with supercritical carbon dioxide in the extraction of caffeine from coffee bean. This process is demonstrated in Figure 11.2. In the process, caffeine will be extracted from coffee bean by liquid carbon dioxide in the extractor. After the extraction, caffeine will be washed from liquid carbon dioxide using water in the scrubber. Liquid carbon dioxide will be recycled back to the extractor. Caffeine will be recovered from water in the separator and water is recycled back to the scrubber.

Another application of SCFE is the replacement of organic solvent in spray painting with carbon dioxide. In the spraying process, unlike organic solvents, carbon dioxide flashes, leaving no organic contaminates in the atmosphere. Propane

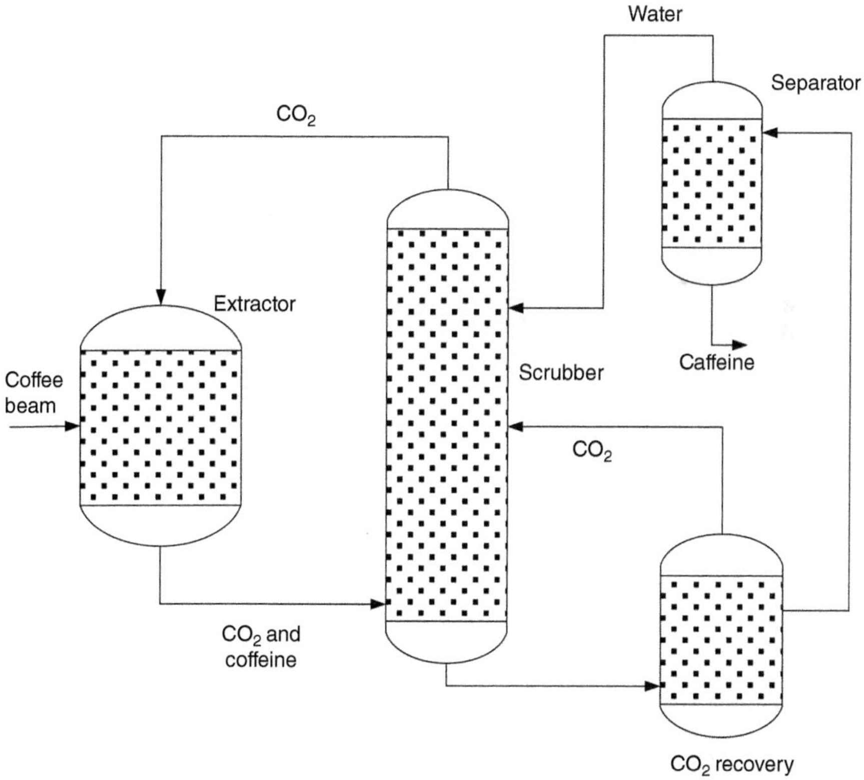

Figure 11.2 Supercritical extraction of caffeine from coffee beans using carbon dioxide solvent.

Table 11.1 Characteristics: liquid–liquid extraction vs. supercritical fluid extraction.

		LLE	SCFE
1	Mass transfer between solute (feed) and solvent	Slow	Fast
2	Solvent recovery	By conventional such distillation	By simply dropping pressure or increasing temperature
3	Operation pressure	Low	High: 1000–5000 psig
4	Capital cost	Low	High: only high-value products will bear the cost
5	Operating procedure	Simpler	Complicated

is another supercritical solvent used in SCFE to recover benzene, toluene, and xylene (BTX) from paraffins and naphthenes, recover lubricating oils from residue, and remove oils from sledges.

11.2.3 Liquid–Liquid Extraction vs. Supercritical Fluid Extraction

LLE has advantages and disadvantages compared to SCFE. For example, SCFE system offers faster mass transfer rate for components to be separated and easier solvent recovering than the LLE system. However, LLE system is less complicated to operate, and needs lower operating pressure and capital cost. Table 11.1 summarizes these features.

11.3 Solvents

The most important component affecting extraction efficiency is the solvent. Therefore, solvent selection is a critical factor in developing an effective extraction system. Criteria in solvent selection are described in this section.

11.3.1 Selectivity

Selectivity is calculated from the ratio of the distribution coefficients at equilibrium of components A and B to be separated. The distribution coefficient of component A is the ratio of concentration A in the **extract** phase to concentration A in the raffinate phase. The distribution coefficient provides a measure of the affinity of component A in the two phases. For the two-component separation, A and B, the separation factor or the selectivity of A to B can be calculated as:

$$\text{Separation factor or selectivity of A to B} = \frac{K_\text{a}}{K_\text{b}}$$

$$K_\text{a} = \frac{X_\text{a}}{Y_\text{a}} = \text{distribution coefficient of A at equilibrium}$$

$K_b = \frac{X_b}{Y_b}$ = distribution coefficient of B at equilibrium

X_a = amount of component A in extract phase

Y_a = amount of component A in raffinate phase

X_b = amount of component B in extract phase

Y_b = amount of component B in raffinate phase

11.3.2 Recoverability

After the extraction, the ability to recover solvent from the product is critical to the efficiency of the process. Typically, this can be done by distillation. Therefore, when selecting the solvent, the boiling point and thermal stability of the solvent need to be taken into consideration when designing the system.

11.3.3 Immiscible

An extraction unit, a liquid stream (feed) containing the component(s) to be recovered, is fed into an extractor, where it contacts a solvent. The solvent must be immiscible with feed liquid stream. This allows solvent to be dispersed as droplets in the continued feed liquid phase.

11.3.4 Density

Mass transfer occurs between the droplets (solvent dispersed phase) and the surrounding liquid (continuous-feed phase). For the two liquids to be subsequently separated, solvent selected must have different density from the feed liquid stream. The droplets then accumulate above or below the continuous phase, depending on the liquids' relative densities.

11.3.5 Chemical Stability

The selected solvent should be physically and chemically stable. The solvent should have no chemical reaction with any of the components to be separated.

11.3.6 Loading Capacity (L_a)

To achieve high extraction capability, the solvent selected should have a high capacity to retain the solute

$$L_a = \frac{(\text{Solute})}{(\text{Solvent})} = \text{mole ratio of solute to solvent.}$$

11.4 Extractors

There are two types of liquid–liquid extractors: simple extractors and mechanical extractors. Examples of simple extractors include spray, sieve trays, and packed column. These extractors do not have any moving components. On the other hand, mechanical extractors, such as mixer-settlers, rotary-disc columns, reciprocating-plate columns (Karr), and centrifugal extractors, achieve separation by mechanical agitation. Table 11.2 summarizes the several types of liquid–liquid extractors.

11.4.1 Simple Extractor

Figure 11.3 shows the fundamental designs of three simple extractors: sieve-tray column, spray column, and packed column. The dispersion of simple extractors is achieved by vertical countercurrent flow of two phases: heavy and light under the influence of gravity.

11.4.1.1 Sieve-tray Column

Sieve-tray columns have a simple geometry design and are used because of their reliability. The columns are used widely in LLE as well as in supercritical extraction. The simple geometry of sieve-tray columns is illustrated in Figure 11.4.

As shown in Figure 11.4, sieve-tray extractor columns are like distillation columns except that weirs are not needed with sieve-tray extraction. The continuous-feed phase moves across the tray and contacts the droplets of dispersed solvent phase, which are generated at the tray perforations. The droplets of dispersed solvent move vertically through the compartments between the trays. The mass transfer between the continuous-feed phase and dispersed solvent-phase droplets to separate the feed components occurs between the trays.

Table 11.2 Liquid–liquid extractors.

Extractor	Characteristics
Simple extractors	
• Spray, sieve, packed column	Simple dispersion mechanism; suitable for only a few stages of separation; low capital cost
Mechanical extractors	
• reciprocating-plate column (Karr)	Used with high viscosity liquids; system becomes complex and costly for higher stage of separation
• centrifugal extractor	Used with liquids containing suspended solids; easy-to-disperse components
	Used with emulsified materials and systems with low-density component differences; short contact time for unstable materials; high capital cost

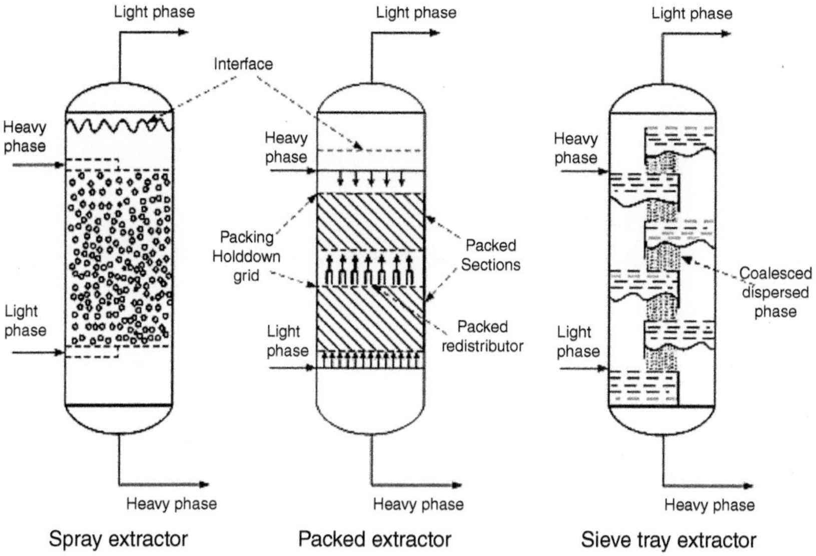

Figure 11.3 Simple extractor.

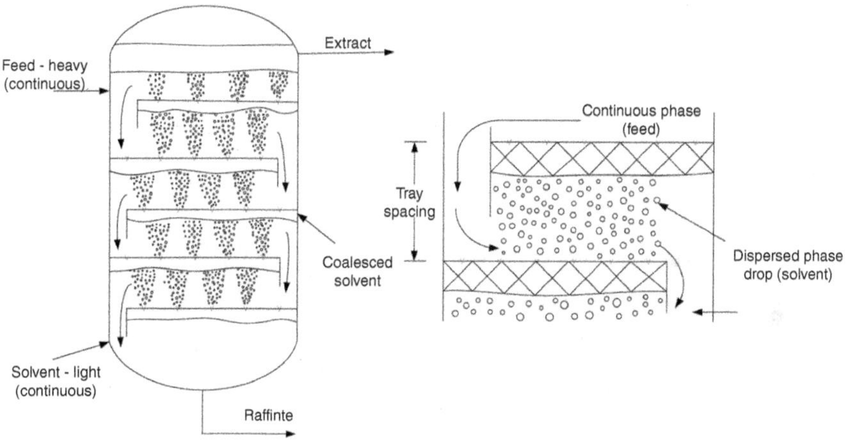

Figure 11.4 Sieve-tray extractor.

In the extraction mechanism described above, the smaller the solvent droplets generated, the higher the surface area of the dispersed solvent phase. The smaller droplets result in better mass transfer between the feed and solvent. Mass transfer efficiency of sieve trays can be predicted using published models and correlations [1].

11.4.1.2 Packed Column

This type of column is used when only a few equilibrium stages are needed. The packing enhances mass transfer by increasing droplets' residence time and

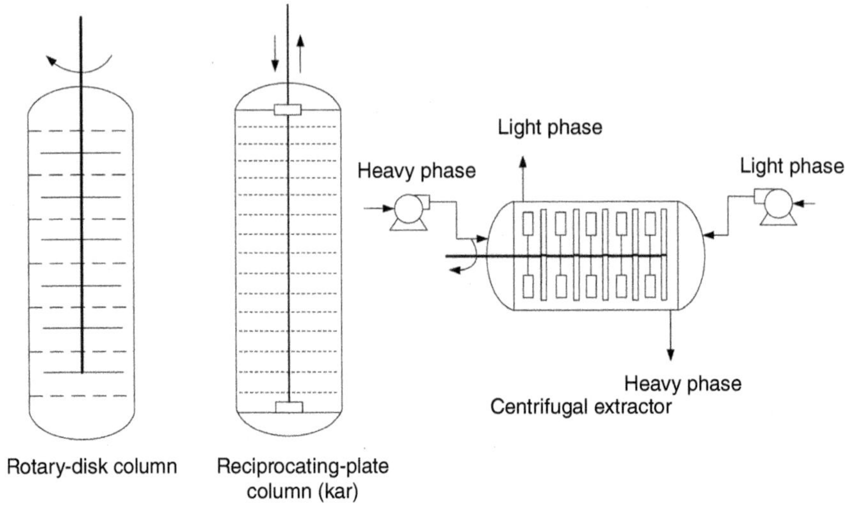

Figure 11.5 Mechanical added extractors.

dispersed-phase holdup. Packing also reduces back mixing of the continuous phase by promoting the breakup of dispersed-phase droplets.

11.4.2 Mechanical Extractors

Mass transfer between the continuous-feed phase and solvent dispersed phase can be aided by the use of mechanical agitation. Example schematics of mechanically aided extractors are given in Figure 11.5: rotary-agitated columns, reciprocating columns, and centrifugal extractor.

11.4.2.1 Rotary and Reciprocal Extractors

Advantages of mechanical extractors are that dispersion and interfacial area can be significantly increased by mechanically aided agitation. Baffling is necessary to prevent back mixing. Agitation in the rotating disk column is achieved by rotary motion combined with baffling. In the reciprocating plate column, agitation is generated by the vertical motion of the plate assembly.

11.4.2.2 Centrifugal Extractors

With the application of centrifugal force instead of gravity, extraction residence time can be reduced, and phase separation enhanced. Because of their complex construction design, the capital cost of centrifugal extractors tends to be higher than that of other types of extractors. Centrifugal extractors also have greater maintenance requirement and therefore higher operational costs. However, they are compact and offer relatively high throughput. These extractors are useful in applications where contact time must be short. For example, in chemically unstable systems (e.g. extraction of antibiotics) when product inventory must be kept at a minimum, liquids tend to emulsify, and components are difficult to separate.

Table 11.3 Features between simple and mechanically added extractors.

	Features	Simple extractors	Mechanical extractors
1	Mass transfer	Low	High
2	Capacity	High	Low (except for centrifugal extractor)
3	Efficiency	Low	High
4	Energy consumption	Low	High
5	Capital cost	Low	High
6	Service cost	Low	High

11.4.3 Simple Extractors vs. Mechanical Extractors

Mechanically aided extractors have many advantages and disadvantages compared to simple extractors. Table 11.3 compares the features of simple and mechanically aided extractors.

11.5 Industrial Extraction

LLE is widely used in industry for hydrometallurgy and hydrocarbon extraction. Applications in gas–liquid extraction called absorption are used commercially to extract carbon dioxide and sulfur-species (acid gas) from natural gas. Absorption was covered in Chapter 10. In this next section, the impact of LLE will be demonstrated with the following applications:

(1) Hydrometallurgy such as extraction of Zr from Hf, and U from Pu; and
(2) Hydrocarbon extraction such as aromatics from nonaromatic, *m*-xylene from other xylene isomers, mercaptan removal, and SCFE of caffeine from coffee.

11.5.1 Hydrometallurgy Extraction

Hydrometallurgy extraction is a process used to recover metals from aqueous ore leachates. However, because of the thermodynamic favorability of ionic species with the polar aqueous phase, simple extraction of metal cation or anion from aqueous phase into a nonpolar organic phase is unlikely to occur. For this reason, additional chemicals into solvent are required to form complexes with the metal ions during the contact of solute and solvent phases. These metal-complex formations enhance the metal to be extracted from an aqueous phase to the nonpolar phase. More specifically, properly designed chemical complexation reactions for specific metal cations or anions enable them to be separated from a mixture of other species. Description of both metal extraction fundamentals and specific applications have been documented and are referenced at the end of this chapter [2].

Table 11.4 Types of extraction solvents.

Types of extraction solvents	Reaction mechanism
Acid extractants: 1. Phosphoric acid (D2EPA) 2. Di(2-ethylhexylphosphoric acid) (D2EPHA) 3. Carboxylic acid (Versatic) 4. Sulfonic acid (DNNSA)	Complexation formation
Basic extractants: 1. Primary amines/1,1,3,3,5,5,7,7,9,9-decamethyl decamine 2. Secondary amines/ditridecylamine 3. Tertiary amines/trioctylamine (TOA), triisooctylamine (TIOA) 4. Quaternary ammonium salts/trioctylmethylammonium chloride	Ion-pair formation
Chelating extractants: 1. Hydroxyoximes (LIX 64) 2. Beta-Diketones (LIX 54) 3. 8-hydroxyquinolines (LIX 34)	Chelate formation
Solvation: 1. Tributyl phosphate (TBP) 2. Phosphine oxide (TOPO)	Solvate formation

In the hydrometallurgy reaction mechanism described above, metal extractions rely on an extractant solvent compound. The solvent compound consists of a diluent that provides the bulk physical properties and at least one other chemical capable of reversibly complexing the solute. The reaction between the solute, in this case the anion or cation metal, and the extractant is usually stoichiometric. Types of extraction solvents for metal extraction are summarized in Table 11.4. These extraction solvents form compounds with specific metals reversibly. This reversible reaction–enhanced extraction is used as the basis for recovery of most metals from leachate. There are four types of extraction solvents: acidic, basic, chelating, and solvation extractants.

11.5.1.1 Zirconium/Hafnium Separation

Zirconium is an ideal material for use in nuclear reactors due to its low absorption for thermal neutrons. Hafnium, on the other hand, with its strong neutron-absorption properties is often a contaminant that can decrease the performance of the zirconium. Because of the need to purify the zirconium, tributyl phosphate (TBP) solvent was developed to extract hafnium from zirconium [3]. Separation of zirconium from hafnium is demonstrated in a typical solvent extraction cycle in Figure 11.6.

In Figure 11.6, zirconium is separated from hafnium by using a solvent composed of TBP in a hydrocarbon diluent such as dodecane. The feed from an acid ore leachate includes both +4 cations of zirconium and hafnium in aqueous nitric acid

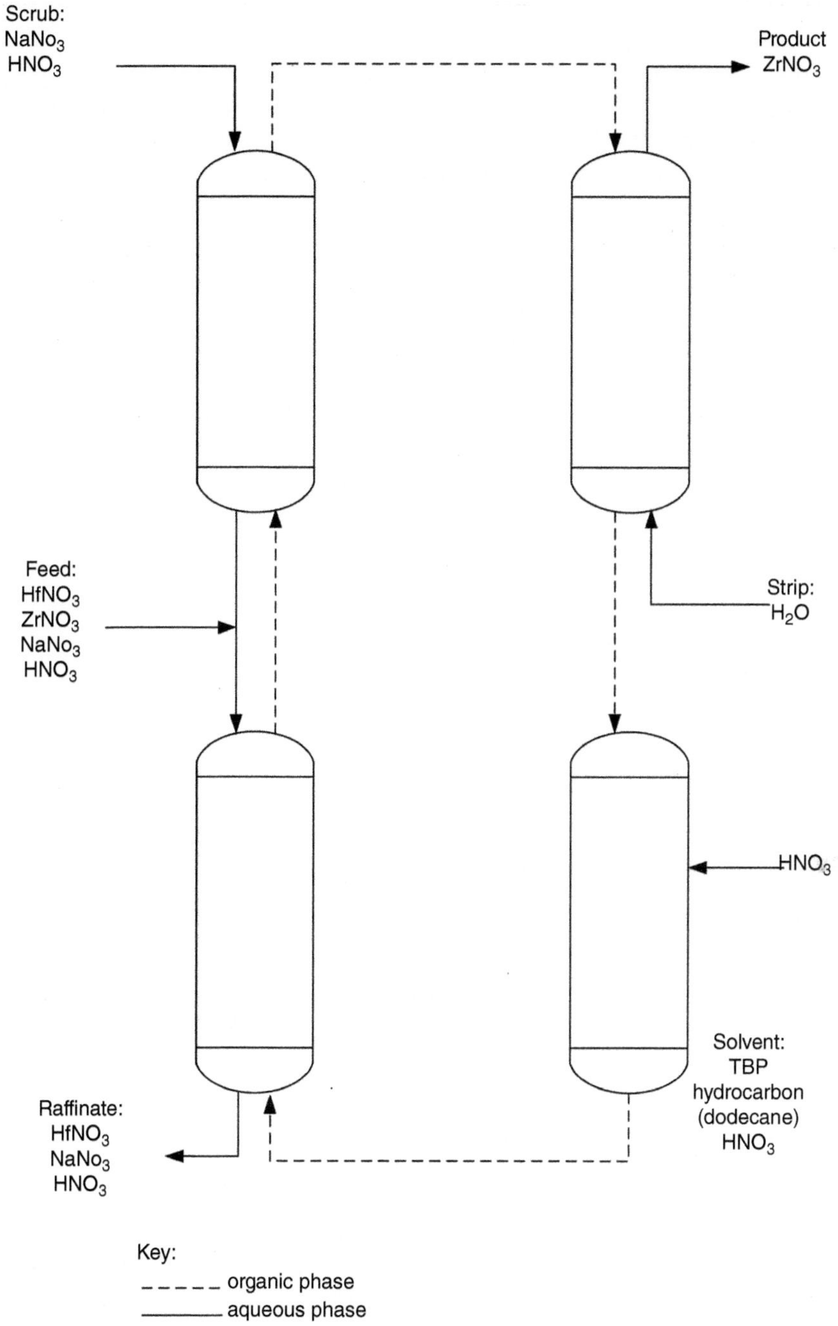

Figure 11.6 Hafnium/zirconium separation utilizing extraction, scrubbing, stripping, and regeneration.

and sodium nitrate. The feed is separated into a raffinate stream containing almost all the hafnium and an extract stream that contains almost all the zirconium. In the extraction unit, most of the zirconium is extracted into the solvent by complexation with the TBP solvent. The hafnium is then back extracted from the solvent into the aqueous phase in the scrubbing section. The scrubbing section is analogous to the rectification section in a distillation column, and the scrub solution is analogous to the distillation column's reflux. The zirconium solute is removed from the solvent in the stripping section by using a water strip solution. Finally, the solvent is purified in the solvent regeneration section and then returned to the extraction unit.

11.5.1.2 Uranium/Plutonium Separation

Metals in nuclear waste can exist in the oxidation states of +1 to +6. More specifically, plutonium and uranium exist in the +3 or +4 state and +4 or higher oxidation state, respectively. In the separation of uranium and plutonium, TBP is used as a solvent as it only forms complexes with +4 and higher oxidation state metals [4]. With this exclusive reaction between TBP and metal cations, uranium can be separated from plutonium by utilizing the difference between the extractability of +3 and +4 cations in TBP.

The process for uranium and plutonium separation in nuclear waste is shown in Figure 11.7. In the process, the solvent TBP is first used to extract +4 plutonium and uranium while rejecting other +1, +2, +3 oxidation state metals from the feed. The +4 plutonium is then reduced to +3 by adding an irreversible aqueous-phase reductant such as hydroxylamine nitrate (HAN) or ferrous sulfamate (FeSAM). This then allows the separation of plutonium from uranium by once again using TBP as the solvent.

11.5.2 Hydrocarbon Extraction

Hydrocarbon extraction is based on the relative solubility of the components to be separated in the solvent phase. The commercial examples that will be described next are *m*-xylene separation from other xylenes using HF–BF_3 solvent, aromatics separation from nonaromatic using sulfolane solvent (Sulfolane process), and mercaptan removal (Merox process).

11.5.2.1 *m*-Xylene/Xylene Isomer Separation

m-Xylene is an aromatic hydrocarbon. It is one of the three isomers of dimethyl-benzene collectively referred to as xylene. The other two isomers are *o*-xylene and *p*-xylene. The major use of *m*-xylene is in the production of iso-phthalic acid, which is used as a copolymerizing monomer to alter the properties of polyethylene terephthalate (PET). Other uses of *m*-xylene are in the production of unsaturated polyester resins. *m*-Xylene and other xylene isomers can be produced using heavy naphtha feed in the reforming process. Because of the closed boiling point of the xylene isomer, *m*-xylene cannot be separated from other xylene isomers using distillation. Development of the *m*-xylene extraction described below was the first technology developed for *m*-xylene separation.

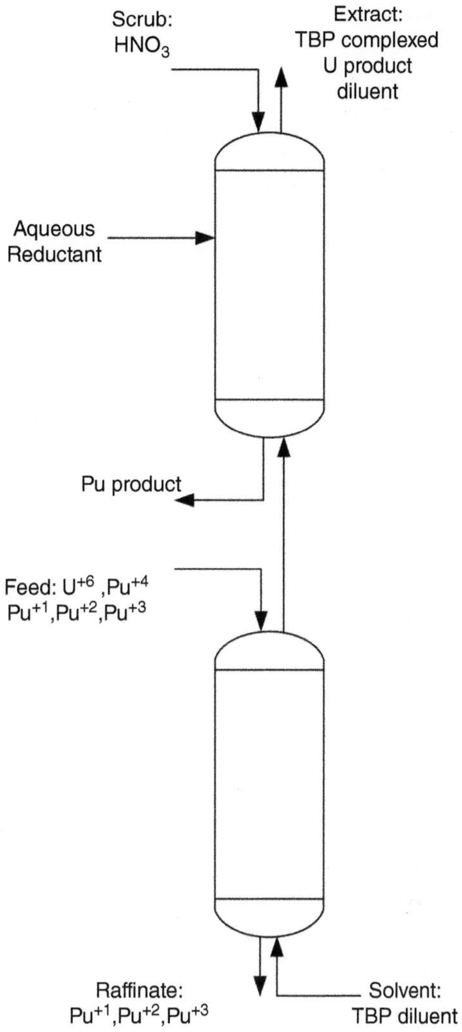

Figure 11.7 U/Pu separation utilizing the difference between the extractability of +4 and +3 cations in the TBP solvent.

Herman Pines [5] reports that the basicity of aromatics increases as the number of alkyl groups bound to the benzene ring increases. Pines' finding is summarized in Table 11.5, which shows the relative basicity of methylbenzenes in $HF–BF_3$ solvent. It is clear all the xylenes are more basic than toluene. Furthermore, among the xylenes, *m*-xylene is the most basic.

Based on the basicity of the different xylene isomers, Japan Gas–Chemical developed a LLE process using $HF–BF_3$ solvent to extract *m*-xylene from another xylene isomer [6, 7]. In this process, *m*-xylene (MX) reacts reversibly to form complexes with boron trifluorides (BF_3) in the presence of liquid hydrogen fluoride (HF). MX complex ($MXHBF_4$) is soluble in HF solvent and then can be separated from the other feed stream isomers. The MX complex formation with $HF–BF_3$ can be expressed by

Table 11.5 Base strengths of methylbenzenes relative to $HF-BF_3$.

Methylbenzene	Relative basicity in $HF-BF_3$
Toluene	0.63
p-Xylene	1
o-Xylene	2
m-Xylene	20
Pseudocumene (1,2,4-)	40
Hemimellitene (1,2,3-)	40
Durene (1,2,4,5-)	120
Prehnitene (1,2,3,4-)	170
Mesitylene (1,3,5-)	2 800
Isodurene (1,2,3,5-)	6 500
Pentamethylbenzene	8 700
Hexamethylbenzene	89 000

Source: Pines [5].

the equation below:

$$MX + HF\,(\text{liquid}) + BF_3(\text{gas}) \rightarrow MXHBF_4 \tag{11.1}$$

To illustrate the extraction of MX through the formation of the $MXHBF_4$ complex, a simple laboratory experiment is introduced. Hydrocarbon-phase xylene isomers are added into aqueous-phase HF. HF has a higher density and will stay at a lower layer. BF_3 gas is then charged into the extractor under mechanical agitation. After the $MXHBF_4$ complex is formed, the hydrocarbon and HF layers are completely separated from each other. Most importantly, the $MXHBF_4$ complex is soluble in HF and remains in the HF phase. The MX in the HF phase is recovered by the reversibility of reaction (11.1), the decomposition of the complex $MXHBF_4$ to $MX + HF + BF_3$.

One of the disadvantages of the MX extraction process is that the $HF–BF_3$ is a corrosive solvent. In practice, the MX LLE process is unfriendly and not economically attractive. With this, a friendlier adsorptive separation process, MX Sorbex [8, 9], was developed that replaced the MX extraction process just described. The MX Sorbex development was detailed in Chapter 7.

The MX extraction process demonstrates that aromatics can be extracted from other isomers based on their differences in basicity. Therefore, it should be expected that other isomers in Table 11.5 can be separated using the same theory. For example, isodurene should be able to be extracted from durene due to their basicity difference.

11.5.2.2 Sulfolane Process

p-Xylene is a critical petrochemical raw material for PTE production. *p*-Xylene (PX) is one of the aromatics that is co-produced using catalytic reforming technology.

Besides aromatic, nonaromatic is also generated. Sulfolane is a process to recover aromatics from Reformate. After sulfolane, aromatics product that contains BTX isomer will be fractionated to BTX isomer. High-purity PX will then be recovered from xylene isomer for the PTE application.

Sulfolane is a process used to recover high-purity aromatics from hydrocarbon mixtures such as reformed petroleum naphtha (reformate). The process is also used to recover aromatics from pyrolysis gasoline (pygas) and coke-oven light oil. The Sulfolane process is named after the solvent used, sulfolane or tetrahydrothiophene-1,1-dioxide. Sulfolane is a five-membered ring structure containing one atom of sulfur and four atoms of carbon with two oxygen atoms bonded to the sulfur atom in the ring. Sulfolane was developed by Shell in the early 1960s as an industrial solvent and is used in aromatics extraction. Universal oil products (UOP) has been the exclusive licensor of Sulfolane since 1965.

Sulfolane is used for extraction aromatics from nonaromatics because of its unique properties in high aromatic/nonaromatic selectivity and hydrocarbons solubility. A study of hydrocarbon solubility in sulfolane highlights its properties:

(1) With hydrocarbons containing the same number of carbon atoms, the solubility decreases in the following order: aromatics > naphthenes > olefins > paraffins.
(2) Within the same hydrocarbon homologous series, the solubility decreases as molecular weight increases.

Due to the unique sulfolane selectivity and solubility properties for hydrocarbons, the Sulfolane process combines both LLE and extractive distillation (ED) in the same process unit. This combination of technologies results in the effective processing of feed stocks with a much broader boiling range than would be possible by either extraction process alone. This combined extraction process has advantages for aromatic recovery in:

(1) LLE systems where light nonaromatic components are more soluble in the solvent than heavy nonaromatics. Thus, LLE is more effective in separating aromatics from the heavy contaminants than the light ones.
(2) ED where light nonaromatic components are more readily stripped from the solvent than heavy nonaromatics. Thus, ED is more effective in separating aromatics from light contaminants than heavy ones.

A simple commercial Sulfolane process [10, 11] flow diagram is shown in Figure 11.8. Feed reformate enters the extractor and flows upward while the Sulfolane solvent flows counter currently downward. The raffinate stream, which is low in aromatics, is withdrawn from the top of the extractor. At the other end of the unit, the Sulfolane solvent loaded with aromatics is withdrawn from the bottom of the extractor and enters the extractive distillation column. In the ED unit, the light nonaromatic components, having volatilities higher than aromatics, are separated from the Sulfolane solvent. The overhead stream from the ED, light nonaromatics, is recycled to the extractor to remove the heavy nonaromatics from the solvent phase.

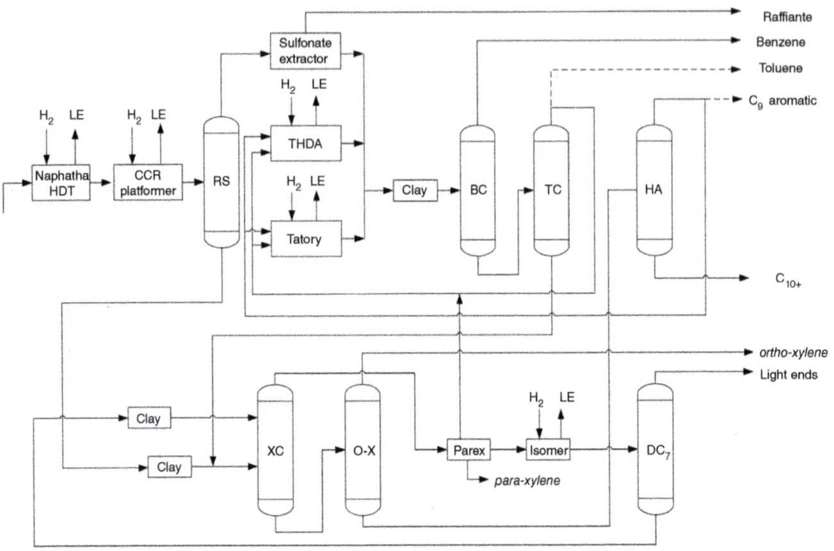

Figure 11.8 UOP aromatics complex flowsheet.

The bottoms stream, which is substantially free of nonaromatic impurities from the extractive distillation column, is sent to the aromatic's recovery column, where the aromatic products are separated from the solvent. This separation is accomplished with minimal energy input, because of the large difference in boiling point between the Sulfolane solvent and the heaviest aromatic components. To minimize temperature exposures, the recovery column is operated under vacuum. The Sulfolane solvent from the bottom of the recovery column is then returned to the extractor. Next, the extract is recovered overhead and sent to distillation columns downstream for recovery of the individual aromatic: BTX products.

To complete the process, the raffinate stream exits the top of the extractor and enters the water wash column, where the dissolved solvent is removed with water. The water in the solvent-rich phase is then vaporized in the water stripper. Accumulated solvent from the bottom of the water stripper is pumped back to the recovery column as the raffinate product exits the top of the raffinate wash column.

As described above, Sulfolane is a process to recover aromatics such as BTX isomer from Reformate. *p*-Xylene, a critical petrochemical raw material for PTE production, is then separated from xylene isomer by the Parex process. Because of the high PX demand, other processes such as isomar and Tatoray are introduced to enhance the PX product. To demonstrate the critical demand of PX, as of 2014, UOP has licensed over 100 aromatic complexes and more than 700 individual process units to produce aromatics. This includes more than 300 continuous catalyst regeneration (CCR) Platforming™, 155 Sulfolane™, 80 Isomar™, 65 Tatoray™, and 100 Parex™ process units. Figure 11.9 shows UOP's aromatic complex flow sheet. In the complex, aromatics will be produced using the CCR Platformer. After CCR Platformer, the stream is fractionated into the C_8 overhead. The C_8 overhead is composed of

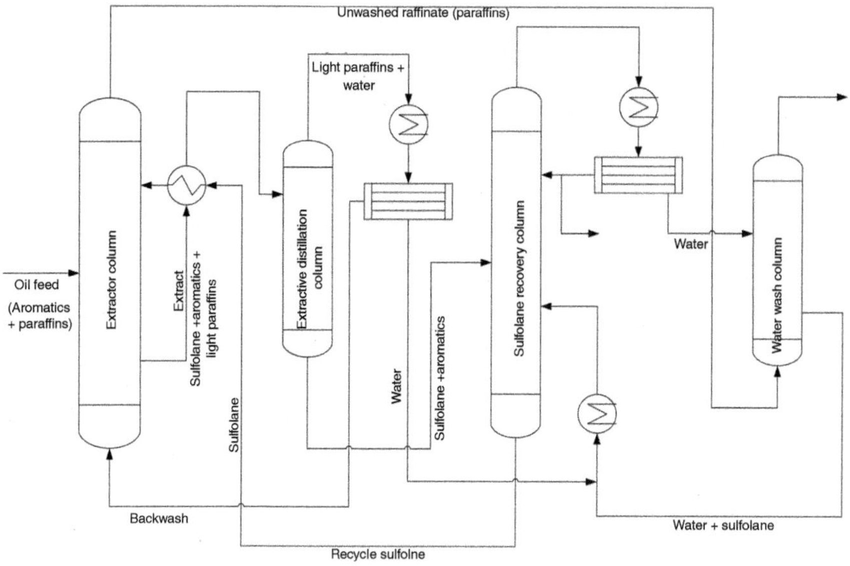

Figure 11.9 Sulfolane process flowsheet.

aromatics and nonaromatic components, which are separated using the sulfolane extraction process.

11.5.2.3 Merox Process

Global sulfur specifications for gasoline have been tightening over the last decade. For example, European Union fuels for on-road and non-road vehicles have been limited to 10 ppm sulfur since 2009 and 2011, respectively. In the United States, ultralow sulfur diesel with a maximum of 15 ppm sulfur has been the requirement since 2006 for on-road vehicles, 2010 for non-road vehicles, and 2012 for locomotives and marine vessels.

Refiners rely heavily on sulfur extraction and sweetening process units, such as the Merox process to meet gasoline and diesel product specifications. Merox is an acronym for mercaptan oxidation. The technology was introduced to the refining industry more than 50 years ago. The process is a proprietary catalytic chemical process developed by UOP used in oil refineries and natural gas processing plants to remove mercaptans from liquefied petroleum gas (LPG), propane, butanes, light naphtha, kerosene, and jet fuel by converting them to liquid hydrocarbon disulfides. The chemistry of catalytic oxidation of mercaptans (RSH) to disulfides (RSSR) in an alkaline environment can be described as follows:

$$\text{Extraction:}\quad \mathrm{RSH + NaOH \rightarrow NaRS + H_2O}$$

$$\text{Caustic regeneration:}\quad \mathrm{NaRS + \frac{1}{4}O_2 + \frac{1}{2}H_2O \rightarrow NaOH + \frac{1}{2}RSSR}$$

$$\text{Overall reaction:}\quad \mathrm{2RSH + \frac{1}{2}O_2 \rightarrow RSSR + H_2O}$$

Table 11.6 Solubility of mercaptans in water and their distribution between isooctane and 0.5-N NaOH aqueous phase at 20 °C

Mercaptans	K_q – Overall extraction constant
Ethyl-mercaptan	77.9
n-Propyl-mercaptan	15.6
n-Butyl-mercaptan	3.59, 3.56
Tert-Butyl-mercaptan	2.46
n-Amyl-mercaptan	0.753
Tert-Amyl-mercaptan	0.362
n-Heptyl-mercaptan	0.0342

The extraction efficiency of the Merox process can be characterized by the extraction coefficient (K_q) in the equilibrium reaction and extraction equation below:

$$\text{RSH (oil)} \leftrightarrow \text{RSH (aqueous)} \leftrightarrow \text{RS}^-\text{(aqueous)}$$

$$K_q = \frac{(\text{RSH (aq)} + \text{RS}^- \text{(aq)})}{\text{RSH (oil)}}$$

The extraction coefficients (K_q) of different mercaptans are summarized in Table 11.6 [12]. The table shows that the lighter mercaptans such as C_3, C_4, LPG, and naphtha gases have higher K_q values. This indicates that the lighter mercaptans can be extracted with a caustic solution. The resultant disulfides from the mercaptan oxidation are separated and the caustic reused for extraction. Hence, the sulfur content of the extracted hydrocarbon feed is reduced.

Heavier hydrocarbon fractions such as naphtha, kerosene, jet fuel, and diesel containing mercaptans with low K_q values cannot be readily extracted from hydrocarbons. Therefore, a sweetening Merox process is used where the mercaptan oxidation reaction takes place in the presence of air and caustic solution. In the environment, mercaptans are converted to disulfides, which remain in the sweetened hydrocarbon product. The overall sulfur content, therefore, remains the same.

Figure 11.10 shows the Merox process applications for different types of hydrocarbons [13]. It is confirmed that over 700 extraction Merox units are being applied to the light mercaptan removal. It is also showed that over 700 sweetening Merox units are being used to convert heavier mercaptans to disulfides. Merox LLE and sweetening Merox will be described in the next sections.

Merox Liquid–Liquid Extraction

The flow diagram for the Merox LLE unit is shown in Figure 11.11. The light naphtha feed flows up the extractor column and is counter-currently extracted with a circulating caustic solution. If hydrogen sulfide (H_2S) is present in the feed, H_2S

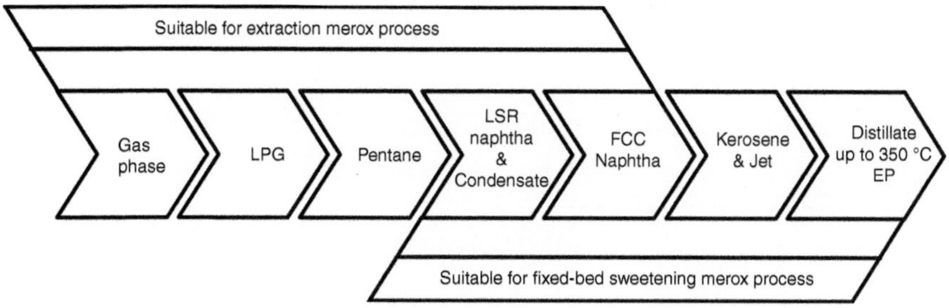

Figure 11.10 Merox process applications.

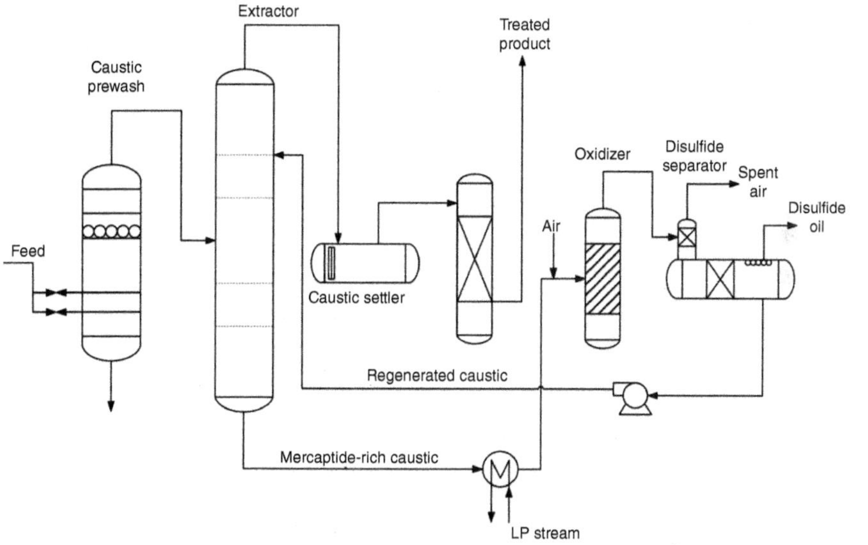

Figure 11.11 Extraction Merox process.

is removed by passing it through a caustic prewash extractor. Removal of the H_2S increases process efficiency by minimizing waste of the circulating caustic. After the extractor column, the product passes to the caustic settler to further reduce caustic carryover. The treated naphtha then flows through a sand filter coalescing system to reduce the sodium content of the naphtha to <1 wt-ppm Na^+.

The Merox unit also contains a caustic regeneration section where the mercaptan-rich caustic solution from the bottom of the extraction column flows to an oxidizer column. Here, the extracted mercaptans are oxidized to disulfides by air injected upstream of the oxidizer. The caustic stream leaving the disulfide separator is then circulated back to the extraction column.

Sweetening Merox Process

The sweetening Merox process is used with pentane and heavier hydrocarbon feed streams (Figure 11.10). As an example, the Minalk*TM* fixed-bed sweetening Merox process is shown in Figure 11.12. The primary application for the Minalk process

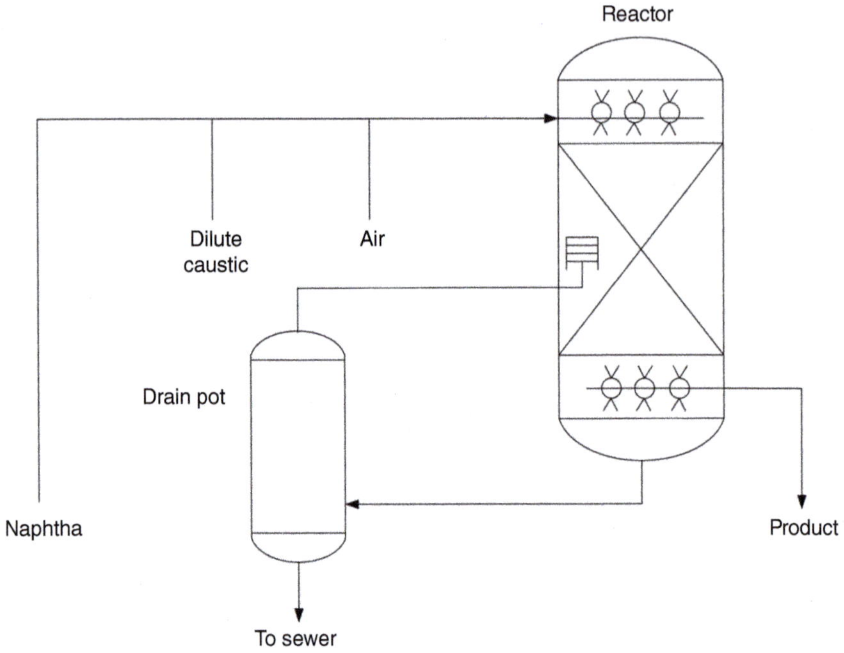

Figure 11.12 Sweetening Merox process.

is for treating naphtha. The Minalk process consists of a single reactor filled with a fixed bed of activated charcoal impregnated with a Merox proprietary catalyst. Dilute caustic is continuously injected into the naphtha feed stream along with air. The naphtha is then passed downward through the fixed bed where the mercaptan is oxidized to disulfide. The disulfide remains in the naphtha-phase product.

11.6 Summary

LLE, or solvent extraction, is one of the oldest and widely used techniques in industrial separation. Supercritical extraction was developed in the later days with more uses being developed. This chapter summarized the key components of the process, including how to choose solvents and type of extractors. In addition to the basic principles of extraction, the chapter highlighted some of the key commercial extraction processes in metal extraction and hydrocarbon separation.

Acknowledgments

The author would like to thank Dr. Ames Kulprathipanja for reviewing the chapter. Thanks also to Dr. Katipot Inkong and Ms. Orawee Lamoonkit of Chulalongkorn University in drawing some of the figures.

References

1 Rocha, J.A., Humphrey, J.L., and Fair, J.R. (1986). Mass transfer efficiency of sieve tray extractors. *Industrial and Engineering Chemistry Process Design and Development* 25: 286.

2 Ritcey, G.M. and Ashbrook, A.W. (1984). *Solvent Extraction, Principles and Applications to Process Metallurgy*. New York: Elsevier.

3 Cox, R.P., Peterson, H.C., and Beyer, G.H. (1958). Separating hafnium from zirconium. Solvent extraction with tributyl phosphate. *Industrial and Engineering Chemistry* 50 (2).

4 Schultz, W.W. and Navratil, J.D. (1984). *Science and Technology of Tributyl Phosphate*. Boca Raton, FL: CRC Press.

5 Pines, H. (1981). *The Chemistry of Catalytic Hydrocarbon Conversions*, 26. Academic Press.

6 Ueno, T. (1970). *Japan Gas-Chemical Xylene Separation Process*, vol. 12. Bulletin of The Japan Petroleum Institute.

7 Talbot, J.L. (1956). Separation of xylene isomers. US Patent 2,738,372.

8 Neuzil, R.W. (1982). US Patent 4,326,092.

9 Kulprathipanja, S. (1999). Process for adsorptive separation of metaxylene from xylene mixtures. US Patent 5,900,523.

10 Kelly, M.F., and Uitti, K.D. (1970). Extractive distillation of aromatics with sulfolane solvent. US Patent 3,551,327.

11 Meyers, R.A. (2016). *Handbook of Petroleum Refining Processes*, Fourthe. UOP Sulfolane Process: McGraw-Hill Professional.

12 Yabroff, D.L. Extraction of mercaptans with alkaline solution. *Industrial and Engineering Chemistry* 32 (2).

13 Sullivan D., *The Role of the Merox Process in the Era of Ultra Low Sulfur Transportation Fuels*. Fifth EMEA Catalyst Technology Conference, 3 and 4 March 2004

Index

Modern Petrochemical Technology: Methods, Manufacturing and Applications, First Edition.
Santi Kulprathipanja, James E. Rekoske, Daniel Wei, Robert V. Slone, Trung Pham, and Chunqing Liu.

b

n

q

r